Psychiatric Controversies in Epilepsy

Psychiatric Controversies in Epilepsy

Andres M. Kanner
Steven C. Schachter

AMSTERDAM • BOSTON • HEIDELBERG • LONDON • NEW YORK • OXFORD
PARIS • SAN DIEGO • SAN FRANCISCO • SINGAPORE • SYDNEY • TOKYO
Academic Press is an imprint of Elsevier

Academic Press is an imprint of Elsevier
525 B Street, Suite 1900, San Diego, CA 92101-4495, USA
30 Corporate Drive, Suite 400, Burlington, MA 01803, USA
32, Jamestown Road, London NW1 7BY, UK
Radarweg 29, PO Box 211, 1000 AE Amsterdam, The Netherlands

First edition 2008

Library of Congress Cataloging-in-Publication Data
A catalog record for this book available from the Library of Congress

British Library Cataloguing in Publication Data
A catalogue record for this book is available from the British Library

ISBN: 978-0-12-374006-9

For information on all Academic Press publications
visit our website at books.elsevier.com

Dedicated to my wife Hilary and my daughters Lesley and Lauren
— Andres M. Kanner

Dedicated to the memory of Richard Bonasera
— Steven C. Schachter

Contents

Preface

The four major psychiatric disorders – mood, anxiety, psychotic and attention deficit hyperactivity disorders – are relatively frequent comorbidities of epilepsy. Traditionally, these psychiatric entities have been considered to be a complication of the seizure disorder; yet, a bidirectional relation has now been demonstrated between epilepsy and at least two of these conditions, mainly mood and attention deficit disorders. Therefore, not only are patients with epilepsy at greater risk of developing these psychiatric disorders, but patients with mood and attention deficit disorders have a significantly greater risk of developing epilepsy.

These bidirectional relationships suggest the existence of common pathogenic mechanisms operant in epilepsy and the major psychiatric disorders. Thus, the identification of these underlying pathogenic mechanisms may shed light on the neurobiological bases of these psychiatric disorders as well as epilepsy. The primary aims of this book are to focus on these mechanisms and to elucidate many of the frequently encountered controversial problems associated with the psychiatric aspects of epilepsy.

The book begins with a review of the relatively high prevalence rates of psychiatric comorbidities in epilepsy using data from population-based studies, which are remarkable considering that the latter are globally under-recognized and under-treated. This contradiction constitutes one of the biggest paradoxes in the practice of medicine and is addressed in three chapters of the book. The first of these chapters reviews the widespread lack of communication between psychiatrists and neurologists from the perspectives of a neurologist and a psychiatrist, who also outline possible reasons and offer solutions. They both agree that the limited (or often non-existent) training in psychiatry for neurologists and vice versa is one of the important causes behind the lack of communication between the two disciplines, which is therefore addressed in the two chapters that follow.

Three chapters address the reported clinical differences of mood and psychotic disorders in patients with epilepsy relative to patients with primary mood and psychotic disorders, which have been the source of much controversy among experts. Six chapters then explore the neurobiological aspects of mood, anxiety, psychotic and attention deficit disorders, raising the question of whether these four psychiatric disorders are in fact neurologic entities with psychological symptoms. One of these chapters discusses the thought-provoking question of whether psychogenic non-epileptic seizures may have a neurological component, while another chapter reviews data on the potentially negative impact that mood disorders can have on the course of the major neurologic disorders, including epilepsy, Parkinson's disease, Alzheimer's dementia and multiple sclerosis.

Controversies on the type and timing of psychiatric evaluations of patients with epilepsy abound. The debate includes questions of whether psychiatric evaluations should be carried out at the time of diagnosis of the seizure disorder, particularly in the pediatric population, given the significant percentage of patients who already present a comorbid psychiatric disorder at the time of the first seizure. The next question pertains to whether psychiatric evaluations should be carried out in every candidate for epilepsy surgery, given that this group of patients has among the highest prevalence rates of psychiatric disorders and are therefore at significant risk of post-surgical psychiatric complications. Yet, this practice is only carried out in a minority of the major American surgical epilepsy centers while the majority of centers instead rely on neuropsychological evaluations. The controversial aspects of such practices are discussed in two chapters of the book.

There is also no lack of controversy regarding the treatment of comorbid psychiatric disorders, whether with psychotropic drugs or psychotherapy, and the impact that these therapies have on seizure occurrence. Three chapters review these important issues. Finally, the psychotropic properties of vagus nerve stimulation and an overview of psychological therapies are presented.

We hope that this book fosters a better understanding of the shared neurobiological mechanisms of epilepsy and related psychiatric conditions, and leads to progress in resolving the most pressing controversial issues in the clinical evaluation and care of patients with epilepsy and comorbid psychiatric disorders.

Andres M. Kanner, M.D.
Steven C. Schachter, M.D.

Contributors

Kenneth Alper Departments of Psychiatry and Neurology, NYU Comprehensive Epilepsy Center, New York University School of Medicine, New York, NY 10016, USA

G. A. Baker Division of Neurological Science, Clinical Sciences Centre for Research and Education, Lower Lane, Liverpool L9 7LJ, UK

John Barry Department of Psychiatry, Stanford University School of Medicine, Palo Alto, CA, USA

David W. Dunn Department of Psychiatry and Neurology, Indiana University School of Medicine, 702 Barnhill Drive, ROC 4300, Indianapolis, IN 46202, USA

Joanne Eatock Division of Neurological Science, Clinical Sciences Centre for Research and Education, Lower Lane, Liverpool L9 7LJ, UK

Eric J. Foltz Department of Neurology, The University of Arizona, 1501 North Campbell Avenue, Room 6205 Tucson, Arizona 85724

Marlis Frey Rush Epilepsy Center, Rush University Medical Center, 1725 W. Harrison Street, Chicago, IL 60612, USA

Frank G. Gilliam The Neurological Institute-7th Floor, Columbia University, 710 West 168th Street, New York, NY 10032, USA

Bruce P. Hermann Matthews Neuropsychology Laboratory, Department of Neurology, Health Sciences Learning Center, University of Wisconsin School of Medicine and Public Health, 750 Highland Avenue Madison, WI 53705, USA

Sean T. Hwang The Neurological Institute-7th Floor, Columbia University, 710 West 168th Street, New York, NY 10032, USA

Jana E. Jones University of Wisconsin School of Medicine and Public Health, Department of Neurology, H4/665 CSC, 600 Highland Ave, Madison, WI 53792-6180, USA

Kousuke Kanemoto Department of Neuropsychiatry, Aichi Medical University, 21 Yazako-Karimata, Nagakute, Aichi 480-1195, Japan

Andres M. Kanner Department of Neurological Sciences and Rush Epilepsy Center, Rush University Medical Center, 1653 West Congress Parkway, Chicago, IL 60612, USA

E. S. Krishnamoorthy The Institute of Neurological Sciences – VHS Hospital, Taramani, Chennai – 600113, India

William G. Kronenberger Section of Psychology, Department of Psychiatry, Indiana University School of Medicine, 702 Barnhill Drive, ROC 4300, Indianapolis, IN 46202, USA

David M. Labiner Associate Head, Department of Neurology, The University of Arizona, 1501 North Campbell Avenue, Room 6205, Tucson, AZ 85724, USA

MaryBeth Lake Northwestern University's Feinberg School of Medicine, Children's Memorial Hospital, 2300 Children's Plaza, PO Box 10, Chicago, IL 60614-3394, USA

David W. Loring Departments of Neurology and Clinical and Health Psychology, University of Florida, L3-100 McKnight Brain Institute, PO Box 100236, Newell Drive, Gainesville, FL 32610-0236, USA

Marco Mula The Neuropsychiatry Research Group, Department of Neurology, Amedeo Avogadro University, C.so Mazzini, 18-28100 Novara, Italy

Tomohiro Oshima Department of Neuropsychiatry, Aichi Medical University, 21 Yazako-Karimata, Nagakute, Aichi 480-1195, Japan

Sigita Plioplys Northwestern University's Feinberg School of Medicine, Children's Memorial Hospital, 2300 Children's Plaza, PO Box 10, Chicago, IL 60614-3363, USA

Markus Reuber Department of Neurology, Royal Hallamshire Hospital, Glossop Road, Sheffield, S10 2JF, UK

Edgar A. Samaniego Department of Neurology, University of Wisconsin-Madison, 600 Highland Ave-H6/574 CSC, Madison, WI 53792-5132, USA

Steven C. Schachter Departments of Neurology Harvard Medical School and Beth Israel Deaconess Medical Center, Kirstein 478, 330 Brookline Avenue, Boston, MA, USA

R. Seethalakshmi The Institute of Neurological Sciences Voluntary Health Services, Mahadev, No. 6, Bhaskarapuram, Mylapore, Chennai, India

Linda M. Selwa Department of Neurology, University of Michigan, 1500 E. Medical Center Drive, 1914 Taubman, Ann Arbor, MI 48109, USA

Raj D. Sheth Department of Neurology, University of Wisconsin-Madison, 600 Highland Ave, H6/574 CSC, Madison, WI 53792-5132, USA

Yukari Tadokoro Department of Neuropsychiatry, Aichi Medical University, 21 Yazako-Karimata, Nagakute, Aichi 480-1195, Japan

José F. Tellez-Zenteno Division of Neurology, Royal University Hospital, 103 Hospital Drive, PO Box 26, Room 1622, Saskatoon, Canada SK S7N OW8

Michael R. Trimble Institute of Neurology, University College London, Queen Square, London WC1N3BG, UK

Samuel Wiebe Division of Neurology, Foothills Medical Centre, 1403 – 29 St NW, Calgary, Alberta, Canada T2N 2T9

Prevalence of Psychiatric Disorders in Patients with Epilepsy: What We Think We Know and What We Know

José F. Tellez-Zenteno and Samuel Wiebe

INTRODUCTION

It is well established that psychiatric conditions coexist with epilepsy. In antiquity and even today in some cultures, the notion of psychiatric comorbidity and the clinical expression of seizures have been viewed as moral and paranormal attributes. The "epileptic personality," a well intended but misguided 20th century concept, bears testimony to the notion of psychiatric comorbidity. In the last two decades, there has been an explosion of efforts to measure the frequency with which psychiatric conditions occur in epilepsy. Most large-scale efforts have focused on the coexistence of these conditions at some point in time (prevalence), rather than on the new onset of psychiatric disorders in epilepsy (incidence). The reasons for this preference are simple. Studies of prevalence are cross-sectional, simpler and cheaper to do than studies of incidence, which require carefully assembled, prospective cohorts.

Readers attempting to make sense of the epidemiological literature on psychiatric comorbidity in epilepsy may find themselves in a maze of conflicting data. As with most situations of between-study variation in clinical research, the reasons for discordant data in psychiatric comorbidity often reduce to differences in definitions, populations, measurement tools and scientific rigor. Using a comprehensive review of the pertinent literature, we aim to shed some light into these differences. The analysis is not methodologically exhaustive, and yet it yields a plethora of data and closely related concepts. This is necessary to illustrate how coherent results can be gleaned from comorbidity studies after accounting for methodological differences.

Originally coined by Feinstein, the term comorbidity is now used to refer to the greater than coincidental association of two conditions in the same individual

(Feinstein, 1970). The psychiatric comorbidities in patients with epilepsy have important clinical and therapeutic implications. Psychiatric disorders associated with epilepsy can precede, cooccur with or follow the diagnosis of epilepsy (Gaitatzis *et al.*, 2004a), which raises important questions about etiology and causation. The most frequent psychiatric diagnoses reported in people with epilepsy include psychoses, neuroses, mood disorders (DSM-IV axis I disorders), personality disorders (DSM-IV axis II disorders) and behavioral problems (Gaitatzis *et al.*, 2004a). Psychiatric symptoms can be classified according to their temporal relationship with seizure occurrence. They can be divided into ictal symptoms (related to the seizure itself) or interictal symptoms (independent of individual seizures).

There are few truly population-based studies evaluating the prevalence of psychiatric disorders in people with epilepsy. Most studies involve selected groups of patients from tertiary centers or specialized clinics. Psychiatric pathology may be overrepresented in selected populations, such as in patients with temporal lobe epilepsy or with refractory seizures. Broad-ranging general population studies using validated psychiatric diagnostic tools are essential to understand the entire spectrum of mental health problems in people suffering from epilepsy.

WHAT IS PREVALENCE?

Prevalence, often expressed as a percentage, refers to the total number of cases of a disease in a given population at a specific time (Hauser and Hesdorffer, 1990). However, prevalence can be measured in different ways, which can explain some of the variation seen in different reports. A very common measure is point prevalence, which refers to the number of people who have the disease in question at a specific time. This measure is useful to describe the burden of the disease on the population and to estimate health care needs (Hauser and Hesdorffer, 1990). Another commonly used measure in psychiatric comorbidity is the lifetime prevalence, which refers to the number of people who ever had the disorder in question divided by the point of midyear population. Because this is a cumulative measure, it is higher than point prevalence. Another commonly used measure in psychiatric comorbidity is the 1-year prevalence, which reflects the number of people who had the disease in question during the previous year (Hauser and Hesdorffer, 1990). Finally, prevalence can be specific to groups of interest, such as age and sex specific prevalence, and these can also be expressed as point, 1-year or lifetime prevalence (Hauser and Hesdorffer, 1990).

WHO HAS A PSYCHIATRIC DISORDER?

Defining "caseness," that is, which individuals have the condition of interest, is extremely important in obtaining accurate estimates of prevalence and incidence in psychiatric comorbidity. The gold standard is the standardized psychiatric interview, which yields DSM IV diagnoses (Manchanda *et al.*, 1996; Davies *et al.*,

2003; Tellez-Zenteno *et al.*, 2005). This is difficult to apply to large populations, and it is more often used in studies of selected populations. A commonly used method, attractive for its simplicity, relies on self-reported symptoms (Kobau *et al.*, 2006) obtained through validated screening questionnaires, for example the Beck Depression Inventory (Grabowska-Grzyb *et al.*, 2006), the Hospital Anxiety and Depression Scale (HADS) (Mensah *et al.*, 2006) and others (Strine *et al.*, 2005). These methods are less accurate, often overestimate the prevalence of psychiatric conditions, and should be explicitly considered as measures of symptom endorsement only. Not all psychiatric symptoms necessarily represent psychiatric disorders. The concept of a disorder requires symptoms to be present, but also imposes additional requirements such as persistence, and associated distress and dysfunction. Some studies have used non-standardized interviews to identify psychiatric disorders. This lack of standardization may under or overestimate prevalence depending on the type of interview, but it often yields higher estimates as compared with studies using standardized interviews (Gudmundsson, 1966). Another method of ascertainment used in recent years relies on coded administrative data using the World Health Organization International Classification of Diseases (ICD) (Gaitatzis *et al.*, 2004b). This method is very attractive because large databases already in existence can be readily accessed and analyzed. However, the sensitivity and specificity of ICD coding is highly variable among psychiatric conditions and also among different studies, due to variations in ICD coding practices.

GENERAL PREVALENCE OF PSYCHIATRIC DISORDERS

The lifetime prevalence of psychiatric disorders in the general population is high, ranging from 41% to 50% (Kessler *et al.*, 1994; Offord *et al.*, 1996; Bijl *et al.*, 1998). On the other hand, the 1-year prevalence ranges between 18.6% and 30% (Kessler *et al.*, 1994; Offord *et al.*, 1996; Bijl *et al.*, 1998).

Few population-based studies have evaluated the prevalence of psychiatric disorders in epilepsy, and different study methods produce variability in results. Table 1.1 shows the studies evaluating psychiatric comorbidity in non-selected populations with epilepsy (general population or practice-based). Three of these studies used structured interviews to obtain DSM-IV diagnoses and found similar prevalence rates of mental health conditions in people with epilepsy, that is, 37% (Davis *et al.*, 2003), 29% (Pond and Bidwell, 1960) and 23.5% (Tellez-Zenteno *et al.*, 2005). These prevalence rates are all higher than in the general population. On the other hand, studies that did not use standardized interviews, such as those by Gudmundsson (1966) and Pond and Bidwell (1960), found even higher prevalence rates of psychiatric diagnoses in people with epilepsy (54%). Prevalence rates in studies using ICD codes from administrative data (Jalava and Sillanpaa, 1996; Bredkjaer *et al.*, 1998; Hackett *et al.*, 1998; Stefansson *et al.*, 1998; Gaitatzis *et al.*, 2004b) are highly variable, reflecting the issues in accuracy described above. For example, Stefansson reported a prevalence of 35.3%, Hackett 23.1%, Gaitatzis 41%, Bredkjaer 16.8% and Jalava 24%. Interestingly

TABLE 1.1 Non-selected populations

Author	N	Ascertainment method of psychiatric conditions	Type of population	Use of controls	Prevalence of psychiatric conditions (%)	Prevalence of depression (%)	Prevalence of anxiety disorder (%)	Prevalence of schizophrenia (%)	Prevalence of psychosis (%)	Prevalence of personality disorders (%)	Prevalence of alcohol dependence (%)
Pond and Bidwell (1960), UK	245 PE	Psychiatric interview not based on DSM-IV	Open population (health survey) Children	Only explored psychiatric comorbidity in patients with epilepsy	29[a]	NS	NS	NS	NS	NS	NS
Gudmundsson (1966), Iceland	654 PE	Interview with physician (no standardized interview)	Open population (health survey) Children and adults	Only explored psychiatric comorbidity in patients with epilepsy	54.5[a]	NE	NE	NE	7[a]	NE	NE
Graham and Rutter (1970), Isle of Wight	63 PE 144 PWE	Psychiatric interview (non-standardized interview)	Open population (PBS) Children	Controlled (healthy children)	28.6/6.8[a]	NE	NE	NE	NE	NE	NE
Edeh and Toone (1987), London	88 PE	CIS	Adults with epilepsy from GP practice	NCOE	48[a]	22[a]	15[a]	1[a]	3.4[a]	2.2[a]	NS
Havlova (1990), Prague	225 PE	Chart review	Cohort of children from a hospital with a neurology program in Prague	NCOE	6.7[c]	NE	NE	NE	NE	NE	NE
Forsgren (1992), Sweden	713 PE	Chart review	Open population (health survey) Adults	NCOE	5.9[a]	NS	NS	0.8	0.7	0.7	NS

Study	Sample	Method	Population	Control							
Jalava and Sillanpaa (1996), Finland	94 PE 199 PWE	Chart review and ICD-9	Adults with epilepsy from different sources	Controlled (random healthy residents of Finland and employee controls of a printing house)	24/0.7[b]	NS	NS	NS	3.1/0[b]	NS	NS
Bredkjaer et al. (1998), Denmark	67,116 PWE	ICD-8	National patient register	NCOE	16.8	NE	NE	NE	NE	NE	NE
Stefansson et al. (1998)	241 PE 482 PWE	ICD-9	Adult patients receiving disability benefits	Controlled (same list as PWE)	35.3/29.7[a]	NE	NE	1.2/0.4[a]	6.2/2.3[a]	18.3/21[a]	5/2.3[a]
Hackett et al. (1998), India	26 PE 1377 PWE	ICD-10	Open population (health survey) Adults	Controlled (healthy adults)	23.1/8.1[a]	NE	NE	NE	NE	NE	NE
Davies et al., 2003), UK	67 PE 10249 PWE	Psychiatric interview based on DSM-IV criteria	Open population (health survey) Children	Controlled (non-epileptic patients)	37/9[a]	NE	NE	NE	NE	NE	NE
Ettinger et al. (2004), USA	775 PE	CES-D	Open population (health survey) Adults	All had epilepsy	NE	36.5[a]	NE	NE	NE	NE	NE
Gaitatzis et al. (2004a, b), UK	5834 PE	ICD-9	Patients from a database generated from GP. All age groups	Controlled (patients without epilepsy)	41 in PE[a]	18.2/9.2[a]	11.1/5.6[a]	0.7/0.1a	9/2[a]	NE	2.4/0.4[a]
Strine et al. (2005), USA	427 PE 30018 PWE	Kessler 6 scale (depression and anxiety symptoms)	Open population (Health survey) Adults	Controlled (non-epileptic patients)	NE	32.6/15.5[a]	14.4/6.8[a]	NE	NE	NE	NE

(Continued)

TABLE 1.1 (*Continued*)

Author	N	Ascertainment method of psychiatric conditions	Type of population	Use of controls	Prevalence of psychiatric conditions (%)	Prevalence of depression (%)	Prevalence of anxiety disorder (%)	Prevalence of schizophrenia (%)	Prevalence of psychosis (%)	Prevalence of personality disorders (%)	Prevalence of alcohol dependence (%)
Kobau et al. (2006), USA	131 PE 4154 PWE	Self reported prevalence	Open population (Health survey) Children and adults	Controlled (non-epileptic patients)	NE	39/15[a]	39/15[a]	NE	NE	NE	NE
Mensah et al. (2006), UK	499 PE	HADS	Adults with epilepsy from GP practice	NCOE	NS	11.2[a]	NS	NS	NS	NS	NS
Our study	253 PE, 36,984 PWE	CIDI	Open population (health survey) Adults	Controlled (non-epileptic patients)	23.5/10.9[d]	17.4/10.8[d]	12.8/4.7[d]	NE	NE	NE	NE

PE, patients with epilepsy; PWE, patients without epilepsy; GP, general practitioners; PBS, population–based study; NCOE, non–controlled, only patients with epilepsy; NS, not stated; NE, not examined. [a]Point prevalence; [b]Prevalence during a follow–up of 35 years; [c]Lifetime prevalence; [d]12–month prevalence; [e]Combined anxiety and depression symptoms.

the study of Havlova (1990), where a review of charts was done (similar principle to ICD), the prevalence of psychiatric disorders was lower than studies using ICD codes. Studies using general practitioners' registries such as those by Edeh and Toone (1987) and Gaitatzis et al. (2004b) report higher prevalence rates of psychiatric conditions (see Table 1.1). The same effect is observed in the study by Stefansson et al. (1998), based on a list of patients with disability. The prevalence of psychiatric conditions in these types of studies could be higher compared with other methodologies because they are biased toward individuals seeking medical attention, for example, sicker populations. A study of 88 adult patients with epilepsy from general practices in the South of London reported a prevalence of psychosis of 4%. The ascertainment method in this study was performed using the clinical interview schedule (Edeh and Toone, 1987).

Studies of selected populations report considerably higher prevalence rates. Perini et al. (1996) performed a controlled study including patients with epilepsy, juvenile myoclonic epilepsy and diabetes; the corresponding rates of psychiatric disorders were 80%, 22% and 10%, respectively. Using standardized instruments, Silberman et al. (1994) evaluated 21 patients at an epilepsy center in the United States, and found a prevalence of psychiatric disorders of 71%. In candidates for epilepsy surgery, and using DSM-III criteria to identify psychiatric diagnoses, Manchanda et al. (1996) found psychiatric disorders in 47.3% of patients. In a study of patients with temporal lobe epilepsy and generalized epilepsy, Shukla et al. (1979) reported a prevalence of psychiatric disorders of 79% and 47%, respectively. Glosser et al. (2000) explored psychiatric disorders in patients with temporal lobe epilepsy who had epilepsy surgery; the prevalence was 65% before and after the resection. Blumer et al. (1998) explored psychiatric comorbidity before and after temporal resections and the prevalence was 57% and 39%, respectively. Finally, Wrench et al. (2004) evaluated the prevalence of psychiatric disorders before and after the surgery in 43 temporal and 17 extratemporal cases; the corresponding prevalences were 54% and 65% before the surgery and 54% and 33% after the surgery.

In summary, psychiatric comorbidity in epilepsy is very high regardless of the method of assessment. The prevalence of psychiatric conditions in people with epilepsy is at least double in studies of selected epilepsy populations as compared with those in non-selected and in the general population. Important variations among studies probably explain most of this variation, in particular the method of ascertainment and the study population.

SPECIFIC PSYCHIATRIC COMORBIDITIES

Depression

Using DSM-IV criteria in the general population, the lifetime prevalence of depression ranges from 12.2% to 16.2%, and the 1-year prevalence ranges from

4.8% to 5.2% (Kessler *et al.*, 2003; Hasin *et al.*, 2005; Patten *et al.*, 2006). Mood disorders are the most common psychiatric condition found in people with epilepsy.

Studies in non-selected populations of people with epilepsy show a higher prevalence of depression than that in the general population. The lifetime prevalence of depression among people with epilepsy in the Canadian general population was estimated at 18.2%, using the Composite International Interview for psychiatric diagnoses, which yields DSM-IV diagnoses (Tellez-Zenteno *et al.*, 2005). The reported point prevalence of depression varies considerably among studies involving non-selected populations. Using the Center for Epidemiology Studies-Depression Scale (CES-D), Ettinger *et al.* (2004) found a point prevalence of 36.5% in people with epilepsy in the community. Mensah *et al.* (2006) studied adult patients with epilepsy from a list of general practitioners. Using the HADS, the prevalence rate was 11.2%. Edeh (Edeh and Toone, 1987) used the CIS (Clinical Interview Scale) in a sample of adults attending with general practitioners, and found a prevalence of 22%. Finally, Jacoby *et al.* (1996) used the HADS in people with epilepsy from primary physician practices and found a prevalence of 9%. Importantly, these studies refer to self-reported symptoms of depression, not to diagnosed depression. Using other methodologies in non-selected epilepsy populations, the rates of depression are consistently higher than those in the general population. For example, Gaitatzis *et al.* (2004b) found a prevalence of depression of 18.2% using the ICD-9 from a data base of general practitioners. Other researchers report similar findings (Strine *et al.*, 2005; Tellez-Zenteno *et al.*, 2005; Kobau *et al.*, 2006).

Several studies have assessed the prevalence of depression in selected epilepsy populations, mostly in temporal lobe epilepsy. The prevalence of depression in these studies is much higher than that in studies of non-selected epilepsy populations. Grabowska-Grzyb *et al.* (2006) found a prevalence of depression of 49.2% in 203 patients with intractable epilepsy, ascertained with the Beck Depression Inventory and the Hamilton Depression Rating Scale. Wrench *et al.* (2004) evaluated 43 patients with temporal lobe epilepsy and 17 with extratemporal epilepsy before and after surgery. Depression was diagnosed in 33% and 53%, respectively, using the Austin CEP interview before surgery. After surgery the corresponding prevalences were 30% and 17%. Victoroff *et al.* (1994) found lifetime and point prevalence of depression of 62% and 30%, respectively, using structured interviews in 53 patients with intractable complex partial seizures. Using similar methodology in patients with temporal lobe epilepsy, Briellmann *et al.* (2007) and Ring *et al.* (1998) found prevalence rates of 44% and 45%.

In summary, depression represents the most frequent interictal psychiatric condition in patients with epilepsy (Kanner, 2005; Schmitz, 2005). The rates are highest in selected populations with intractable epilepsy (about 40–60%), and intermediate in people with epilepsy in the general population (about 20%). These rates are much higher than those reported in the general population (12–16%). The timely identification and treatment of depression in epilepsy is increasingly recognized as an area requiring attention (Gilliam *et al.*, 2006).

Anxiety

The lifetime prevalence of anxiety disorder in the general population in large epidemiological studies ranges from 1.9% to 5.1% (Weissman and Merikangas, 1986; Wittchen et al., 1994; Hunt et al., 2002; Wittchen, 2002; Wober-Bingol et al., 2004; Lieb et al., 2005).

The prevalence of anxiety disorders in people with epilepsy in general population studies using ICD and DSM-IV case ascertainment ranges from 11% to 15%. Gaitatzis et al. (2004b) obtained a point prevalence of 11.1% in the National Practice Study, using ICD coded data. Tellez-Zenteno et al. (2005) reported a lifetime prevalence of 12.8% in a Canadian general population study using structured psychiatric interviews (Composite International Diagnostic Interview (CIDI)). Edeh reported a point prevalence of 15% in a register of patients with epilepsy from general practitioners in the United Kingdom. Strine et al. (2005) reported a point prevalence of 14.4% from a health survey where anxiety was ascertained with the Kessler 6 scale. On the other hand, studies using self-reported methods report higher prevalence rates, for example, 39% in Kobau et al.'s (2006) study.

Devinsky et al. (2005) studied 360 patients with refractory epilepsy before temporal lobe surgery. Using the Beck Anxiety Inventory and the CIDI the rate of anxiety disorders was 24.7%. This study followed patients up to 24 months and the postoperative rates of anxiety and depression improved after surgery. In Manchanda et al.'s (1996) study of 300 consecutive candidates for epilepsy surgery, 10.7% fulfilled DSM-IV criteria for anxiety disorders. A small pediatric study of epilepsy from outpatient clinics found depression in 16% of patients using the Revised Children's Manifest Scale (Ettinger et al., 1998). Another pediatric study evaluated the prevalence of anxiety in 171 patients with epilepsy compared with 93 healthy children; the rates were 33% vs. 6%, respectively (Caplan et al., 2005). Finally, the rate of anxiety disorders among patients with intractable epilepsy seems higher in those with temporal lobe epilepsy (23%) than in those with extratemporal epilepsy (18%), and surgery did not seem to lower the rates substantially in either (24% and 17%, respectively) (Wrench et al., 2004).

The prevalence of anxiety disorders in epilepsy follows the pattern seen in other psychiatric comorbidities. It is higher in selected populations of intractable epilepsy (twofold higher than in non-selected epilepsy populations). Interestingly the overall rates of depression and anxiety are not modified in patients with intractable epilepsy that underwent epilepsy surgery.

Personality disorders

The prevalence of personality disorders in the general population ranges between 6% and 13% (Cohen et al., 1994; Samuels et al., 1994; Jackson and Burgess, 2000; Torgersen et al., 2001; Lenzenweger et al., 2007).

Few epidemiological studies of personality disorders have been done in non-selected populations of people with epilepsy. Forsgren (1992) found a prevalence of 0.7% in a cohort of 713 patients with epilepsy from Sweden (see Table 1.1). In patients from general practices assessed with the Clinical Interview Schedule in London, Edeh and Toone (1987) reported a prevalence of 2.2%. Finally, population-based studies using the ICD as a method of ascertainment found a prevalence of 18.3% in people with epilepsy (Stefansson et al., 1998). The broad difference between the first two studies and the last could be explained by the method of ascertainment, or by the study population. For example, Stefansson et al. (1998) could have overestimated the prevalence because they focused on a population receiving disability benefits.

The rates of personality disorders in selected populations differ broadly from the rates found in non-selected populations. Perini et al. (1996) found a prevalence of 40% evaluating 20 patients with temporal lobe epilepsy using structured interviews. Manchanda et al. (1996) found DSM-IV diagnoses of personality disorders in 18%. Lopez-Rodriguez et al. (1999) evaluated 52 patients with medically refractory epilepsy using the DSM-III criteria and found a prevalence of 21%. These and other studies have shown that the prevalence could range from 18% to 40% in populations of patients with refractory epilepsy (Gaitatzis et al., 2004a; Swinkels et al., 2005). The specific personality disorders that have been described in people with epilepsy include antisocial personality (Koch-Weser et al., 1988), avoidant (Lopez-Rodriguez et al., 1999), obsessive-compulsive (Monaco et al., 2005), schizoid (Naylor et al., 1994), schizotypal (Victoroff et al., 1994), and dependent and dissociative (Naylor et al., 1994).

Interestingly some studies have shown that epilepsy surgery does not modify the prevalence of personality disorders in patients with intractable epilepsy (Glosser et al., 2000; Wrench et al., 2004).

Taken as a whole the prevalence of personality disorders in non-selected populations of people with epilepsy is close to that in the general population, but the rates in selected epilepsy populations (patients with refractory epilepsy with over-representation of temporal lobe cases) is almost twice as high.

Psychosis

The prevalence of psychosis in the general population ranges between 1% and 2% (Johns and van Os, 2001). People with epilepsy are at risk to develop psychosis, which may be ictal, postictal or chronic interictal (Olfson et al., 1996; van Os et al., 2001).

The prevalence of interictal psychosis in non-selected epilepsy populations studies varies from 3.1% to 9% (Edeh and Toone, 1987; Forsgren, 1992; Jalava and Sillanpaa, 1996; Stefansson et al., 1998; Gaitatzis et al., 2004b). In a population-based study performed in Iceland using a non-standardized interview, the prevalence of psychosis was 6–9% for males and females, respectively (Gudmundsson, 1966). Three population or primary practice-based studies using ICD coded data have found a prevalence of psychosis two to three times higher than in the general population, but less than 10%. Gaitatzis et al. (2004b) reported a prevalence of 9%, Stefansson et al. (1998) of 6.2%, and Jalava and Sillanpaa (1996) of 3.1%. Using ICD-8 coded data, a national Danish study evaluated the new onset of psychiatric disorders in a cohort of patients with epilepsy in a 15-year follow-up period after the diagnosis of epilepsy (Bredkjaer et al., 1998). The standardized incidence ratio of non-organic affective psychosis developing after a diagnosis of epilepsy was 3 (Bredkjaer et al., 1998). That is, people with epilepsy were three times more likely to develop psychosis than controls. Finally, Forsgren (1992) performed a practice-based study in Sweden reviewing charts of 713 patients. The prevalence of psychosis was 0.7%. This outstanding low prevalence in the category of population or practice-based studies may relate to a low sensitivity of their case-finding method (see Table 1.1).

The prevalence of interictal psychosis in selected epilepsy populations is higher than in studies of non-selected populations cited above. In studies of patients with temporal lobe or refractory epilepsy, the prevalence varies from 10% to 19%, which is almost double compared with non-selected populations (Taylor, 1972; Lindsay et al., 1979; Shukla et al., 1979; Sherwin et al., 1982; Victoroff et al., 1994; Umbricht et al., 1995). Finally, the prevalence of psychosis in temporal lobe epilepsy is reportedly higher (19%) compared with that in primary generalized epilepsy (10%) (Shukla et al., 2003).

A common question in clinical practice pertains to the risk of psychosis in patients who had epilepsy surgery. The reports are variable, but all are in the range of 1–8% (Kanner, 2000). For example, Ney et al. (1998) reported an incidence rate of 3%, Manchanda et al. (1993) of 1.3%, Leinonen et al. (1994) of 8.8%, Koch-Weser et al. (1988) of 8% and Christodoulou et al. (2002) of 1%. Most of these data refer to cumulative, rather than annual, incidence. This is important, because any events occurring after a patient has surgery automatically become "postsurgical," and the longer patients are followed, the more likely comorbidities are to occur (Ferguson et al., 1993). Therefore, a clear statement of the period of follow-up, and the adjusted annual incidence are crucial to make sense of these data. The majority of the reported studies have been done in temporal lobe epilepsy. There is a tendency for a higher frequency in patients with right temporal lobe epilepsy (Manchanda et al., 1993), and in those with foreign tissue lesions (Taylor, 1975).

Finally, there are not many epidemiological studies evaluating ictal psychosis, considered a rare phenomenon (Kanner, 2000). This form of psychosis usually is an expression of non-convulsive status epilepticus, including simple partial status, complex partial status and absence status (Wolf and Trimble, 1985; Kanner, 2000).

As observed with personality disorders, depression, and psychiatric disorders in general, the prevalence of interictal psychosis is considerably higher than that of psychosis in the general population.

Suicide in epilepsy

Approximately 1.1–4.6% of the general population will make a suicide attempt during their lifetime (Kessler *et al.*, 1999; Weissman *et al.*, 1999). Suicide in epilepsy is a major cause of death and it is usually underestimated (Tomson *et al.*, 2004). Suicide as the primary outcome is rarely the main focus of studies and usually is a secondary outcome reported in studies evaluating mortality. There is a great variation of the reported rate of suicide, from 0% to 33% in some studies (Hitiris *et al.*, 2007; Jones *et al.*, 2003). The lifetime prevalence of suicide in people with epilepsy has been estimated in large population-based studies and is between 1.1% and 1.2% (Jones *et al.*, 2003). Recent meta-analyses have compared the rates of suicide in the general population and in people with epilepsy. In a meta-analysis of 29 cohorts the mean number of suicides in epilepsy was 112 ± 172 $(0 - 833)$ per 100,000, compared with 13.2 ± 6 $(0.8 - 25.9)$ in the general population (Pompili *et al.*, 2005). Yet, the rates were highly variable and some studies even showed lower rates of suicide in people with epilepsy. The study concludes that the rate of suicide is increased in epilepsy (Pompili *et al.*, 2005). A subsequent meta-analysis calculated the expected rate of suicide and deaths in people with epilepsy using 30 available cohorts of mortality in people with epilepsy (Pompili *et al.*, 2006a). This meta-analysis concludes that the rate of suicides in epilepsy is under-reported in cohorts studying mortality considering that the expected rates of suicide are similar to the overall rate of mortality in people with epilepsy (Pompili *et al.*, 2006a). Finally, a third meta-analysis compared the rate of suicide in cohorts of patients who had epilepsy surgery (Pompili *et al.*, 2006b) with the general population. This study included 11 cohorts, comprising 2,425 patients, 24 of whom committed suicide, and concludes that the rate of suicide is high in people with epilepsy surgery compared with the general population (Pompili *et al.*, 2006b). This meta-analysis suggests that surgery does not normalize the increased mortality related to epilepsy, including suicide, as other authors have suggested (Guldvog *et al.*, 1991; Hennessy *et al.*, 1999). Some studies have shown that epilepsy surgery can decrease mortality, especially in seizure free patients (Salanova *et al.*, 2002), but this has not been proven in other studies (Ryvlin and Kahane, 2003), and the lack of studies of mortality with long-term outcome precludes strong assertions (Tellez-Zenteno *et al.*, 2007).

Illicit and abused drugs

In a US national household survey on drug abuse (NHSDA) performed in 1994, 10% of people older than 12 years reported non-medical use of prescription drugs, 38% had used an illicit drug, and 85% had drunk an alcoholic beverage in their lifetime (Rouse, 1996). Little systematic information exists in this regard in patients with epilepsy (Devinsky, 2003). One of the current debates pertains to the therapeutic use of marijuana in patients with seizures. A survey performed in a tertiary center in Canada showed that 21% of patients had smoked marijuana in the last year in order to decrease the frequency of seizures, the majority of whom reported benefits. Further, 24% of the entire sample believed it to be beneficial (Gross et al., 2004). The effects of marijuana on seizure activity are not fully understood, and some studies suggest some beneficial effect on seizures (Consroe et al., 1975; Wada et al., 1975; Gross et al., 2004), but other reports state the opposite (Brust et al., 1992; Tilelli and Spack, 2006). There is no epidemiological information regarding the use of cocaine in patients with epilepsy, and this is an area requiring further investigation.

The US national probability survey of drug use (NHSDA) explored the prevalence of alcohol dependence in the general population using the DSM-IV criteria. Among current male drinkers, 4.6% of adolescents (aged 12–17 years) met criteria for the past year dependence, and the rate increased to 8.5% in the 18–23 age group. Approximately, 5% of the population reported heavy drinking, defined as 5 or more drinks per day for at least 5 days a month (Rouse, 1996). Currently, it is not clear if alcohol dependence is more frequent in patients with epilepsy and the lack of studies precludes strong conclusions. In a review performed by Chan (1985), the prevalence of alcoholism ranged from 12% to 36% in patients with seizures. The validity of this report is limited by a lack of distinction in the majority of the studies between alcohol withdrawal seizures and epilepsy. This is likely to result in an overestimation of the comorbidity of alcoholism and epilepsy (as opposed to seizures). A more recent study compared the rates of alcohol consumption using the Alcohol Use Disorders Identification Test in different neurological conditions. The prevalence was 18% in patients with stroke, 12% in epilepsy outpatients and 13% in healthy controls (Brathen et al., 2000). This supports the notion that alcohol use is overestimated in epilepsy, and is likely due to the common occurrence of alcohol-related acute provoked seizures (Rathlev et al., 2006).

In summary, after accounting for major methodological differences, a relatively coherent picture of psychiatric comorbidity in epilepsy emerges. Yet, many studies lack appropriate definitions and descriptions, and there is no standardization of study methods. When referring to the frequency of psychiatric conditions in epilepsy, one must look for explanatory qualifiers describing at the very least the type of patient population (e.g., select groups, general population), the type of measure (e.g., type of prevalence or incidence), how cases were ascertained (e.g., structured interview, coded data, self-reports), and how accurate the results are (e.g., confidence intervals). One must also resist the temptation to report or interpret cases of positive self-reports in screening instruments as actually diagnosed psychiatric conditions. Finally, a mere statement about coexistence of medical conditions does little to further our understanding of cause and effect, and to help us develop interventions to modify risk factors and therapies. This is the next frontier.

REFERENCES

Bijl, R. V., Ravelli, A., van Zessen, G. (1998). Prevalence of psychiatric disorder in the general population: results of The Netherlands Mental Health Survey and Incidence Study (NEMESIS). *Soc Psychiatry Psychiatr Epidemiol* **33**, 587–595.

Blumer, D., Wakhlu, S., Davies, K., Hermann, B. (1998). Psychiatric outcome of temporal lobectomy for epilepsy: incidence and treatment of psychiatric complications. *Epilepsia* **39**, 478–486.

Brathen, G., Brodtkorb, E., Sand, T., Helde, G., Bovim, G. (2000). Weekday distribution of alcohol consumption in Norway: influence on the occurrence of epileptic seizures and stroke? *Eur J Neurol* **7**, 413–421.

Bredkjaer, S. R., Mortensen, P. B., Parnas, J. (1998). Epilepsy and non-organic non-affective psychosis. National epidemiologic study. *Br J Psychiatry* **172**, 235–238.

Briellmann, R. S., Hopwood, M. J., Jackson, G. D. (2007). Major depression in temporal lobe epilepsy with hippocampal sclerosis: clinical and imaging correlates. *J Neurol Neurosurg Psychiatry* **78**, 1226–1230.

Brust, J. C., Ng, S. K., Hauser, A. W., Susser, M. (1992). Marijuana use and the risk of new onset seizures. *Trans Am Clin Climatol Assoc* **103**, 176–181.

Caplan, R., Siddarth, P., Gurbani, S., Hanson, R., Sankar, R., Shields, W. D. (2005). Depression and anxiety disorders in pediatric epilepsy. *Epilepsia* **46**, 720–730.

Chan, A. W. (1985). Alcoholism and epilepsy. *Epilepsia* **26**, 323–333.

Christodoulou, C., Koutroumanidis, M., Hennessy, M. J., Elwes, R. D., Polkey, C. E., Toone, B. K. (2002). Postictal psychosis after temporal lobectomy. *Neurology* **59**, 1432–1435.

Cohen, B. J., Nestadt, G., Samuels, J. F., Romanoski, A. J., McHugh, P. R., Rabins, P. V. (1994). Personality disorder in later life: a community study. *Br J Psychiatry* **165**, 493–499.

Consroe, P. F., Wood, G. C., Buchsbaum, H. (1975). Anticonvulsant nature of marijuana smoking. *JAMA* **234**, 306–307.

Davies, S., Heyman, I., Goodman, R. (2003). A population survey of mental health problems in children with epilepsy. *Dev Med Child Neurol* **45**, 292–295.

Devinsky, O. (2003). Psychiatric comorbidity in patients with epilepsy: implications for diagnosis and treatment. *Epilepsy Behav* **4**(Suppl 4), S2–10.

Devinsky, O., Barr, W. B., Vickrey, B. G., Berg, A. T., Bazil, C. W., Pacia, S. V., Langfitt, J. T., Walczak, T. S., Sperling, M. R., Shinnar, S., Spencer, S. S. (2005). Changes in depression and anxiety after resective surgery for epilepsy. *Neurology* **65**, 1744–1749.

Edeh, J., Toone, B. (1987). Relationship between interictal psychopathology and the type of epilepsy. Results of a survey in general practice. *Br J Psychiatry* **151**, 95–101.

Ettinger, A. B., Weisbrot, D. M., Nolan, E. E., Gadow, K. D., Vitale, S. A., Andriola, M. R., Lenn, N. J., Novak, G. P., Hermann, B. P. (1998). Symptoms of depression and anxiety in pediatric epilepsy patients. *Epilepsia* **39**, 595–599.

Ettinger, A., Reed, M., Cramer, J. (2004). Depression and comorbidity in community-based patients with epilepsy or asthma. *Neurology* **63**, 1008–1014.

Feinstein, A. R. (1970). The pretherapeutic classification of comorbidity in chronic disease. *J Chronic Dis* **23**, 455–468.

Ferguson, S. M., Rayport, M., Blumer, D. P., Fenwick, F. B. C., Taylor, D. C. (1993). Postoperative psychiatric changes. In *Surgical treatment of the epilepsies* (J. J. Engel, ed.), pp. 649–661. New York: Raven Press, Ltd.

Forsgren, L. (1992). Prevalence of epilepsy in adults in northern Sweden. *Epilepsia* **33**, 450–458.

Gaitatzis, A., Trimble, M. R., Sander, J. W. (2004a). The psychiatric comorbidity of epilepsy. *Acta Neurol Scand* **110**, 207–220.

Gaitatzis, A., Carroll, K., Majeed, A., Sander, W. (2004b). The epidemiology of the comorbidity of epilepsy in the general population. *Epilepsia* **45**, 1613–1622.

Gilliam, F. G., Barry, J. J., Hermann, B. P., Meador, K. J., Vahle, V., Kanner, A. M. (2006). Rapid detection of major depression in epilepsy: a multicentre study. *Lancet Neurol* **5**, 399–405.

Glosser, G., Zwil, A. S., Glosser, D. S., O'Connor, M. J., Sperling, M. R. (2000). Psychiatric aspects of temporal lobe epilepsy before and after anterior temporal lobectomy. *J Neurol Neurosurg Psychiatry* **68**, 53–58.

Grabowska-Grzyb, A., Jedrzejczak, J., Naganska, E., Fiszer, U. (2006). Risk factors for depression in patients with epilepsy. *Epilepsy Behav* **8**, 411–417.

Graham, P., Rutter, M. (1970). Organic brain dysfunction and child psychiatric disorder. *Br Med J* **3**(5620), 695–700.

Gross, D. W., Hamm, J., Ashworth, N. L., Quigley, D. (2004). Marijuana use and epilepsy: prevalence in patients of a tertiary care epilepsy center. *Neurology* **62**, 2095–2097.

Gudmundsson, G. (1966). Epilepsy in Iceland. A clinical and epidemiological investigation. *Acta Neurol Scand* **43**(Suppl 124).

Guldvog, B., Loyning, Y., Hauglie-Hanssen, E., Flood, S., Bjornaes, H. (1991). Surgical versus medical treatment for epilepsy. I. Outcome related to survival, seizures, and neurologic deficit. *Epilepsia* **32**, 375–388.

Hackett, R., Hackett, L., Bhakta, P. (1998). Psychiatric disorder and cognitive function in children with epilepsy in Kerala, South India. *Seizure* **7**, 321–324.

Hasin, D. S., Goodwin, R. D., Stinson, F. S., Grant, B. F. (2005). Epidemiology of major depressive disorder: results from the National Epidemiologic Survey on Alcoholism and Related Conditions. *Arch Gen Psychiatry* **62**, 1097–1106.

Hauser, W. A., Hesdorffer, D. C. (1990). Incidence and prevalence. In *Epilepsy: frequency, causes and consequences* (W. A. Hauser, D. C. Hesdorffer, eds), pp. 1–51. New York: Demos.

Havlová, M. (1990). Prognosis in childhood epilepsy. *Acta Univ Carol Med Monogr* **135**, 1–105.

Hennessy, M. J., Langan, Y., Elwes, R. D., Binnie, C. D., Polkey, C. E., Nashef, L. (1999). A study of mortality after temporal lobe epilepsy surgery. *Neurology* **53**, 1276–1283.

Hitiris, N., Mohanraj, R., Norrie, J., Brodie, M. J. (2007). Mortality in epilepsy. *Epilepsy Behav* **10**(3), 363–376.

Hunt, C., Issakidis, C., Andrews, G. (2002). DSM-IV generalized anxiety disorder in the Australian National Survey of Mental Health and Well-Being. *Psychol Med* **32**, 649–659.

Jackson, H. J., Burgess, P. M. (2000). Personality disorders in the community: a report from the Australian National Survey of Mental Health and Well-Being. *Soc Psychiatry Psychiatr Epidemiol* **35**, 531–538.

Jacoby, A., Baker, G. A., Steen, N., Potts, P., Chadwick, D. W. (1996). The clinical course of epilepsy and its psychosocial correlates: findings from a UK community study. *Epilepsia* **37**, 148–161.

Jalava, M., Sillanpaa, M. (1996). Concurrent illnesses in adults with childhood-onset epilepsy: a population-based 35-year follow-up study. *Epilepsia* **37**, 1155–1163.

Johns, L. C., van Os, J. (2001). The continuity of psychotic experiences in the general population. *Clin Psychol Rev* **21**, 1125–1141.

Jones, J. E., Hermann, B. P., Barry, J. J., Gilliam, F. G., Kanner, A. M., Meador, K. J. (2003). Rates and risk factors for suicide, suicidal ideation, and suicide attempts in chronic epilepsy. *Epilepsy Behav* **4**(Suppl 3), S31–S38.

Kanner, A. M. (2000). Psychosis of epilepsy: A neurologist's perspective. *Epilepsy Behav* **1**, 219–227.

Kanner, A. M. (2005). Depression in epilepsy: a neurobiologic perspective. *Epilepsy Curr* **5**, 21–27.

Kessler, R. C., McGonagle, K. A., Zhao, S., Nelson, C. B., Hughes, M., Eshleman, S., Wittchen, H. U., Kendler, K. S. (1994). Lifetime and 12-month prevalence of DSM-III-R psychiatric disorders in the United States. Results from the National Comorbidity Survey. *Arch Gen Psychiatry* **51**, 8–19.

Kessler, R. C., Borges, G., Walters, E. E. (1999). Prevalence of and risk factors for lifetime suicide attempts in the National Comorbidity Survey. *Arch Gen Psychiatry* **56**, 617–626.

Kessler, R. C., Berglund, P., Demler, O., Jin, R., Koretz, D., Merikangas, K. R., Rush, A. J., Walters, E. E., Wang, P. S. (2003). The epidemiology of major depressive disorder: results from the National Comorbidity Survey Replication (NCS-R). *JAMA* **289**, 3095–3105.

Kobau, R., Gilliam, F., Thurman, D. J. (2006). Prevalence of self-reported epilepsy or seizure disorder and its associations with self-reported depression and anxiety: results from the 2004 HealthStyles Survey. *Epilepsia* **47**(11), 1915–1921.

Koch-Weser, M., Garron, D. C., Gilley, D. W., Bergen, D., Bleck, T. P., Morrell, F., Ristanovic, R., Whisler, W. W., Jr. (1988). Prevalence of psychologic disorders after surgical treatment of seizures. *Arch Neurol* **45**, 1308–1311.

Leinonen, E., Tuunainen, A., Lepola, U. (1994). Postoperative psychoses in epileptic patients after temporal lobectomy. *Acta Neurol Scand* **90**, 394–399.

Lenzenweger, M. F., Lane, M. C., Loranger, A. W., Kessler, R. C. (2007). *DSM-IV* personality disorders in the national comorbidity survey replication. *Biol Psychiatry* **62**(6), 553–564.

Lieb, R., Becker, E., Altamira, C. (2005). The epidemiology of generalized anxiety disorder in Europe. *Eur Neuropsychopharmacol* **15**, 445–452.

Lindsay, J., Ounsted, C., Richards, P. (1979). Long-term outcome in children with temporal lobe seizures. III: Psychiatric aspects in childhood and adult life. *Dev Med Child Neurol* **21**, 630–636.

Lopez-Rodriguez, F., Altshuler, L., Kay, J., Delarhim, S., Mendez, M., Engel, J., Jr. (1999). Personality disorders among medically refractory epileptic patients. *J Neuropsychiatry Clin Neurosci* **11**, 464–469.

Manchanda, R., Miller, H., McLachlan, R. S. (1993). Post-ictal psychosis after right temporal lobectomy. *J Neurol Neurosurg Psychiatry* **56**, 277–279.

Manchanda, R., Schaefer, B., McLachlan, R. S., Blume, W. T., Wiebe, S., Girvin, J. P., Parrent, A., Derry, P. A. (1996). Psychiatric disorders in candidates for surgery for epilepsy. *J Neurol Neurosurg Psychiatry* **61**, 82–89.

Mensah, S. A., Beavis, J. M., Thapar, A. K., Kerr, M. (2006). The presence and clinical implications of depression in a community population of adults with epilepsy. *Epilepsy Behav* **8**, 213–219.

Monaco, F., Cavanna, A., Magli, E., Barbagli, D., Collimedaglia, L., Cantello, R., Mula, M. (2005). Obsessionality, obsessive-compulsive disorder, and temporal lobe epilepsy. *Epilepsy Behav* **7**(3), 491–496.

Naylor, A. S., Rogvi-Hansen, B., Kessing, L., Kruse-Larsen, C. (1994). Psychiatric morbidity after surgery for epilepsy: short term follow up of patients undergoing amygdalohippocampectomy. *J Neurol Neurosurg Psychiatry* **57**, 1375–1381.

Ney, G. C., Barr, W. B., Napolitano, C., Decker, R., Schaul, N. (1998). New-onset psychogenic seizures after surgery for epilepsy. *Arch Neurol* **55**, 726–730.

Offord, D. R., Boyle, M. H., Campbell, D., Goering, P., Lin, E., Wong, M., Racine, Y. A. (1996). One-year prevalence of psychiatric disorder in Ontarians 15 to 64 years of age. *Can J Psychiatry* **41**, 559–563.

Olfson, M., Weissman, M. M., Leon, A. C., Farber, L., Sheehan, D. V. (1996). Psychotic symptoms in primary care. *J Fam Pract* **43**, 481–488.

Patten, S. B., Wang, J. L., Williams, J. V., Currie, S., Beck, C. A., Maxwell, C. J., El Guebaly, N. (2006). Descriptive epidemiology of major depression in Canada. *Can J Psychiatry* **51**, 84–90.

Perini, G. I., Tosin, C., Carrazo, C., Bernasconi, G., Canevini, M. P., Canger, R., Pellegrini, A., Testa, G. (1996). Interictal mood and personality disorders in temporal lobe epilepsy and juvenile myoclonic epilepsy. *J Neurol Neurosurg Psychiatry* **61**, 601–605.

Pompili, M., Girardi, P., Ruberto, A., Tatarelli, R. (2005). Suicide in the epilepsies: a meta-analytic investigation of 29 cohorts. *Epilepsy Behav* **7**, 305–310.

Pompili, M., Girardi, P., Tatarelli, R. (2006a). Death from suicide versus mortality from epilepsy in the epilepsies: a meta-analysis. *Epilepsy Behav* **9**, 641–648.

Pompili, M., Girardi, P., Tatarelli, G., Angeletti, G., Tatarelli, R. (2006b). Suicide after surgical treatment in patients with epilepsy: a meta-analytic investigation. *Psychol Rep* **98**, 323–338.

Pond, D. A., Bidwell, B. H. (1960). A survey of epilepsy in fourteen general practices. II. Social and psychological aspects. *Epilepsia* **1**, 285–299.

Rathlev, N. K., Ulrich, A. S., Delanty, N., D'Onofrio, G. (2006). Alcohol-related seizures. *J Emerg Med* **31**, 157–163.

Ring, H. A., Moriarty, J., Trimble, M. R. (1998). A prospective study of the early postsurgical psychiatric associations of epilepsy surgery. *J Neurol Neurosurg Psychiatry* **64**, 601–604.

Rouse, B. A. (1996). Epidemiology of illicit and abused drugs in the general population, emergency department drug-related episodes, and arrestees. *Clin Chem* **42**, 1330–1336.

Ryvlin, P., Kahane, P. (2003). Does epilepsy surgery lower the mortality of drug-resistant epilepsy? *Epilepsy Res* **56**, 105–120.

Salanova, V., Markand, O., Worth, R. (2002). Temporal lobe epilepsy surgery: outcome, complications, and late mortality rate in 215 patients. *Epilepsia* **43**, 170–174.

Samuels, J. F., Nestadt, G., Romanoski, A. J., Folstein, M. F., McHugh, P. R. (1994). DSM-III personality disorders in the community. *Am J Psychiatry* **151**, 1055–1062.

Schmitz, B. (2005). Depression and mania in patients with epilepsy. *Epilepsia* **46**(Suppl 4), 45–49.

Sherwin, I., Peron-Magnan, P., Bancaud, J., Bonis, A., Talairach, J. (1982). Prevalence of psychosis in epilepsy as a function of the laterality of the epileptogenic lesion. *Arch Neurol* **39**, 621–625.

Shukla, G. D., Srivastava, O. N., Katiyar, B. C., Joshi, V., Mohan, P. K. (1979). Psychiatric manifestations in temporal lobe epilepsy: a controlled study. *Br J Psychiatry* **135**, 411–417.

Shukla, G. D., Bhatia, M., Singh, V. P., Jaiswal, A., Tripathi, M., Gaikwad, S., Bal, C. S., Sarker, C., Jain, S. (2003). Successful selection of patients with intractable extratemporal epilepsy using non-invasive investigations. *Seizure* **12**, 573–576.

Silberman, E. K., Sussman, N., Skillings, G., Callanan, M. (1994). Aura phenomena and psychopathology: a pilot investigation. *Epilepsia* **35**, 778–784.

Stefansson, S. B., Olafsson, E., Hauser, W. A. (1998). Psychiatric morbidity in epilepsy: a case controlled study of adults receiving disability benefits. *J Neurol Neurosurg Psychiatry* **64**, 238–241.

Strine, T. W., Kobau, R., Chapman, D. P., Thurman, D. J., Price, P., Balluz, L. S. (2005). Psychological distress, comorbidities, and health behaviors among US adults with seizures: results from the 2002 National Health Interview Survey. *Epilepsia* **46**, 1133–1139.

Swinkels, W. A., Kuyk, J., van Dyck, R., Spinhoven, P. (2005). Psychiatric comorbidity in epilepsy. *Epilepsy Behav* **7**, 37–50.

Taylor, D. C. (1972). Mental state and temporal lobe epilepsy. A correlative account of 100 patients treated surgically. *Epilepsia* **13**, 727–765.

Taylor, D. C. (1975). Factors influencing the occurrence of schizophrenia-like psychosis in patients with temporal lobe epilepsy. *Psychol Med* **5**(3), 249–254.

Tellez-Zenteno, J. F., Wiebe, S., Patten, S. B. (2005). Psychiatric comorbidity in epilepsy: a population based analysis. *Epilepsia* **46**, 264–265.

Tellez-Zenteno, J. F., Dhar, R., Hernandez-Ronquillo, L., Wiebe, S. (2007). Long-term outcomes in epilepsy surgery: antiepileptic drugs, mortality, cognitive and psychosocial aspects. *Brain* **130**, 334–345.

Tilelli, J. A., Spack, L. D. (2006). Marijuana intoxication presenting as seizure – comment. *Pediatr Emerg Care* **22**, 141.

Tomson, T., Beghi, E., Sundqvist, A., Johannessen, S. I. (2004). Medical risks in epilepsy: a review with focus on physical injuries, mortality, traffic accidents and their prevention. *Epilepsy Res* **60**, 1–16.

Torgersen, S., Kringlen, E., Cramer, V. (2001). The prevalence of personality disorders in a community sample. *Arch Gen Psychiatry* **58**, 590–596.

Umbricht, D., Degreef, G., Barr, W. B., Lieberman, J. A., Pollack, S., Schaul, N. (1995). Postictal and chronic psychoses in patients with temporal lobe epilepsy. *Am J Psychiatry* **152**, 224–231.

van Os, J., Anisen, M., Bijl, R. V., Vollebergh, W. (2001). Prevalence of psychotic disorder and community level of psychotic symptoms: an urban-rural comparison. *Arch Gen Psychiatry* **58**, 663–668.

Victoroff, J. I., Benson, F., Grafton, S. T., Engel, J., Jr, Mazziotta, J. C. (1994). Depression in complex partial seizures. Electroencephalography and cerebral metabolic correlates. *Arch Neurol* **51**, 155–163.

Wada, J. A., Wake, A., Sato, M., Corcoran, M. E. (1975). Antiepileptic and prophylactic effects of tetrahydrocannabinols in amygdaloid kindled cats. *Epilepsia* **16**, 503–510.

Weissman, M. M., Merikangas, K. R. (1986). The epidemiology of anxiety and panic disorders: an update. *J Clin Psychiatry* **47**(Suppl), 11–17.

Weissman, M. M., Bland, R. C., Canino, G. J., Greenwald, S., Hwu, H. G., Joyce, P. R., Karam, E. G., Lee, C. K., Lellouch, J., Lepine, J. P., Newman, S. C., Rubio-Stipec, M., Wells, J. E., Wickramaratne, P. J., Wittchen, H. U., Yeh, E. K. (1999). Prevalence of suicide ideation and suicide attempts in nine countries. *Psychol Med* **29**, 9–17.

Wittchen, H. U. (2002). Generalized anxiety disorder: prevalence, burden, and cost to society. *Depress Anxiety* **16**, 162–171.

Wittchen, H. U., Zhao, S., Kessler, R. C., Eaton, W. W. (1994). DSM-III-R generalized anxiety disorder in the National Comorbidity Survey. *Arch Gen Psychiatry* **51**, 355–364.

Wober-Bingol, C., Wober, C., Karwautz, A., Auterith, A., Serim, M., Zebenholzer, K., Aydinkoc, K., Kienbacher, C., Wanner, C., Wessely, P. (2004). Clinical features of migraine: a cross-sectional study in patients aged three to sixty-nine. *Cephalalgia* **24**, 12–17.

Wolf, P., Trimble, M. R. (1985). Biological antagonism and epileptic psychosis. *Br J Psychiatry* **146**, 272–276.

Wrench, J., Wilson, S. J., Bladin, P. F. (2004). Mood disturbance before and after seizure surgery: a comparison of temporal and extratemporal resections. *Epilepsia* **45**, 534–543.

CHAPTER 2

Why Do Neurologists and Psychiatrists Not Talk to Each Other?

Andres M. Kanner and John Barry

INTRODUCTION

The close and complex relationships between psychiatric and neurologic disorders have been identified for a long time. Not surprisingly, any clinician or neuroscientist who has some understanding of the two disciplines would conclude that there is a pressing need for close collaboration between neurologists and psychiatrists, both in the clinical and research domains. In fact, there are now six programs in the United States that offer a combined neurology and psychiatry residency program. And yet, the sad reality is that psychiatrists and neurologists do not talk to one another! The purpose of this chapter is to analyze this peculiar phenomenon, try to understand its roots from the perspectives of a neurologist (Andres M. Kanner) and a psychiatrist (John Barry), who both work closely with patients who suffer from comorbid neurologic and psychiatric disorders, and to offer some potential solutions.

WHY SHOULD NEUROLOGISTS AND PSYCHIATRISTS TALK TO ONE ANOTHER?

During the 19th and first half of the 20th centuries, psychiatrists and neurologists collaborated very closely, and it was expected that neurologists be well versed in psychiatry and vice versa. The work of Gowers, Charcot, Freud and Gastaut, to name a few, exemplifies the fruits of such close collaboration between the two disciplines. Up until 50 years ago, neurology residents would rotate for 1 year on the psychiatry service and vice versa. Paradoxically, while such is not the case

Psychiatric Controversies in Epilepsy
Copyright © 2008 by Elsevier Inc. All rights of reproduction in any form reserved.

any longer, the need for a closer collaboration between the two disciplines has become even more obvious. Indeed, an increasing body of literature demonstrates the relatively high prevalence of psychiatric comorbidity in neurological disorders. For example, prevalence rates of post-stroke depression have been reported to range between 30% and 50% (Robinson, 2003), the lifetime prevalence of a major depression in patients with multiple sclerosis has been found to vary from 10% to 60% (Minden and Schiffer, 1990; Feinstein, 1999), and approximately 46% of patients with Parkinson's disease have been found to suffer from depressive disorders (Gotham et al., 1986). In patients with epilepsy, lifetime and cross-sectional prevalence rates of depressive, anxiety, and attention deficit disorders as well as psychosis are several-fold higher than in the general population.

By the same token, there is an increasing body of literature demonstrating a bidirectional relationship between some neurological and psychiatric disorders. For example, patients with a history of depression have been found to have a 4- to 7-fold greater risk of developing seizure disorders (Forsgren and Nystrom, 1990; Hesdorffer et al., 2000; Hesdorffer et al., 2006) (see also Chapter 12). Likewise, population-based studies have shown that patients with depression have a 2.5-fold higher risk of developing a stroke, even after controlling for major risk factors (e.g., hypertension, diabetes, obesity, nicotine addiction, etc.) (Colantonio et al., 1992; Everson et al., 1998; Jonas and Mussolino, 2000; Larson et al., 2001; May et al., 2002), and a 2.2-fold higher risk of developing Parkinson's disease (Schuurman et al., 2002). Finally, children with attention deficit hyperactivity disorder (ADHD) of the inattentive type have a 3-fold higher risk of developing seizure disorders (Hesdorffer et al., 2004). These data do not suggest that neurologic disorders are the cause of psychiatric disorders or vice versa. Rather, they suggest the existence of common pathogenic mechanisms operant in neurologic and psychiatric disorders that facilitate the development of one condition in the presence of the other, regardless of the sequence with which they clinically present.

If these data are not sufficient to convince neurologists and psychiatrists to collaborate with each other, here is some additional evidence: post-stroke depression impacts on the quality of life of stroke patients with respect to worse recovery of cognitive impairments and activities of daily living and is associated with higher mortality risks (Parikh et al., 1990; Robinson et al., 2000; Robinson, 2003). For example, data from one study of 140 patients demonstrated that post-stroke major depression was associated with greater cognitive impairment 2 years after a stroke (Robinson, 2003). In another study, responders to treatment with antidepressant medication had higher scores in the mini-mental state exam compared to nonresponders (Kimura et al., 2000). Also, post-stroke depression has been shown to have a negative impact on the recovery of activities of daily living. For example, one study showed that in-hospital post-stroke depression was the most important variable that predicted poor recovery of activities of daily living over a 2-year period (Parikh et al., 1990) while another demonstrated that successful treatment of post-stroke depression with nortriptyline was associated with significant recovery in the activities of daily living (Chemerinski *et al.*, 2001).

By the same token, the negative impact of these psychiatric comorbid disorders on the quality of life of neurologic patients has been demonstrated by multiple investigators. For example, in one study of 226 patients with Parkinson's disease, depression was found to be the factor most closely related to quality of life (Kuopio *et al.*, 2000), while stage and duration of illness and cognitive ability were of lesser importance. These findings were confirmed in a separate study of 97 patients with Parkinson's disease (Schrag *et al.*, 2000) and the Global Parkinson's Disease Survey Steering Committee reached similar conclusions after reviewing the data of a multicenter study of 2,020 patients (The Global Parkinson's Disease Survey Steering Committee, 2002).

Of greater concern has been the data indicating a significant relationship between the presence of post-stroke depression and a higher mortality risk following stroke. Indeed, in a study of 976 patients followed for 1 year, patients with post-stroke depression had 50% higher mortality than those without (Robinson, 2003). By the same token, in another study, patients treated with fluoxetine or nortriptyline had an increased survival probability at 6 years (61%) compared to patients given placebo (34%) (Robinson *et al.*, 2000). Furthermore, treatment with antidepressant medication was an independent predictor of increased survival (Robinson, 2003). And yet, despite all of these data, *none* of the major studies that have investigated the outcome of stroke have factored in the existence of post-stroke depression as a variable.

In patients with epilepsy, the negative impact of depressive disorders on their quality of life has been demonstrated in five separate studies. For example, depression was the single strongest predictor for each domain of a health-related quality-of-life instrument in a study of 56 patients carried out in Germany (Lehrner *et al.*, 1999). The significant association of depression with health-related quality-of-life measures persisted after controlling for seizure frequency, seizure severity and other psychosocial variables. In a US-based study of 125 patients more than 1 year after temporal lobe surgery, mood status was the strongest clinical predictor of the patients' assessments of their own health status (Gilliam *et al.*, 1999). In a separate cohort of 194 epilepsy clinic patients, the authors also found that a depressed mood and neurotoxicity from antiepileptic drugs were the only variables with a significant correlation with poorer self-reported health status (Gilliam, 2002).

Suicidality in neurologic disorders is a relatively common complication that has been identified, particularly in patients with migraines with aura, epilepsy, stroke, multiple sclerosis, Huntington's disease, traumatic brain injury, early stages of dementia and amyotrophic lateral sclerosis. Furthermore, in the case of epilepsy, a population-based study showed that a history of attempted suicide increased the risk of developing epilepsy by a factor of 5.1 (95% confidence interval, 2.2–11.5) (Hesdorffer *et al.*, 2006). Attempted suicide increased seizure risk even after adjusting for age, sex, cumulative alcohol intake, and major depression or number of symptoms of depression. Furthermore, the increased suicidal risk of patients with epilepsy relative to the general population has been demonstrated in several epidemiologic studies. For example, the suicide rate in depressed patients with epilepsy is 9–25 times higher in patients with partial seizures of temporal lobe origin than expected in the overall population (Robertson, 1997, Rafnsson *et al.*, 2001, Christensen *et al.*, 2007). A review of 17 studies pertaining to mortality in epilepsy found that suicide was 10 times more frequent than in the general population (Robertson, 1997). A population-based incidence cohort study in patients with epilepsy from Iceland showed that suicide had the highest standard mortality rate (5.8) of all causes of death (Rafnsson *et al.*, 2001). Furthermore, in a recent population-based study, patients with epilepsy had a 3-fold higher risk of committing suicide, which persisted after controlling for concomitant psychiatric disease and socio-economic factors (Christensen *et al.*, 2007); though in patients with comorbid psychiatric disease, the relative risk was almost 14-fold higher than in controls (RR: 13.7, 95% CI, 11.8–16.0). Of note, the highest risk was identified during the first year after the diagnosis of epilepsy was made and it was particularly high (RR: 29.2, 95% CI, 16.4–51.9) in those with a psychiatric diagnosis.

Patients with stroke appear as well to have a higher suicidal risk. To address this question, the role of cerebrovascular risk factors in late-life suicide was investigated (Chan *et al.*, 2007). The cerebrovascular risk factor score was calculated, based on the American Heart Association Criteria, for each case and these scores were compared with the scores of subjects from a psychological autopsy study of suicide among community-dwelling adults over age 50 years. The cerebrovascular risk factor scores were significantly higher in suicide cases than in community-dwelling comparison subjects after accounting for age, sex, depression diagnosis, and functional status.

Patients with multiple sclerosis have a significantly greater suicidal risk as well. For example, the suicide risk among Danish citizens with multiple sclerosis was compared with that of the general population (Brønnum-Hansen *et al.*, 2005). The study was based on linkage of the Danish Multiple Sclerosis Registry to the Cause of Death Registry. It comprised all 10,174 persons in whom multiple sclerosis was diagnosed in the period 1953–1996. In all, suicides had occurred in 115 persons with multiple sclerosis (63 men, 52 women), whereas the expected number of suicides was 54.2 (29.1 men, 25.1 women). Thus, the suicide risk among persons with multiple sclerosis was more than twice that of the general population (standard mortality risk = 2.12). The increased risk was particularly high during the first year after diagnosis (standardized mortality ratio (SMR) = 3.15).

Given these data, it is only logical to expect first a very close collaboration between neurologists and psychiatrists in the evaluation and management of their neurologic patients, and second, that patients with neurologic disorders would be screened for psychiatric comorbidity with the aim of incorporating the appropriate psychiatric treatment into their overall management. Sadly, this appears to be the exception rather than the rule. For example, in a study of 226 patients seen at a neurology clinic in which depressive disorders were identified in 88 (40%) patients, including 54 (26%) who had major depression, 69 (78%) patients continued to display symptoms of depression 8 months later (Carson *et al.*, 2003), and 46 (85%) of the 54 patients with major depression continued suffering from major depression. Clearly, this study demonstrates the fact that psychiatric comorbidities remain unrecognized and untreated in a large percentage of neurologic patients.

CONSEQUENCES

The lack of communication between the two disciplines has serious implications in the management of these patients. Here are some examples: one out of every four to five patients referred to video-electroencephalogram (video-EEG) monitoring units with a diagnosis of refractory epilepsy do not suffer from epileptic seizures. The majority of these non-epileptic events are of psychogenic origin and are referred as psychogenic non-epileptic seizures (PNES). Ideally, after the diagnosis of PNES is established patients are expected to be enrolled in some type of psychiatric treatment. A study of 174 patients with PNES showed that more often than not, this does not happen: almost 82% of patients were readmitted to a neurologic ward and 41% continued on antiepileptic drugs after the diagnosis was reached (this percentage excluded patients who had epileptic seizures in addition to psychogenic events) (Reuber *et al.*, 2003). Failure to institute the proper treatment can have dire consequences for these patients. Besides not being started on appropriate psychotropic therapy, their repeated readmissions to emergency rooms often result in misdiagnosis of epileptic seizures, including status epilepticus, which in turn leads to admissions to intensive care units where patients are intubated, placed on supportive respiratory therapy and treated with aggressive status protocols including induced coma with midazolam and propofol, which are known to have significant morbidity and mortality risks. Indeed, several studies have found that 30–50% of patients with PNES are admitted to intensive care units.

To understand some of the obstacles that preclude patients from being started on psychiatric treatment, a survey was given to a group of neurologists and psychiatrists (and their trainees) regarding their standard evaluation and management of PNES (Harden *et al.*, 2003). About 75 psychiatrists and 50 neurologists were asked about the diagnostic significance of video-EEG data in the evaluation of patients with PNES. Only 18% of psychiatrists stated that video-EEG is an accurate diagnostic method "most of the time" for patients with PNES, in contrast to 70% of neurologists. This difference in opinion was also found between psychiatry and neurology trainees. These data speak volumes about the total lack of communication between neurologists (or epileptologists) and psychiatrists! Conversely, the advantage of a close collaboration between both disciplines can be clearly appreciated in one study that evaluated the outcome of PNES patients: those who were seen by a counselor affiliated with the epilepsy program had a superior overall outcome with seizure cessation/reduction rates of 68% vs. 48% cessation/reduction among those who were seen by a counselor outside the program (Aboukasem *et al.*, 1998).

Another (relatively common) circumstance in which close communication between neurologists and psychiatrists is of the essence is illustrated in the case of patients with PNES and comorbid epileptic seizures that may go unsuspected or undetected at the time of the video-EEG monitoring study in which PNES are recorded. In the course of psychiatric treatment, such patients may report "seizures" different "than the ones that I had in the hospital" in which case a repeat monitoring study may be required, but patients may be told these "different" seizures are also PNES. Thus, the failed collaboration between neurologists and psychiatrists frequently results in the communication of contradictory (and very confusing) messages to patients and their families.

Furthermore, the misdiagnosis of epileptic seizures that mimic PNES is becoming more frequent as primary care physicians and neurologists are starting to become aware of the entity of PNES (Parra *et al.*, 1999), particularly as it pertains to frontal lobe epilepsy (Williamson *et al.*, 1985). Clearly, psychiatrists must be aware of the clinical similarities (and differences!) between the two types of disorders and be acquainted with the diagnostic strategies used to reach a correct diagnosis in the course of a video-EEG monitoring study in order to identify those patients that were erroneously diagnosed. Yet, the poor agreement between psychiatrists and neurologists on the diagnostic value of video-EEG data shown by the data from Harden's study (Harden *et al.*, 2003) attests to the reasons for the obstacles at play in the delivery of optimal treatment for these patients.

Another example! As already stated above, the prevalence of psychiatric comorbid disorders is relatively high among patients with epilepsy, and it is even higher among patients with refractory epilepsy. Thus, various series have identified depressive and anxiety disorders in up to 50% of patients who are considered for epilepsy surgery (Blumer, 1997). Various case series have identified the development of *de-novo* psychiatric complications after surgery, including major depressive episodes complicated by suicidal attempts or psychotic disorders. In addition, up to 30% of patients can experience a transient exacerbation of pre-surgical depressive episodes during the 3–12 months after surgery. Clearly, identification of patients at risk for post-surgical psychiatric complications is expected to be part of a pre-surgical evaluation. Or is it? In fact, in a survey of 47 major epilepsy centers, a pre-surgical psychiatric evaluation was performed in every surgical candidate in only 10 (21%) centers. In fact, in many epilepsy centers, psychiatrists have been replaced by neuropsychologists! This serious problem and the results of this survey are reviewed in greater detail in Chapter 16.

Misconceptions about the impact of psychotropic drugs on the seizure threshold have resulted in one of the more frequent causes of under-treatment of psychiatric disorders in patients with epilepsy. For example, ADHD has been recognized in approximately 30% of children with epilepsy (Dunn, 2001). The use of central nervous system stimulant drugs like methylphenidate has been proven to be safe and effective in the treatment of ADHD in epilepsy (Dunn, 2001). Yet, there is a wide misconception that these drugs lower the seizure threshold, and hence many clinicians refuse to prescribe them to these children. A similar concern exists with respect to the use of antidepressant drugs. While tricyclic antidepressant drugs can lower the seizure threshold at toxic doses in non-epileptic patients, such is not the case for antidepressants belonging to the selective serotonin reuptake inhibitors (SSRI) (or at worse they are minimal (Kanner *et al.*, 2000)), selective serotonin norepinephrine reuptake inhibitors, and monoamine-oxidase-inhibitors, all of which appear to be safe in epilepsy patients. These two important problems are reviewed in detail in Chapter 17, respectively. Better communication between neurologists and psychiatrists would eliminate such misconceptions.

Finally, the lack of collaboration between the two disciplines has had a negative impact on the training of psychiatrists on the neurobiologic bases of psychiatric disorders and on the training of neurology residents on the psychiatric aspects of neurologic disorders. These points are reviewed in greater detail in Chapters 3 and 4. Thus, while neuropsychiatric research relies on sophisticated neurophysiologic and neuroimaging techniques such as positron emission tomography, magnetic resonance spectroscopy, and volumetric measurements of frontal and temporal lobe structures on magnetic resonance imaging, the average psychiatry residency program lacks adequate training on the use and limitations of neuroimaging and neurophysiologic diagnostic tests. A common result is the tendency of psychiatrists to order EEG studies in patients with mood disorders, psychosis or ADHD in the absence of any evidence of clinical seizures, or the assumption that psychopathology in the presence of epileptiform activity must be the expression of a seizure disorder, even in the absence of any evidence of clinical seizures. The total misuse of quantitative EEG technology for diagnosis of psychiatric disorders is one of the more blatant consequences resulting from the lack of communication between neurophysiologists and psychiatrists.

Conversely, the absence of any training of neurology residents in psychiatry precludes them from placing psychiatric phenomena in the proper perspective, even in neurologic disorders in which the psychiatric phenomena are among the cardinal symptoms (e.g., in the early stages of Wilson's disease, Huntington's disease and porphyria). It is not unusual for neurology residents (and neurologists) to view psychiatric phenomena in a very simplistic manner. This is exemplified by the common misperception that depressive and anxiety disorders are "normal" reactive processes to the presence of neurologic diseases that do not require any specific treatment, and that referral for evaluation and treatment is only necessary in the advanced stages of the psychiatric condition when patients begin displaying suicidal ideation or lose their ability to function.

WHAT CAUSES THIS LACK OF COMMUNICATION?

Given all of the examples cited in the previous sections of this chapter, poor communication between the two disciplines is a bizarre phenomenon that is hard to understand on logical bases. After all, neurologists and psychiatrists share the same professional board (The American Board of Psychiatry and Neurology) and the written examination taken by neurologists includes a section with questions about psychiatry and vice versa. Furthermore, the creation of neuropsychiatric professional associations such as the American or International Neuropsychiatric Associations (ANPA and INA, respectively) attest to the recognition of the need to understand psychiatric and neurologic disorders from the perspectives of both disciplines. A review of the index of titles of any textbook of neuropsychiatry confirms these observations and, as stated before, in recent years six residency programs have been created in the United States offering dual training in both disciplines in a 6-year period with eligibility for board certification in both specialties. So how can we explain this schism?

THE NEUROLOGIST'S PERSPECTIVE

Various reasons can be considered:

(1) The poor education of neurologists on the diagnosis, clinical implications and treatment of psychiatric comorbidities of neurologic disorders and vice versa, and the poor education that psychiatry residents get on the latter as well as on neurologic complications of psychiatric disorders. In addition, psychiatry residents are still not trained appropriately on the diagnostic yield, or on appropriate indications and limitations of neurophysiologic and neuroimaging studies.

For the last 30 years, psychiatry residents have had very limited training in neurology while in the last 15–20 years neurology residents have not been required to have any formal training in psychiatry. Clearly, there is a disconnect between the amount of training and the expectations of knowledge at the end of the respective residency programs by the authorities in charge of regulating the curricula of these two specialties. Otherwise, why would the Psychiatry Board exam include a section of neurology questions and vice versa?

(2) The poor communication of scientific advances between the two disciplines, both in the peer-reviewed journals and in scientific meetings of the respective professional societies. For example, there is no liaison between the major neurologic and psychiatric professional societies, such as the American Academy of Neurology (AAN), the American Neurological Association (ANA), on the one hand, and the American Psychiatric Association (APA) on the other. The absence of a formal liaison is evident between psychiatric and neurologic professional societies (like the American Epilepsy Society (AES)) that deal with neurologic disorders in which psychiatric comorbidities are prominent. Nonetheless, attempts

to address this problem can be seen in the establishment of professional journals like *Epilepsy & Behavior*, with both neurologists and psychiatrists on the Editorial Board.

(3) The neurologists' conceptions (or misconceptions) of the "poor quality" of psychiatric evaluations of patients with neurological disorders. This view is supported from the data obtained in the survey cited above (see also Chapter 16) in which 75% of respondents from 47 epilepsy centers opined that "it is hard to find a psychiatrist interested in performing in-depth evaluations of patients with epilepsy."

(4) The neurologists' perception of the unavailability of psychiatrists to evaluate patients with neurological disorders.

The lack of proper training of both neurologists and psychiatrists is a major culprit. Yet, we cannot attribute the lack of communication between the two disciplines to the lack of education only, as this was not always an issue. As stated above, in the first half of the 20th century, neurology and psychiatry residents spent 6 months–1 year training in the other specialty, paradoxically during a time period when there was little understanding of the neurobiologic bases of psychiatric disorders and when recognition of psychiatric comorbidity of neurologic disorders was limited. Of note, in neurology and psychiatry residency programs in Europe, there are currently longer rotations in neurology and psychiatry services, respectively. Thus, did psychiatrists and neurologists in the first half of the 20th century or do they in Europe today have greater communication than in North America? Not really! Thus, the other variables cited above are also operant.

THE PSYCHIATRIST'S PERSPECTIVE

I do not view this as a problem of lack of interest. It has been this author's (John Barry) experience that there is a considerable amount of interest on the part of both psychiatrists and neurologists to understand both disciplines. The schism that has taken place started with Freud but in recent years there has been a merging of the two professions. In 2004, the APA's annual meeting was entitled, "Dissolving the Mind–Brain Barrier." Presently, psychiatrists are frequently talking about neurosurgical procedures for treatment resistant depression including vagus nerve and deep brain stimulation. The fact that a neurologist/psychiatrist and a psychiatrist specializing in neuropsychiatry are writing this article together attests to that fact.

Much of the difficulty that arises between psychiatrists and neurologists may stem from fear and discomfort. As mentioned above, psychiatrists are not familiar enough with medicine in general, and specifically neurology. The same holds true for neurologists dealing with psychiatric problems in their patients. As a result, instead of admitting difficulty, which has its own set of problems, the practitioner may deny the existence of the problem or avoid asking questions that would put him or her in an unfamiliar situation. If a ready conduit can be provided to either psychiatric consultation for the neurologist or neurologic consultation for the psychiatrist, many seemingly insurmountable barriers could be overcome. The more available the consultant, the more the practitioner will take chances and venture forth into unfamiliar territory with the knowledge that support is readily available.

In my opinion, one of the major issues is financial. As Holtz (1992) and Cavanaugh and Milne (1995) noted, Consultation-Liaison (CL) services have increased financial pressures. Cavanaugh *et al.* surveyed 119 institutions with 76 responding; 57% noted that funding issues were a major problem associated with understaffing in 51%. About 70% of the programs had their services stopped or reduced. The factors involved were increasing acuity, shorter hospital stay and older populations. Another factor is the need for authorization for psychiatric services that often times is not obtained because of patient acuity resulting in poor collection rates.

Attempts have been made to creatively expand services and in some instances, institutions have provided funding for CL services (Bourgeoid *et al.*, 2003). Cost effectiveness of these procedures has been documented by some but there are very little data to help support such a move (Andreoli *et al.*, 2003). In addition, decreasing psychiatric input can increase overall costs (Horn, 2003). This author's experience has been extremely positive with both services relying on the other and with the opening of many doors of communication that has benefited our patient care and research opportunities.

Perhaps as a result of this issue, another very salient problem has become lack of availability of psychiatrists for referrals. In addition, time pressures experienced by physicians in general, and neurologists in particular, have become an increasing problem. Asking patients psychiatric questions requires that there be a ready availability of psychiatrists when a positive answer is obtained or the questions will not be asked. In addition, referrals to outpatient psychiatry have become increasingly difficult because of availability but also because of insurance issues. Those patients with the most difficult psychiatric problems often have poor insurance, and as a result, referrals can be difficult at best. Additionally, physician time pressures often limit the number of problems and complexity of issues that can be dealt with efficiently at any one office visit. Until there is more funding for overlap services between psychiatry and neurology and health care in general improves, especially for the seriously and chronically medically and psychiatrically ill, lack of recognition and poor care will continue.

The wide variety of psychiatric problems seen in neurological patients is challenging but potentially exciting. For example, in patients with epilepsy, not only are depression and anxiety disorders frequent, but psychosis and personality issues may be present as well. A potential challenge for the practitioner is that these patients may not meet familiar diagnostic criteria according to the Diagnosis and Statistical Manual (DSM) classification. In such cases, it is necessary to think "outside the box" and be extremely liberal with diagnostic categories. Although typical DSM characteristics can be seen in patients with epilepsy and psychiatric disorders, presentations are sometimes "atypical" and that can be extremely challenging, especially for the new resident (Kanner and Barry, 2001). Again, avoidance may result from unfamiliarity.

WHERE DO WE GO FROM HERE?

As suggested by the data reviewed in this chapter, there is no excuse for psychiatrists and neurologists to maintain poor communication, as it is yielding a very negative impact on patient management. Clearly, the first step must include closer interaction between the two disciplines, starting at the medical school level where students should be exposed to the close relations between the two disciplines. This effort must be followed at the residency training level, whereby neurology residents should spend a minimum of a 3-month (ideally 6 months) rotation on the psychiatry service and vice versa. Finally, annual meetings of the respective professional societies serving the needs of neurologists and psychiatrists must start including full-day courses, symposia and workshops to educate both disciplines on the close relation between psychiatric and neurologic disorders.

Are neuropsychiatry professional societies solving the problems discussed above? While the intention of these societies has been to bring the two disciplines closer, their impact on clinical practice has been very limited to date. The relatively sparse involvement of neurologists accounts in part for the limited success of these societies. Yet, given their mission, these societies are in an ideal position to facilitate the education of neurologists and psychiatrists. Indeed, the establishment of formal liaisons between these (ANPA, INA) and the major psychiatric (APA) and neurologic societies (AAN, ANA, AES) can facilitate a rapprochement between the two disciplines by planning educational programs during their respective annual meetings.

Sadly, the lack of trust between neurologists and psychiatrists, including the old misconceptions and prejudices, remains the biggest obstacle for improving the close collaboration between the two disciplines, in the authors' opinions. At some point it will be necessary to openly confront such misconceptions with frank discussions that are not inhibited by a need to be "politically correct." Yet, at the end of such an open dialog, the collaboration between the two disciplines will change with better education at all levels and finding a solution for the economic problems that limit access of neurologic patients to psychiatric care. In fact, the chances of solving this latter problem may be more likely if neurologists and psychiatrists joined efforts!

REFERENCES

Aboukasem, A., Mahr, G., Gahry, B. R., et al. (1998). Retrospective analysis of the effects of psychotherapeutic intervention on outcomes of psychogenic seizures. *Epilepsia* **39**, 470–473.

Andreoli, P. B. A., Citero, V. A., Mari, J. J. (2003). A systematic review of studies of the cost-effectiveness of mental health consultation-liaison interventions in general hospitals. *Psychosomatics* **44**, 499–507.

Blumer, D. (1997). Antidepressant and double antidepressant treatment for the affective disorder of epilepsy. *J Clin Psychiatry* **58**, 3–11.

Bourgeoid, J. A., Hilty, D. M., Klein, S. C., et al. (2003). Expansion of the consultation-liaison psychiatry paradigm at a university medical center: integration of diversified clinical and funding models. *Gen Hosp Psychiatry* **25**(4), 262–268.

Brønnum-Hansen, H., Stenager, E., Nylev Stenager, E., Koch-Henriksen, N. (2005). Suicide among Danes with multiple sclerosis. *J Neurol Neurosurg Psychiatry* **76**(10), 1457–1459.

Carson, A. J., Postma, K., Stone, J., Warlow, C., Sharpe, M. (2003). The outcome of depressive disorders in neurology patients: a prospective cohort study. *J Neurol Neurosurg Psychiatry* **74**, 893–896.

Cavanaugh, S., Milne, J. M. (1995). Recent changes in consultation-liaison psychiatry, a blueprint for the future. *Psychosomatics* **36**, 95–102.

Chan, S. S., Lyness, J. M., Conwell, Y. (2007). Do cerebrovascular risk factors confer risk for suicide in later life? A case-control study. *Am J Geriatr Psychiatry* **15**(6), 541–544.

Chemerinski, E., Robinson, R. G., Arndt, S., Kosier, J. T. (2001). The effect of remission of poststroke depression on activities of daily living in a doubleblind randomized treatment study. *J Nerv Ment Dis* **89**, 421–425.

Christensen, J., Vestergaard, M., Mortensen, P. B., Sidenius, P., Agerbo, E. (2007). Epilepsy and risk of suicide: a population-based case-control study. *Lancet Neurol.* **6**, 693–698.

Colantonio, A., Kasi, S. V., Ostfeld, A. M. (1992). Depressive symptoms and other psychosocial factors as predictors of stroke in the elderly. *Am J Epidemiol* **136**, 884–894.

Dunn, D. W. (2001). Attention-deficit hyperactivity disorder, oppositional defiant disorder, and conduct disorder. In *Psychiatric Issues in Epilepsy: A Practical Guide to Diagnosis and Treatment* (A. B. Ettinger, A. M. Kanner, eds), pp. 111–126. Philadelphia, PA: Lippincott, Williams and Wilkins.

Everson, S. A., Roberts, R. E., Goldberg, D. E., Kaplan, G. A. (1998). Depressive symptoms and increased risk of stroke mortality over a 29-year period. *Arch Intern Med* **158**, 1133–1138.

Feinstein, A. (1999). Multiple sclerosis and depression. In *The Clinical Neuropsychiatry of Multiple Sclerosis* (A. Feinstein, ed.), pp. 26–50. Cambridge, UK: Cambridge University Press.

Forsgren, L., Nystrom, L. (1990). An incident case-referent study of epileptic seizures in adults. *Epilepsy Res* **6**, 66–81.

Gilliam, F. (2002). Optimizing health outcomes in active epilepsy. *Neurology* **58**(Suppl 5), S9–S19.

Gilliam, F., Kuzniecky, R., Meador, K., *et al.* (1999). Patient-oriented outcome assessment after temporal lobectomy for refractory epilepsy. *Neurology* **53**(4), 687–694.

Gotham, A. M., Brown, R. G., Marsden, C. D. (1986). Depression in Parkinson's disease: a quantitative and qualitative analysis. *J Neurol Neurosurg Psychiatry* **49**, 381–389.

Harden, C. L., Tuna Burgut, F., Kanner, A. M. (2003). The diagnostic significance of video-EEG monitoring findings on pseudoseizure patients differ between neurologists and psychiatrists. *Epilepsia* **44**, 453–456.

Hesdorffer, D. C., Hauser, W. A., Annegers, J. F., *et al.* (2000). Major depression is a risk factor for seizures in older adults. *Ann Neurol* **47**, 246–249.

Hesdorffer, D. C., Ludvigsson, P., Olafsson, E., Gudmundsson, G., Kjartansson, O., Hauser, W. A. (2004). ADHD as a risk factor for incident unprovoked seizures and epilepsy in children. *Arch Gen Psychiatry* **61**(7), 731–736.

Hesdorffer, D. C., Hauser, W. A., Olafsson, E., Ludvigsson, P., Kjartansson, O. (2006). Depression and attempted suicide as risk factors for incident unprovoked seizures and epilepsy. *Ann Neurol* **59**(1), 35–41.

Holtz, J. L. (1992). Making a consultation service work, an organizational commentary. *Psychosomatics* **33**(3), 324–328.

Horn, S. D. (2003). Limiting access to psychiatric services can increase total health care costs. *J Clin Psychiatry* **64**(suppl 17), 23–28.

Jonas, B. S., Mussolino, M. E. (2000). Symptoms of depression as a prospective risk factor for stroke. *Psychosom Med* **62**, 463–471.

Kanner, A. M., Kozak, A. M., Frey, M. (2000). The use of sertraline in patients with epilepsy: is it safe? *Epilepsy Behav* **1**, 100–105.

Kanner, A. M., Barry, J. J. (2001). Depression and psychotic disorders associated with epilepsy – are they unique? *Epilepsy Behav* **2**, 170–186.

Kimura, M., Robinson, R. G., Kosier, T. (2000). Treatment of cognitive impairment after poststroke depression. *Stroke* **31**, 1482–1486.

Kuopio, A. M., Marttila, R. J., Helenius, H., *et al.* (2000). The quality of life in Parkinson's disease. *Mov Disord* **15**, 216–223.

Larson, S. L., Owens, P. L., Ford, D., Eaton, W. (2001). Depressive disorder, dysthymia, and risk of stroke. Thirteen-year follow-up from the Baltimore epidemiological catchment area study. *Stroke* **32**, 1979–1983.

Lehrner, J., Kalchmayr, R., Serles, W., *et al.* (1999). Health-related quality of life (HRQOL), activity of daily living (ADL) and depressive mood disorder in temporal lobe epilepsy patients. *Seizure* **8**(2), 88–92.

May, M., McCarron, P., Stansfeld, S., *et al.* (2002). Does psychological distress predict the risk of ischemic stroke and transient ischemic attack? The Caerphilly Study. *Stroke* **33**, 7–12.

Minden, S. L., Schiffer, R. B. (1990). Affective disorders in multiple sclerosis, review and recommendations for clinical research. *Arch Neurol* **47**, 98–104.

Parikh, R. M., Robinson, R. G., Lipsey, J. R., Starkstein, S. E., Fedoroff, J. P., Price, T. R. (1990). The impact of post-stroke depression on recovery in activities of daily living over two-year follow-up. *Arch Neurol* **47**, 785–789.

Parra, J., Iriarte, J., Kanner, A. M. (1999). Are we overusing the diagnosis of psychogenic non-epileptic events? *Seizure* **8**(4), 223–227.

Rafnsson, V., Olafsson, E., Hauser, W. A., *et al.* (2001). Cause-specific mortality in adults with unprovoked seizures. A population-based incidence cohort study. *Neuroepidemiology* **20**(4), 232–236.

Reuber, M., Pukrop, R., Bauer, J., Helmstaedter, C., Tessendorf, N., Elger, C. E. (2003). Outcome in psychogenic non-epileptic seizures: 4 to 10-year follow-up in 164 patients. *Ann Neurol* **53**, 305–311.

Robertson, M. M. (1997). Suicide, parasuicide, and epilepsy. In *Epilepsy: A Comprehensive Textbook* (J. Engel, T. A. Pedley, eds), 2141–2151. Philadelphia, PA: Lippincott-Raven.

Robinson, R. G. (2003). Poststroke depression: prevalence, diagnosis, treatment and disease progression. *Biol Psychiatry* **54**, 376–387.

Robinson, R. G., Schultz, S. K., Castillo, C., Kopel, T., Kosier, T. (2000). Nortriptyline versus fluoxetine in the treatment of depression and in short term recovery after stroke: a placebo controlled, double-blind study. *Am J Psychiatry* **157**, 351–359.

Schrag, A., Jahanshahi, M., Quinn, N. (2000). What contributes to quality of life in patients with Parkinson's disease? *J Neurol Neurosurg Psychiatry* **69**, 308–312.

Schuurman, A. G., Van Den Akker, M., Ensinck, K. T., *et al.* (2002). Increased risk of Parkinson's disease after depression: a retrospective cohort study. *Neurology* **58**, 1501–1504.

The Global Parkinson's Disease Survey Steering Committee (2002). Factors impacting on the quality of life in Parkinson's disease: results from an international survey. *Mov Disord* **17**, 60–67.

Williamson, P. D., Spencer, D. D., Spencer, S. S., Novelly, R. A., Mattson, R. H. (1985). Complex partial seizures of frontal lobe origin. *Ann Neurol* **18**(4), 497–504.

CHAPTER 3

Are Neurologists Trained to Recognize and Treat the Psychiatric Comorbidities of Epilepsy?

Linda M. Selwa

INTRODUCTION

The primary providers of medical care for many patients with refractory epilepsy are neurologists, often with special additional subspecialty training as epileptologists. The American Epilepsy Society (AES) registered 3,035 subspecialty members in 2006 (AES report, 2007), and there are likely to be many more providers of epilepsy care among the nearly 15,000 neurologists listed as members of the American Academy of Neurology (AAN) in their 2004 AAN census (Henry, 2005). For AAN-surveyed neurologists in the United States, epilepsy was listed as the third most common area of practice, after general neurology and headache. Practitioners caring for epilepsy patients encounter a broad spectrum of psychiatric disease. The spectrum of psychiatric comorbidity in refractory epilepsy includes a substantial incidence of major depressive disorder (MDD), anxiety, psychosis, attention deficit hyperactivity disorder (ADHD) and conversion syndromes such as nonepileptic psychogenic seizures. The medications used to control epilepsy may have numerous behavioral, cognitive and psychiatric side effects. Training received by these clinicians in evaluation and management of psychiatric illnesses is limited. In the year 2006, a single month of mandatory psychiatry education was established for neurology residencies, and there are no current recommendations for further psychiatry education in subspecialty training.

In this chapter, we will briefly emphasize the importance of psychiatric issues to the health of epilepsy patients, examine the current level of training available to neurologists and discuss access to psychiatry services. Possible strategies to improve training for neurologists providing epilepsy care will also be explored.

IS RECOGNITION AND TREATMENT OF PSYCHIATRIC DISORDERS IMPORTANT IN THE TREATMENT OF EPILEPSY?

In order to evaluate the depth and importance of the training needed to serve epilepsy patients, it seems useful to briefly review the complexity and incidence of psychiatric disorders, which are discussed in detail in other chapters of this volume. The factors that increase the incidence of these disorders in epilepsy include structural and neurochemical brain disturbances, effects of treatment modalities and the impact of psychosocial stressors (Hermann and Whitman, 1984). Personality disorders and conversion/dissociative disorders have also been linked to various specific types of epilepsy (Swinkels *et al.*, 2005). Each comorbidity has a significant impact on quality of life, compliance, seizure frequency and/or cognitive function in these patients.

Depression

Depression is perhaps the most commonly recognized comorbidity in epilepsy patients. Estimates of prevalence range from 6% to 48% depending on the scale and the population studied, with a median near 30% (Kanner and Nieto, 1999; Hermann, 2000). While depression is noted to be higher in patients with refractory epilepsy than in those whose seizures are well-controlled (Grabowza-Grzyb *et al.*, 2006), it is also true that depression correlates less with seizure frequency, age of onset and duration of epilepsy than with a number of psychosocial variables (Hermann, 2000; Kendler *et al.*, 1999). Structural pathology in frontal or temporal regions is increasingly noted to be a determinant of depression in epilepsy and other populations (Sheline, 2003). The impact of this depression on quality-of-life scales in epilepsy is substantial: depression has been noted in some studies to be the most closely linked variable to quality-of-life scaled scores (Meldolesi *et al.*, 2006). The treatment of depression in epilepsy is complicated by potential drug interactions and effects on seizure thresholds. In some patients cognitive and behavioral therapy may also be useful (Krishnamoorthy, 2003).

Psychosis

Several types of psychotic syndromes have been described in association with epilepsy. Psychosis can be ictal – seizures may present with auditory or visual hallucinations, and nonconvulsive status epilepticus can present as a psychotic state. Interictal psychosis, with a prevalence of between 6% and 24% in patients with refractory epilepsy can be distinguished from schizophrenia because of a lack of negative symptoms and somewhat better prognosis, and may occur more frequently at times when seizures disappear (forced normalization/alternative psychosis) (Toone *et al.*, 1982;

Kanner, 2000). Postictal psychosis, often consisting of a brief period of confusion, delirium and delusional thinking, is a very common manifestation – seen in 7–10% of patients hospitalized for monitoring (Kanner, 2000; Kanner *et al.*, 2000). Other forms of psychosis may be related to medication toxicity or withdrawal, and psychotic syndromes are observed in a small fraction of those who have undergone temporal lobectomy with or without seizure control (Trimble, 1992). A recent Japanese study indicated that the incidence of psychotic symptoms is significantly higher in those with intellectual disabilities (Matsuura *et al.*, 2005).

Anxiety

The prevalence of various forms of anxiety is very high among epilepsy patients, ranging from 19% to 60% (Jones *et al.*, 2005). The types of disorders with increased incidence in epilepsy patients include panic disorder, generalized anxiety disorder, phobias and obsessive compulsive disorders (Beyenburg *et al.*, 2005). Not surprisingly, focal epilepsies, especially those associated with the temporal lobe, have a stronger association than other seizure types. The effects of anxiety on quality of life are substantial and separate from the effects of depression (Cramer *et al.*, 2005). Cognitive and behavioral therapy can be helpful. Medication treatment decisions are often complicated by concerns about drug interactions, as well as the possibility of reducing seizure threshold.

ADHD/personality disorders

Clinical epidemiologic investigations have found the incidence of ADHD to be 3–5 times higher in the population with epilepsy than in the general population (Aldenkamp *et al.*, 2006; Hempel *et al.*, 1995). There is still a significant controversy over whether treatment of the ADHD in these children may reduce the seizure threshold in patients with active epilepsy. The clinical scenario is also clouded by the fact that phenobarbital may exacerbate the symptoms of ADHD in children. The interplay of these disorders is complex: in a series of EEG studies recorded in ADHD patients, one author found that 30% had unequivocal epileptiform activity (Hughes *et al.*, 2000). The incidence of personality disorders in those with epilepsy is also higher than the norm, recently estimated at 4–38% (Swinkels *et al.*, 2005). The types of disorders identified are varied, and further investigation and more clear categorization of the risk and type of coping strategies and personality responses in epilepsy are warranted.

Nonepileptic psychogenic seizures

As many as 30–40% of patients presenting to academic medical centers for CCTV–EEG monitoring of refractory seizure disorders are discovered to have nonepileptic events. The diagnosis is commonly made and discussed by the neurologists reviewing the EEG data, and it is well recognized that the manner of disclosure of the diagnosis may affect outcome (Walzac et al., 1995). These disorders are felt to most closely parallel conversion or dissociative disorders but there is generally a high incidence of comorbid depression, anxiety and personality disorders (Devinsky, 2003). There is a difference in outcome depending on prior risk factors, the semiology of the events and prior degree of psychiatric comorbidity (Kanner and Nieto, 1999; Kanner et al., 1999; Selwa et al., 2000). Between 10% and 25% of these patients suffer from both psychogenic and epileptic seizures, which further complicates diagnosis and treatment.

Effects of medications and surgical treatments

Antiepileptic medications have a striking range of effects on the presence of mood disorders, psychosis, and cognitive and personality parameters and are well reviewed in a recent monograph (Meador et al., 2001). Some drugs with excellent efficacy in epilepsy may have relatively high discontinuation rates related to cognitive and psychiatric effects (Bootsma et al., 2004). Vagus nerve stimulation may have some positive effects on dysphoric syndromes in epilepsy patients (Hoppe et al., 2001). The potential effects of epilepsy surgery on cognition, mood and psychosocial adjustment are substantial. In some cases, depression or psychosis first appear or are significantly exacerbated in the postsurgical setting (Kanner, 2003). Surgical and nonsurgical treatment outcome seems to be linked to the presence of psychiatric comorbidities. In some studies, the rate of seizure freedom after a new diagnosis of epilepsy and the rate of seizure freedom after anterior temporal lobectomy was lower in those with substantial comorbid depression (Anhoury et al., 2000; Mohanraj and Brodie, 2003). Even measures of seizure frequency may well be affected by attention to the psychosocial stressors and psychiatric comorbidities that can influence seizure recurrence.

Given the complexity and prevalence of psychiatric comorbidity in epilepsy, the strong relationship of these factors to patient quality of life and the potential impact of all types of treatment currently available, a thorough education in the diagnosis and treatment of psychiatric disorders would seem to be critical for practitioners providing care to patients with refractory epilepsy.

WHAT TRAINING IS AVAILABLE FOR NEUROLOGISTS OR EPILEPTOLOGISTS IN EVALUATION AND TREATMENT OF PSYCHIATRIC DISORDERS?

Medical school students generally have basic science and clinical exposures in both neurology and psychiatry. A psychiatry clerkship is mandated at most medical schools, while the neurology clerkship is more variable, ranging from no mandatory exposure to 2–4 weeks of neurology, and sometimes including combined time in neurology and psychiatry or neurosurgery. The core curriculum for these clerkships certainly includes some multidisciplinary education, but the level of sophistication of the medical student in the range of presentations and treatment of patients with epilepsy will limit the effectiveness of this early exposure.

After a brief exposure to psychiatry in medical school, the graduate who has chosen residency training in neurology enters a program approved and certified by the American Board of Psychiatry and Neurology (ABPN). This board was established in 1934 at a time of dawning recognition that sciences of the mind and brain would eventually converge, and has continued to supervise the training and certification of both specialties (ABPN.com – history page). At the time the board was formed, there were less than 1,000 neurologists. Currently, the board of the ABPN is comprised of eight neurologists and eight psychiatrists, who work closely with the national Accreditation Council for Graduate Medical Education (ACGME), often serving on their residency review committees (RRCs), which accredit individual programs.

Early in the history of the neurology residency, training included a mandatory 3-month rotation on psychiatry. Even in the early 1960s, the majority of neurology programs required a 3-month exposure in psychiatry. By 1969, only 44/82 neurology programs required any psychiatry, whereas 170 of the 230 psychiatry programs required an average of 3 months of neurology (Rose, 1969). The time neurology residents spent in psychiatry gradually eroded after 1965, when the RRC deleted the requirement for rotations on psychiatry (Price et al., 2000). In the late 1970s, there were few neurology programs requiring any official exposure to psychiatry (Martin, 2002). Meanwhile, in psychiatry, when internships were reinstated in 1978 as a mandatory part of the psychiatry resident experiences, a 2-month requirement was established in neurology. Pediatric neurologists have maintained 1 month of psychiatry training since the accreditation of the specialty. The oral examination of neurologists for certification also originally included a 1 h psychiatric vignette, and that requirement was dropped more than two decades ago. As of 2002, the neurology residency in-training examination had a better positive correlation with the ABPN part one examination in neurology than in psychiatry (Goodman et al., 2002).

In the year 2004, the ABPN defined several new requirements, some of which addressed the issue of psychiatry training (Rosen et al., 2004). A 1-month rotation in psychiatry was mandated, and this change took effect in July 2006. This mandate does not specify the type of exposure, and the programs have instituted various combinations of outpatient and inpatient, largely observational, psychiatry experiences. In addition, a core competency requirement was established that would ultimately include specification of the psychiatry knowledge that was to be acquired. This competency requires supervision by a board certified psychiatrist and currently reads "They must learn about the psychosocial aspects of the physician patient relationship and the importance of personal, social and cultural factors in disease processes and their clinical expression. Residents must learn the principles of psychopathology, psychiatric diagnosis and therapy and the indications for and complications of drugs used in psychiatry" (ACGME.org). These goals are admirable, and would seem difficult to accomplish in a 1-month rotation. After the first 11 months of this required rotation, there is currently no feedback available from neurology residency program directors about the effectiveness of the new psychiatry exposure in achieving these goals.

Should a resident seek more extensive exposure to both psychiatry and neurology, there are currently nine programs in the country offering combined residencies (DeKosky, AAN Education Colloquium, April 30, 2007, Boston, MA) which require a 6-year commitment to mastering both fields of neurology and psychiatry. In the year 2004, only 22 residents were participating in these combined programs. A fellowship in Behavioral Neurology and Neuropsychiatry was approved by the United Council of Neurologic Subspecialties (UCNS) in June of 2004, and development is underway. Fellowships in this area have been available to a subset of neurology-trained residents for several years, but the total number of trainees, although undoubtedly small, is not known.

A clinical neurophysiology (CNP) fellowship was approved by the ACGME in 1992 (Burns et al., 2000), and as of 2004 there were 87 accredited CNP fellowships with a total of 169 filled positions to train future specialists in neurophysiology. By 2004, only 21 of the total graduates of these programs described themselves as trained in psychiatry (Juul et al., 2004). The requirements for CNP subspecialty training include knowledge about "the clinical correlation of autonomic disorders and movement disorders." The only mention of psychiatry in the core competencies is that EEG track fellows should be able to "determine if a patient's symptoms are the result of a disease affecting the CNS or PNS or are of another origin (e.g., of a systemic, psychiatric or psychogenic illness)" by performing appropriate testing. Even in this setting, where psychiatrists and neurologists regularly interact, some basic discrepancies remain in views about diagnosis and treatment of nonepileptic seizures. In one example, 70% of neurologists accepted the CCTV–EEG diagnosis of NES as accurate "most of the time" whereas only 18% of the psychiatrists surveyed had a similar response (Harden et al., 2003). There is no indication in the current suggested core curriculum that psychiatry issues may present important issues in other areas of epilepsy.

Specialists at academic centers may expect that trainees in CNP obtain some interdisciplinary training in psychiatric issues during their fellowship during surgical conferences, and from consultations with their psychiatry colleagues. Andres Kanner has recently finished a survey of 47 epilepsy surgery centers. Only 21% of the centers responding to the survey involved a psychiatrist as a member of the team, and required evaluation for every surgical case. About 45% of the centers did feel that postoperative psychiatric complications were severe enough to warrant a preoperative psychiatric evaluation (presented at Education Colloquium

AAN, April 30, 2007, Boston, MA, submitted data). Clearly education during fellowship remains variable.

A proposal is being reviewed by the AAN to establish a separate ACGME approved Epilepsy fellowship (report to Epilepsy section, Dr Greg Cascino, AAN Epilepsy section meeting, April 30, 2007, Hynes Convention Center Boston, MA). It is not currently known whether this proposal would expand the exposure of subspecialty training in epilepsy to include expanded opportunities for education in psychiatry.

In summary, specific training in psychiatry for neurology epilepsy specialists is currently quite limited. Outspoken critics of the substantial deficits that can exist in multidisciplinary training have commented that "neurologists are not trained to deal with psychiatric patients and little attention is given to psychiatric disorders or their treatment in neurology textbooks or training programs. These patients usually are not offered appropriate comprehensive treatment that includes attention to the psychosocial aspects of their disorders" (Chemali, 2005). For neurologists treating epilepsy and its psychiatric comorbidities, more education would certainly improve the ability to recognize and treat conditions that are closely linked to overall quality of life for these patients.

WHAT ACCESS DO NEUROLOGISTS HAVE FOR CONSULTATION OR JOINT MANAGEMENT OF EPILEPSY PATIENTS WITH PSYCHIATRISTS?

As an alternative to training neurologists in psychiatric issues in epilepsy, some might suggest that consultation or joint management be the rule. If the disciplines continue to function in separate realms, perhaps we can simply try to intersect in areas where the expertise of both disciplines is required to optimize care. Several studies in primary care indicate that a collaborative evaluation with a psychiatrist may result in improved care and patient satisfaction (Katon *et al.*, 1995). Unfortunately, there are several barriers to this solution as well.

Some of the issues limiting access to clinical psychiatrists in training centers are financial. In most centers where a psychiatrist is part of the epilepsy team, a portion of their salary must be paid by the epilepsy surgery group to obtain time for these activities, which do not provide salary support. Many third party payers in the United States do not cover outpatient psychiatric treatment, and those with epilepsy who are most disabled are likely to have access to the poorest reimbursement and personal resources.

The second barrier is the possibility of miscommunication between the specialists, or the inability of neurologists to recognize the seriousness of the psychiatric comorbidity and appropriately convince patients of the value of a referral to a psychiatric specialist.

WHAT POSSIBLE STRATEGIES EXIST FOR THE IMPROVEMENT OF PSYCHIATRIC CARE FOR PATIENTS WITH REFRACTORY EPILEPSY?

At each level of the spectrum of training there are possible improvements that could focus treatment on both abolishing seizures and treating the comorbidities that would facilitate the best overall health. At the resident level, clearly more didactic sessions, interdisciplinary interactions at conferences and in clinics and more exposure to the outpatient management of mood disorders, anxiety, psychosis and ADHD would be helpful in the training of a neurologist. Specifically, more emphasis on the frontal lobe-related and affective portions of the mental status examination would be helpful, and neurobehavioral rounds in an outpatient or inpatient setting can serve to better integrate the disciplines (Matthews *et al.*, 1998). The current moment seems an excellent time to develop a curriculum for our residents in psychiatry: the residents just starting mandatory rotations and their psychiatry mentors seem ideally positioned to comment on what that experience should entail. What rotations provided the most insight? Would more time with psychiatry be helpful?

Fellowship training in epilepsy would seem the ideal time to incorporate some significant exposure to neuropsychiatric syndromes encountered in patients with seizures. A specific competency could be developed that requires some level of didactic exposure and a measure of sensitivity and diagnostic acumen for psychiatric comorbidities. There could be a clear curriculum about when to refer patients, and how to consider or measure quality of life in decisions about medications and surgical treatment.

After graduation from residency and fellowship training, ongoing updates to provide the best interdisciplinary information to psychiatrists and neurologists could be made available through regular courses at national meetings in the primary discipline or in epilepsy (i.e., at American Neurological Association – ANA, American Epilepsy Society – AES, American Clinical Neurophysiology Society – ACNS and/or American Psychiatry Association – APA). CME courses could be provided by both disciplines online or at these meetings as a way to maintain focused attention on the important role that psychiatric management plays in long-term epilepsy outcomes.

In conclusion, the current level of sophistication in training for psychiatric comorbidities of epilepsy is very limited, and there is an important opportunity to strengthen interdisciplinary education, ultimately aiming to improve the health and satisfaction of patients with epilepsy. Our residency and fellowship programs would benefit from more focused attention on improving access for epilepsy patients to the most complete and integrated interventions that will improve quality of life and outcome.

REFERENCES

Aldenkamp, A. P., Arzimangolou, A., Reijs, R., Mil, S. V. (2006). Optimizing therapy of seizures in children and adolescents with ADHD. *Neurology* **67**, S49–S51.

Anhoury, S., Brown, R. J., Krishnamoorthy, E. S., Trimble, M. R. (2000). Psychiatric outcome after temporal lobectomy: a predictive study. *Epilepsia* **41**, 1608–1615.

Annual report of the American Epilepsy Society (accessed May 22, 2007), ⟨http://www.aesnet. org/Members/documents⟩.

Beyenburg, S., Mitchell, A., Schmidt, D., Elger, C., Reuber, M. (2005). Anxiety in patients with epilepsy: systematic review and suggestions for clinical management. *Epilepsy Behav* **7**, 161–171.

Bootsma, H. P. R., Coolen, F., Aldenkamp, A. P., *et al.* (2004). Topirimate in clinical practice: long-term experience in patients with refractory epilepsy referred to a tertiary epilepsy center. *Epilepsy Behav* **5**, 380–387.

Burns, R., Daube, J., Jones, R. (2000). CNP training and certification in the US: 2000. *Neurology* **55**, 1773–1778.

Chemali, Z. (2005). The essentials of neuropsychiatry: teaching residents and fellows the interface between psychiatry and neurology. *Harv Rev Psychiatry* **13**, 312–315.

Cramer, J. A., Bradenburg, N., Xu, X. (2005). Differentiating anxiety and depression symptoms in patients with partial epilepsy. *Epilepsy Behav* **6**(4), 563–569.

Devinsky, O. (2003). Psychiatric comorbidity in patients with epilepsy: implications for diagnosis and treatment. *Epilepsy Behav* **4**, S2–S10.

Goodman, J. C., Juul, D., Westmoreland, B., Burns, R. (2002). RITE performance predicts outcome on the ABPN Part 1 examination. *Neurology* **58**, 1144–1146.

Grabowza-Grzyb, A., Jedrzejczak, a., Nagaska, E., Fiszer, U. (2006). Risk factors for depression in patients with epilepsy. *Epilepsy Behav* **8**(2), 411–417.

Harden, C. L., Burgut, F. T., Kanner, A. M. (2003). The diagnostic significance of video-EEG findings on pseudoseizure patients differs between psychiatrists and neurologists. *Epilepsia* **44**(3), 453–456.

Hempel, A. M., Froest, M. D., Ritter, F. J., Parnham, S. (1995). Factors influencing the incidence of ADHD in pediatric epilepsy patients. *Epilepsia* **36**(Suppl 4), 122.

Henry, K., Lawyer, B. L., Member Demographics Subcommittee of AAN, *Neurologists 2004*, St Paul, American Academy of Neurology 2005 (available at aan.com).

Hermann, B. P., Whitman, S. (1984). Behavioral and personality correlates of epilepsy: a review, methodological critique and conceptual model. *Psychol Bull* **95**, 451–497.

Hermann, B. P., Seidenberg, M., Bell, B. (2000). Psychiatric comorbidity in chronic epilepsy; identification, consequences and treatment of major depression. *Epilepsia* **41**(Suppl 2), S31–S41.

Hoppe, C., Helmstaedter, C., Scherrmann, J., Elger, C. (2001). Self-reported mood changes following 6 months of VNS in epilepsy patients. *Epilepsy Behav* **2**, 235–242.

Hughes, J. R., DeLeo, A. J., Melyn, M. (2000). The EEG in ADHD: emphasis on epileptiform discharges. *Epilepsy Behav* **1**, 271–277.

Jones, J. E., Hermann, B. P., Barry, J. J., *et al.* (2005). Clinical assessment of Axis I psychiatric morbidity in chronic epilepsy: a multicenter investigation. *J Neuropsych Clin Neurosci* **17**(2), 172–179.

Juul, D., Scheiber, S. C., Kramer, T. A. M. (2004). Subspecialty certification by the ABPN. *Acad Psychiatry* **28**(1), 12–17.

Kanner, A. M. (2000). Psychosis of epilepsy: a neurologist's perspective. *Epilepsy Behav* **1**, 219–227.

Kanner, A. M. (2003). Depression in epilepsy: a frequently neglected multifaceted disorder. *Epilepsy Behav* **4**, S11–S19.

Kanner, A., Nieto, J. (1999). Depressive disorders in epilepsy. *Neurology* **53**, 26–32.

Kanner, A. M., Parra, J., Frey, M., *et al.* (1999). Psychiatric and neurologic predictors of pseudoseizure outcome. *Neurology* **53**, 933–938.

Kanner, A. M., Soto, A., Gross-Kanner, H. R. (2000). There is more to epilepsy than seizures: a reassessment of the post-ictal period. *Neurology* **54**(Suppl 3), 352A.

Katon, W., VonKorff, W., Lin, E., *et al.* (1995). Collaborative management to achieve treatment guidelines: impact on depression in primary care. *JAMA* **273**, 1026–1031.

Kendler, K., Karkowski, L., Prescott, C. (1999). Causal relationship between stressful life events and onset of major depression. *Am J Psychiatry* **156**, 837–841.

Krishnamoorthy, E. S. (2003). Treatment of depression in patients with epilepsy: problems, pitfalls and some solutions. *Epilepsy Behav* **4**, S46–S54.

Martin, J. B. (2002). Integration of neurology, psychiatry and neuroscience in the 21st century. *Am J Psychiatry* **159**(5), 695–704.

Matsuura, M., Adachi, N., Muramatsu, R., Kato, M., Onuma, T., Okubo, Y., Oano, Y., Mara, T. (2005). Intellectual disability and psychotic disorders of adult epilepsy. *Epilepsia* **46**(Suppl 1), 11–14.

Matthews, M. K., Koenigsberg, R., Schindler, B., *et al.* (1998). Neurobehavior rounds and interdisciplinary education in neurology and psychiatry. *Med Educ* **32**, 95–99.

Meador, K. J., Gilliam, F. G., Kanner, A. M., Pellock, J. M. (2001). Cognitive and behavioral effects of antiepileptic drugs. *Epilepsy Behav* **2**, S1–S17.

Meldolesi, G. N., Picardi, A., Quarato, P. P., *et al.* (2006). Factors associated with generic and disease-specific quality of life in temporal lobe epilepsy. *Epilepsy Res* **69**(2), 135–146.

Mohanraj, R., Brodie, M. J. (2003). Predicting outcomes in newly diagnosed epilepsy. *Epilepsia* **44**(Suppl 9), 15. (abstract)

Price, B. H., Adams, R. D., Coyle, J. T. (2000). Neurology and psychiatry: closing the great divide. *Neurology* **54**, 8–14.

Rose, A. S. (1969). Presidential address. The current status of education for neurology. *Trans Am Neurol Assoc* **94**, 1–10.

Rosen, N., Daube, J. R., Sulton, L. (2004). Update for the neurology residents and fellows from the NRRC. *Neurology* **63**, 2–3.

Selwa, L. M., Geyer, J., Nikakhtar, N., Brown, M. B., Schuh, L. A., Drury, I. (2000). Nonepileptic seizure outcome varies by type of spell and duration of illness. *Epilepsia* **41**(10), 1330–1334.

Sheline, Y. I. (2003). Neuroimaging studies of mood disorder effects on the brain. *Biol Psychiatry* (54), 338–352.

Swinkels, W. A. M., Kyuk, J., vanDyck, R., Spinhoven, Ph. (2005). Psychiatric comorbidity in epilepsy. *Epilepsy Behav* **7**, 37–50.

Toone, B. K., Garralda, M. E., Ron, M. A. (1982). The psychosis of epilepsy and the functional psychosis: a clinical and phenomenological comparison. *Br J Psychiatry* **141**, 256–261.

Trimble, M. R. (1992). Behavior changes following temporal lobectomy with special reference to psychosis. *J Neurol Neurosurg Psychiatry* **55**(2), 89–91.

Walzac, T. S., Papacoastas, S., Williams, D. T., *et al.* (1995). Outcome after diagnosis of psychogenic nonepileptic seizures. *Epilepsia* **36**, 1131–1137.

CHAPTER 4

The Challenge of Teaching Psychiatry Residents About Psychopathology in Patients with Epilepsy

Sigita Plioplys and MaryBeth Lake

INTRODUCTION

Epilepsy is the third most common neurological disorder in adults after Alzheimer's disease and stroke. In North America, the overall prevalence of epilepsy in adults is about 5–10 per 1,000 (Theodore *et al.*, 2006). In pediatrics, epilepsy is the most common neurologic disorder, affecting up to 1% of children under the age of 16 (Shinnar and Pellock, 2002). Over the past 20 years, population-based longitudinal epidemiological research studies have convincingly demonstrated that epilepsy has a generally positive long-term outcome (Sillanpaa *et al.*, 1998). Approximately 70% of adults with childhood-onset epilepsy achieve stable seizure remission by adulthood whether they stay on or off antiepileptic drugs (AEDs) (Sillanpaa *et al.*, 1998). However, patients with epilepsy have high rates of comorbid psychopathology, poor psychosocial functioning, and require more psychiatric treatment compared to healthy controls. Since psychiatrists inevitably encounter patients with epilepsy in their clinical practices, they should have a solid knowledge of epilepsy, its diagnosis, and treatment.

Enhanced understanding of epilepsy as a *bona fide* neuropsychiatric disorder characterized by seizures, psychopathology, and cognitive and linguistic deficits has been one of the most important developments in contemporary epileptology. It has led to progressive changes in traditional neurologic care of patients with epilepsy and improved collaborations between neurologists and psychiatrists. Undoubtedly, patient care has advanced with the change in emphasis from focused seizure treatment to more comprehensive biopsychosocial management. Such a change has increased the need for psychiatric services in this population – and for psychiatrists trained to address epilepsy and its associated psychiatric, cognitive, developmental, and social problems.

Psychiatric Controversies in Epilepsy

In this chapter, we will discuss issues pertinent to the practical education of psychiatry residents about epilepsy and associated psychopathology.

THE CLINICIAN'S POINT OF VIEW

Dr Sigita Plioplys is a child and adolescent psychiatrist with training in neurology and psychiatry. She is a consultant for the Epilepsy Center at Children's Memorial Hospital in Chicago, Illinois, and teaches a 2-year neuropsychiatry course for child and adolescent psychiatry fellows at Children's Memorial Hospital (Northwestern University's Feinberg School of Medicine).

Epilepsy training models

Epilepsy teaching starts in medical school, where students learn about ictal physiology and pathology, seizure semiology, and diagnostic and treatment methods. After medical school, psychiatry residents further advance their medical/neurologic knowledge about this disorder through encounters with patients during their medical internship and neurology rotation. Formal epilepsy training for psychiatrists often stops here, as they continue with the general psychiatric education.

Perhaps the most effective integrated education on epilepsy and psychopathology takes place at specialized epilepsy treatment centers. Such centers usually function as multidisciplinary epilepsy treatment teams, in which mental health professionals play an integral role (Goldstein *et al.*, 2004). Frequently, psychiatrists working at such programs have combined training in both neurology and psychiatry, and are well equipped to provide the highest quality of neuropsychiatric education. At these programs, psychiatry trainees are directly supervised by psychiatry and neurology/epilepsy specialists who can offer a rich clinical experience with the most interesting and challenging epilepsy patients. Residents are taught to meaningfully interpret and then incorporate objective neurological, radiological, and psychological evaluation data into comprehensive psychiatric formulations. Moreover, learning about functional neuroimaging and neuropsychological assessments contributes to appreciation of higher cortical and neurocognitive functioning of patients with epilepsy. As a result, psychiatry trainees develop integrated understanding of psychopathology and neurocognitive functioning in patients with epilepsy, which allows them to determine effective psychiatric treatment strategies. Additionally, psychiatry trainees in these centers gain clinical experience with the newest AEDs, as well as learn about invasive and alternative seizure treatment strategies. Psychiatry trainees become skilled in recognizing acute problems and appreciating chronic aspects of living with epilepsy and psychopathology.

This multidisciplinary epilepsy teaching model is not widely available nationwide, as tertiary epilepsy centers tend to be located at large academic medical institutions.

In smaller training programs, psychiatry residents may not have access to such integrated epilepsy programs, but they will no doubt encounter patients with epilepsy through psychiatry rotations. On the consultation-liaison service, psychiatry trainees are most often exposed to acute psychopathology or changes in mental status that are associated with delirium, psychosis, mood disorder exacerbations, dangerousness, and medication side effects. Trainees become acquainted with peri-ictal and ictal presentations of psychopathology and the temporal relationship with the course of epilepsy. Also, they become familiar with non-epileptic seizures and underlying psychopathology, such as conversion, malingering, and factitious disorders. However, the consulting team usually has limited clinical contact with individual patients with epilepsy, and thus the learning experience may be fragmented.

Due to the high rates of comorbidity between epilepsy and psychopathology, psychiatry trainees inevitably become familiar with these patients in the inpatient and outpatient psychiatric setting. While managing psychopathology, psychiatry residents should be trained to vigilantly identify associated neurological problems and seek collaboration with neurologists. At this point, integrated learning can be challenged by the level of interaction and communication between psychiatry and neurology services.

Another obstacle to effective teaching about epilepsy and psychopathology is the relative delay in accessing the latest information about the AEDs, their psychiatric side effects, and drug interactions with psychotropic medications. This is especially relevant in training child and adolescent psychiatrists, as children with epilepsy are usually excluded from the already limited number of psychopharmacological trials performed in children.

Future recommendations

The multidisciplinary epilepsy team approach is the most progressive method for integrated teaching about epilepsy and psychopathology. Its driving principle, consistent and effective interdisciplinary communication, should be incorporated in all psychiatric settings. The use of contemporary communication methods, such as teleconferences or videoconferences, may allow colleagues to communicate with each other without constraints of time and distance. Clinical case conferences and interdepartmental grand rounds may also be effective avenues for interdisciplinary learning and teaching about psychopathology in patients with epilepsy.

THE EDUCATOR'S POINT OF VIEW

Dr MaryBeth Lake is the Director of Education and Child and Adolescent Psychiatry Residency Training at Children's Memorial Hospital (Northwestern University's Feinberg School of Medicine).

Current standards for neurology and neuroscience training in psychiatry

Epilepsy is an exemplary disorder model for the integrated teaching of neuropsychiatry given the high rates of psychiatric comorbidities, the necessity for multimodal treatment, the complex pharmacologic interactions, and the necessity for understanding the overall clinical course and long-term prognosis of neurological and psychiatric comorbidities. In addition to its value as a neuropsychiatric training tool, epilepsy should of course be taught for its utility: because epilepsy is so common, practicing psychiatrists will routinely encounter patients with epilepsy in their clinical practice.

The Accreditation Council of Graduate Medical Education (ACGME) does require training programs in psychiatry and the psychiatric subspecialties to provide clinical experience in neurology, but the requirements are quite general, leaving programs to develop neurology rotations without consistent parameters. Rather than tailor neurology rotations to key subspecialty areas, programs are often left to schedule residents into established services, which may or may not include significant exposure to conditions such as epilepsy. While the ACGME's lack of prescription is generally helpful to programs as it allows for flexibility and use of existing programs and teaching faculty, in the case of exposure to epilepsy, it may result in graduates completing training without adequate clinical knowledge base and experience.

In addition to the many ambiguities that surround neurology teaching in psychiatry, there are also questions with regard to neuroscience teaching. ACGME training guidelines for psychiatry and psychiatry subspecialties include no specific mention of "neuroscience" requirements as separate from neurology requirements (although neurobiology and neurophysiology are mentioned). Roffman and colleagues (Roffman et al., 2006) urge more attention to this area, referring to neuroscience as the "glue" to connect mind and brain, which can incorporate both psychotherapy and somatic treatments. Although the emphasis on neuroscience has increased in the last decade, teaching in this area remains variable in psychiatry training programs, and is more likely to take place in a didactic seminar format than in patient-related settings, according to the authors above.

At this point, there does not appear to be a consensus on what neuroscience topics should be taught. Furthermore, Duffy and Camlin's survey (Duffy and Camlin, 1995) highlights that more than a third of responding psychiatry residency training directors did not have a clear conceptualization of what "clinical entity is described by the term neuropsychiatry." Given these data, clear guidelines for specific neuroscience education requirements in psychiatry training, such as epilepsy teaching, could provide more consistency and standardization. This, along with specialized neuropsychiatry clinical experiences, would provide more comprehensive epilepsy education for psychiatrists.

Selwa and colleagues' survey findings indicate that educators consider current neurology training of psychiatry residents during their neurology rotations to be "adequate but not ideal" (Selwa *et al.*, 2006). They propose that interactive training models may lead to more beneficial collaborative learning for rotating psychiatry residents in neurology clinics. By integrating trainees from different disciplines in active outpatient care, and by careful selection of new patients or returning patients for interdisciplinary clinical care and/or teaching purposes, a more comprehensive neurology training experience could be provided.

Given the prevalence of epilepsy, it is surprising that in Selwa *et al.*'s survey, epilepsy was ranked only 8th out of 14 possible content areas identified by psychiatry training directors as important areas for training. Lack of standardized national guidelines and the shortage of neuropsychiatrists available for resident teaching may explain in part why this territory is left relatively neglected amid an ever-crowded curriculum and competing demands for faculty teaching, research, and clinical commitments.

Future recommendations

There is relatively little in the way of research or guidelines to assist psychiatry training directors in the teaching of epilepsy and related disorders. The only 10-year review article on the subject pertains to research and is in press (Plioplys *et al.*, 2007). Of note, the new *Clinical Manual of Pediatric Psychosomatic Medicine* (Shaw and DeMaso, 2006) has limited discussion about epilepsy and psychopathology, despite the high prevalence of epilepsy in pediatric patients seen in child and adolescent psychiatry clinical settings. The establishment of more detailed ACGME training guidelines for psychiatry programs could be immensely useful. Practice parameters or practice guidelines from the American Psychiatric Association (APA) and subspecialty national organizations such as the American Academy of Child and Adolescent Psychiatry (AACAP) on the evaluation and treatment of epilepsy and psychiatric comorbidities would also be of great use.

With the addition of more guidance, portable standardized model curricula could be developed by specialized epilepsy treatment centers and replace the apprenticeship model in this area with a structured evidence-based content to be shared among training programs (Glick and Zisook, 2005). This could be a particularly viable option in this era of limited faculty teaching resources and could be adapted to a specialized area of the psychiatric interface with epilepsy. Clinical competency of trainees could then be naturally evaluated within such a structure. This educational model would complement the current problem-based integrative learning methods used within medical schools that provide the essential framework for more sophisticated education in residency programs.

In addition to standardized model curricula, clinical teaching could be integrated into clinic settings such as neuropsychiatry clinics, as discussed previously, if available. If such a specialized clinic does not exist in a particular program, rather than assigning trainees to general neurology clinics, neurology consult services, or neurology inpatient units, should a specified portion of the neurology rotation in psychiatry training be dedicated specifically to work with epileptologists in clinics, inpatient units, and/or EEG/video-EEG labs?

It would follow that increased emphasis in this area of training should translate to examination on American Board of Psychiatry and Neurology (ABPN) exams. Although Sharp and colleagues point out the limited research to date on the correlation between board certification and clinical outcomes (accepted national standards of care), an increased emphasis on board exams would certainly highlight the importance of training in the area of epilepsy for psychiatry graduates (Sharp *et al.*, 2002).

Increased collaboration between epileptologists and psychiatrists (and trainees in both disciplines) is necessary to beneficially improve clinical care of patients with epilepsy (Kanner, 2003). There is hope that residency training can be shaped in this direction by the clinical needs of our patients.

REFERENCES

Duffy, J. D., Camlin, H. (1995). Neuropsychiatric training in American psychiatric residency programs. *J Neuropsychiatry* **7**(3), 290–294.

Glick, I. D., Zisook, S. (2005). The challenge of teaching psychopharmacology in the new millennium: the role of curricula. *Acad Psychiatry* **29**(2), 134–140.

Goldstein, J., Plioplys, S., Zelko, F., Mass, S., Corns, C., Blaufuss, R., Nordli, D. (2004). Multidisciplinary approach to childhood epilepsy: exploring the scientific rationale and practical aspects of implementation. *J Child Neurol* **19**(5), 362–378.

Kanner, A. M. (2003). When did neurologists and psychiatrists stop talking to each other? *Epilepsy Behav* **4**(6), 597–601.

Plioplys, S., Dunn, D., Caplan, R. (2007). Psychiatric problems in children with epilepsy: a 10-year research review update. *J Am Assoc Child Adol Psychiatry* **46**(11): 1389–1402.

Roffman, J. L., Simon, A. B., Prasad, K. M., Truman, C. J., Morrison, J., Ernst, C. L. (2006). Neuroscience in psychiatry training: how much do residents need to know? *Am J Psychiatry* **163**, 919–926.

Selwa, L. M., Hales, D. J., Kanner, A. M. (2006). What should psychiatry residents be taught about neurology? A survey of psychiatry residency directors. *The Neurologist* **12**(5), 268–270.

Sharp, L. K., Bashook, P. G., Lipsky, M. S., Horowitz, S. D., Miller, S. H. (2002). Specialty board certification and clinical outcomes: the missing link. *Acad Med* **77**(6), 534–542.

Shaw, R. J., DeMaso, D. R. (2006). *Clinical Manual of Pediatric Psychosomatic Medicine: Mental Health Consultation With Physically Ill Children And Adolescents.* Arlington, American Psychiatric Publishing.

Shinnar, S., Pellock, J. M. (2002). Update on the epidemiology and prognosis of pediatric epilepsy. *J Child Neurol* **17**(Suppl 1), S4–S17.

Sillanpaa, M., Jalava, M., Kaleva, O., Shinnar, S. (1998). Long-term prognosis of seizures with onset in childhood. *New Engl J Med* **338**, 1715–1722.

Theodore, W. H., Spencer, S. S., Wiebe, S., Langfitt, J. T., Ali, A., Shafer, P. O., Berg, A. T., Vickrey, B. G. (2006). Epilepsy in North America: a report prepared under the auspices of the Global Campaign Against Epilepsy, the International Bureau for Epilepsy, the International League Against Epilepsy, and the World Health Organization. *Epilepsia* **47**(10), 1700–1722.

CHAPTER 5

What Do We Know about Mood Disorders in Epilepsy?

Marco Mula and Michael R. Trimble

INTRODUCTION

It has been known for a long time that there are some associations between epilepsy and depression. There are several obvious reasons why the two may be closely linked (Kanner, 2003). Epilepsy is a chronic disorder, which brings about much social discrimination. In patients with intractable seizures there is the ever-ready risk of either becoming unconscious, of falling and damaging the self, and in public of social embarrassment. As with any chronic disorder then, epilepsy might be expected to be linked to demoralization and a negative perspective on life, but further to disturbances of affect and reactive depressive features. However, more recent research, while reinforcing this association, has revived ancient ideas regarding a biological contribution to the association based on neuroanatomical and neurochemical principles. Associations between antiepileptic drug prescription and depression have been one focus of attention. In contrast to the associations with depression, bipolar disorder or overt mania in epilepsy was always considered to be much less common, and indeed classic bipolar disorder was considered rare (Wolf, 1982; Mula and Monaco, 2006).

Depression and epilepsy: an overview

Assessing the frequency of epilepsy and depression from selected clinic samples gives a bias toward the more severely affected patients and also those on the most medication. A better understanding of comorbid psychopathology should come from community studies. Edeh and Toone (1987) carried out a general practice study in the United Kingdom and reported that 22% of unselected patients with epilepsy were rated as having a depressive disorder. A Canadian Community Health Survey examined 253 people with epilepsy using a rating scale to identify a history of depression, and noted a lifetime prevalence of depression at 22%: this was compared with 12% in the general population (Tellez-Zenteno et al., 2005).

More recently, Ettinger *et al.* (2004) assessed depression in 775 people with epilepsy, and compared the incidence with patients with asthma, and also healthy controls. In this study, a rating scale assessment was again used (Centers of Epidemiological Studies – Depression Instrument). Symptoms of depression were significantly more frequent in the epilepsy group (36.5%) in comparison with those with asthma (27.8%) and the controls.

Several studies have noted a correlation with seizure frequency. In an epidemiological study, Jacoby *et al.* (1996) noted that depression occurred in 4% of seizure-free patients, in 10% of patients with less than one seizure a month, but at a rate of 21% in patients with higher seizure frequency. O'Donoghue *et al.* (1999) noted that patients with epilepsy with continuing seizures were significantly more likely to suffer from depression than those in remission (33% vs. 6%).

There are a number of studies from selected patient groups, for example tertiary centers, or from those awaiting surgery for epilepsy, which note in these populations an even higher frequency of depression. Victoroff *et al.* (1994) evaluated 60 patients with intractable complex partial seizures using a structured clinical interview for Diagnosis and Statistical Manual (DSM)-IIIR diagnoses, and observed that 58% had histories of depressive disorders. Jones *et al.* (2005a) examined 199 patients from five epilepsy centers, again using structured clinical interview techniques and noted 34% to meet criteria for a mood or anxiety disorder, and 19% to meet criteria for major depression. Ring *et al.* (1998) examined 60 patients awaiting temporal lobe epilepsy surgery, and noted at preoperative assessment that a major depressive disorder was present in 21%. These data suggest that we know that epilepsy and depression are commonly linked together, an association found more frequently than in some other chronic disorders such as asthma. We also know that the relationship of depression is greater in those with higher seizure frequencies and with continuing seizures.

Thus, the presence of depression can be said to be even greater in selected populations that most likely reflects in a large part on the intractability of the seizure disorders.

Another finding, which has been verified, is the association between depressive symptomatology and quality of life in people with epilepsy. Gilliam *et al.* (1997) noted depressive symptoms to be the most important predictor of quality of life, and these were a more powerful predictor than the actual seizure frequency. Perrine *et al.* (1995) and Boylan *et al.* (2004) have reported similar findings.

SYNDROMES, SEIZURES, DRUGS AND MOOD

The relationship of epilepsy syndromes to depression

There has been considerable debate, which is unresolved, as to the association between any particular epilepsy syndrome and depression. It is the case that people with lesional epilepsy from the temporal lobes are more likely to have intractable seizures, and they also are more likely to be taking more extensive medication than those with non-temporal epilepsy, and they may be at increased risk. Thus, some studies have shown patients with temporal lobe epilepsy to be more prone to depression than other groups, but other investigations have failed to confirm this. Examining etiology of epilepsy, Quiske *et al.* (2000) found that patients with temporal lobe epilepsy who had mesial temporal sclerosis were more likely to report depression. In general terms, there is more agreement that patients with complex partial seizures are more likely to have a depressive disorder (Robertson, 1998).

There are also studies linking frontal lobe dysfunction to the depression of epilepsy. The latter have emerged from investigations using brain imaging (positron emission tomography – PET or single-photon emission-computed tomography – SPECT), and neuropsychological batteries. Hermann *et al.* (1991) noted that patients with temporal lobe epilepsy and depression were more likely to perform poorly on frontal lobe neuropsychological tasks, especially with a left-sided seizure focus. Schmitz *et al.* (1997) noted similar frontal changes and localizations using SPECT, and Bromfield *et al.* (1990) using PET reported that patients with temporal lobe epilepsy and associated depression revealed bilateral reductions of frontal lobe metabolism.

Although these studies, of necessity, were on a limited number of patients, the concordance between the findings does support an anatomical association between temporal lobe epilepsy, depression and frontal lobe dysfunction.

With regards to the temporal lobe association with depression, it is of interest that there are a number of studies outside the field of epilepsy which suggest that hippocampal volume loss is associated with affective disorders (Bremner *et al.*, 2000; Frodl *et al.*, 2002). Thus, although further research in this area is needed, neuroimaging studies are revealing an underlying neurocircuitry of depression in psychiatric patients without a neurological disorder, which includes the hippocampus, thus in keeping with the findings in sub-groups of patients with epilepsy.

The relationship of antiepileptic drugs (AEDs) to depression

There is an older literature, which links barbiturate-related drugs to depression, but the more recent data suggests that it is possible to distinguish between AEDs with the potential to have positive effects on mood, such as carbamazepine and valproic acid, from these older compounds.

The role of AEDs in precipitating depression has become of considerable interest following the introduction of a spectrum of molecules referred to as new AEDs (Mula and Sander, 2007). Within this literature the concept of forced normalization has been revived; the phenomenon which describes the sudden switching off of seizures in people with intractable epilepsy who then develop an alternative psychiatric syndrome. Very often this is a psychotic disorder, but depressive symptoms are also reported (Trimble, 1998). This literature is of interest because the psychoses appear to be most usually related to complete cessation of seizures, whereas depression, while often related to control of seizure frequency, is not so often associated with complete cessation (Ring et al., 1993; Thomas et al., 1996; Mula et al., 2008).

The AEDs most associated with this effect seem to be those which act at the benzodiazepine–GABA receptor complex, and include tiagabine, topiramate and vigabatrin. Since in psychiatric practice it is known that benzodiazepines and other γ-aminobutyric acid (GABA) agonists are associated clinically with depression, and that abnormalities of cerebrospinal fluid (CSF) GABA have been reported in patients with depression (Trimble, 1996), the link between sudden cessation of seizures, GABAergic agents and the onset of depression seems reasonably secure. Further, these studies have revealed that patients with epilepsy and a prior history of an affective disorder are the more likely to develop depression in these circumstances.

Bipolar disorder and epilepsy

It used to be confidently stated that bipolar disorder was rare in patients with epilepsy (Wolf, 1982). Such statements were made prior to the use of standardized diagnostic manuals such as the DSM-IV, and were also based on clinical impression rather than being assessed by the use of rating scales. It was accepted however that in the context of the postictal state, patients could develop a postictal psychosis, the features of which were often manic or hypomanic, although more generally the presentation was one of a mixed affective state often with psychotic features. A recent study (Kanner et al., 2004) of the postictal symptoms of 100 patients noted postictal hypomanic symptoms in 22 patients, often with associated psychotic phenomenology. Nishida et al. (2006) recently showed that postictal mania has a distinct position among mental disorders observed in the postictal period. Postictal manic episodes last for a longer period than postictal psychotic episodes. They have a higher frequency of recurrence than postictal psychoses and are associated with an older age at onset, electroencephalogram (EEG) frontal discharges and right hemisphere involvement.

In the study of Ettinger et al. (2005), bipolar symptoms were revealed in 12.2% of the epilepsy patients, which was twice as common as in people with asthma, and seven times as common as in the healthy comparison group. Of those who in the screening process were rated as potential patients with bipolar symptoms, nearly 50% were rated by a physician as having a bipolar disorder.

These data raise some doubts about the previous suggestions that bipolar disorder is rare in people with epilepsy, and raise doubts as to our knowledge of the association between these two disorders. The older discussions related more to classical manic-depressive disorder, as opposed to the concept of the bipolar spectrum, which is currently the focus of psychiatric interest. However, it is clearly evident that more studies are warranted to clarify the prevalence of bipolar disorder in epilepsy and the possible impact on social and quality-of-life measures.

A bi-directional relationship

Another finding which has been replicated is that the link between depression and epilepsy is not necessarily unidirectional, namely that patients with the comorbidity always present with the seizure disorder before the emergence of the depression. Thus, it has been noted in epidemiological studies that having a prior mood disorder can be associated with an increased risk of epilepsy (Forsgren and Nystrom, 1990; Hesdorffer *et al.*, 2000). There may be a number of reasons for this, including perhaps the development of epilepsy following suicidal attempts, following drug abuse or following some other kinds of trauma such as head trauma. However, the findings may reflect on an underlying common pathogenesis, which may relate to some as yet unknown genetic factor, or some link with neurotransmitter function (e.g., related to transmitters that are known to play a role in both epilepsy and depression such as serotonin or GABA).

IS THERE AN EPILEPSY-SPECIFIC AFFECTIVE SYNDROME?

The issue of phenomenology of depression in epilepsy is very much a matter of debate. Betts (1974) reported that patients with epilepsy and depression manifested more endogenous features, while Mulder and Daly (1952) commented on a more reactive nature of depression. The spectrum is likely to be large. In general terms, it is reasonable to hypothesize that patients with epilepsy can experience forms of mood disorders identical to those of patients without epilepsy. However, it is equally reasonable to assume that the underlying brain pathology can influence the expression of mood disorder symptoms, making less evident some aspects or emphasizing others.

The comorbidity between mood and anxiety disorders could be a potential example of such a peculiarity (see Chapter 16). Comorbid anxiety symptoms have been identified in 73% of patients with epilepsy and depression (Robertson *et al.*, 1987; Jones *et al.*, 2005a). However, it is still unknown whether this is an association identical to that described in primary mood disorders or a comorbidity typical of chronic illnesses (where anxiety is present, with irritability and generalized turmoil), or if it is distinctive only of epilepsy and interlinked with the underlying brain pathology. The recognition of comorbid anxiety symptoms is very important clinically, since they may worsen the quality of life of depressed patients and significantly increase the risk of suicide (Kanner, 2006).

One of the main reasons in favor of the existence of an epilepsy-specific affective disorder comes from the clinical observation that the psychopathology often has unique manifestations that are poorly reflected by conventional classification systems such as DSM-IV and ICD-10 (Krishnamoorthy *et al.*, 2007). Mendez *et al.* (1986) investigated the clinical semiology of depression in 175 patients with epilepsy, and reported that 22% could be classified as having atypical features. In particular, classic endogenous-type depressive symptoms, such as feelings of guilt, "*Gefühl der Gefühllosigkeit*," and a circadian pattern of symptom severity were rarely reported. Kanner *et al.* (2000) showed that 71% of patients with refractory epilepsy and depressive episodes, severe enough to need a psychopharmacological treatment, failed to meet criteria for any DSM-IV axis I diagnosis. A study of 199 consecutive patients with epilepsy revealed that 64% failed to meet any DSM-IV criteria using two structured clinical interviews, namely the Structured Clinical Interview for DSM-IV Axis I (SCID-I) and the Mini International Neuropsychiatric Interview (MINI) (Kanner *et al.*, 2004).

In contrast, a recent paper by Jones *et al.* (2005a) showed that a current diagnosis of any mood disorder could be allocated in 24% of patients recruited in five tertiary referral centers for epilepsy in United States, using the MINI. A lifetime diagnosis of major depressive episode was reported in 17.2% of cases, 4% for dysthymic disorder and 2.8% for bipolar disorder. This study, although in line with previous authors, clearly suggests that it is possible to apply standardized criteria of DSM in a not negligible proportion of patients.

Among different potential causes for the atypical features of affective disorders in epilepsy, the peri-ictal cluster of symptoms, to some degree, may account for their features (see Chapter 15) but the possibility that the mood disorders of epilepsy may have unique characteristics has plausibility.

The interictal dysphoric disorder

Kraepelin (1923), and then Bleuler (1949), were the first authors to describe in epilepsy a pleomorphic pattern of symptoms, including affective symptoms with prominent irritability intermixed with euphoric mood, fear, anxiety as well as anergia, pain and insomnia. Gastaut *et al.* (1955) confirmed Kraepelin's and Bleuler's observations, leading Blumer to coin the term interictal dysphoric disorder (IDD) to refer to this type of depressive disorder in epilepsy (Blumer, 2000). Blumer's description of the IDD is particularly intriguing. It is characterized by eight key symptoms grouped in three major categories (Table 5.1): labile depressive symptoms (depressive mood, anergia, pain, insomnia), labile affective symptoms (fear, anxiety), and supposedly "specific" symptoms (paroxysmal irritability, euphoric moods). Blumer preferred the term "dysphoria" to more accurately translate the original definition of Kraepelin "*Verstimmungszustand*" to stress the periodicity of mood changes of the patients and the presence of irritability and outbursts of aggressive behavior as key symptoms. The dysphoric episodes are described as occurring without external triggers and without clouding of consciousness.

TABLE 5.1   Main symptoms clusters of the interictal dysphoric disorder

Labile depressive symptoms	Labile affective symptoms	Specific symptoms
Depressed mood	Fear	Euphoric moods
Anergia	Anxiety	Paroxysmal irritability
Pain		
Insomnia		

They begin and end rapidly and recur fairly regularly in a uniform manner, occurring every few days to every few months and lasting a few hours up to 2 days.

The theoretical framework suggested by Blumer goes beyond a narrow IDD profile, and he speculated that affective symptoms in epilepsy exist along a continuum, from a dysphoric disorder with fleeting symptoms, to a more severe disorder with transient psychotic features, to an even more debilitating disorder with prolonged psychotic states. This scenario is deeply influenced by classic German psychiatry, especially Kraepelin's view of the relationship between manic-depressive illness and schizophrenia (Kraepelin, 1923).

It is obvious that the IDD can be recognized today with features that are different from those described by premodern psychiatry. For example, depressed mood and anergia may be much more evident than before because modern antiepileptic medications may accentuate the dysphoric symptoms. Himmelhoch (1984), and subsequently Kanner and Balabanov (2002), highlighted the chronic course of this state of moderate neurotic depression with symptom-free intervals typical of epilepsy, referring to a dimension very close to dysthymia. In our opinion, IDD patients have several features in common with a specific subset of cyclothymic subjects, where depressive periods and labile-angry-irritable moods dominate the clinical picture. Nonetheless, Blumer reported that patients with IDD benefit from a combined therapy of AEDs and antidepressant drugs (Blumer et al., 2004), a combination extensively used in psychiatry in bipolar depression, suggesting that IDD is close to the bipolar spectrum. It is, therefore, evident that the features of IDD overlap with a variety of affective disorders seen in clinical psychiatric practice, but the concept needs to be further defined in terms of semiology and clinical description (Figure 5.1). In a recent study, we investigated prevalence and psychopathological features of IDD in patients with epilepsy compared with a group of patients with migraine (Mula et al., 2008). Our data suggest that IDD is a robust construct and can be diagnosed in a significant proportion of patients (about 20%), being one of the most frequent mood disorders in epilepsy. Nevertheless, it seems that IDD is not typical only of patients with seizure disorders but can be seen also in other central nervous system (CNS) disorders such as migraine with the same prevalence. This concept was partly suggested by Blumer himself, stating that IDD can occasionally occur in the absence of clinical seizures in patients with

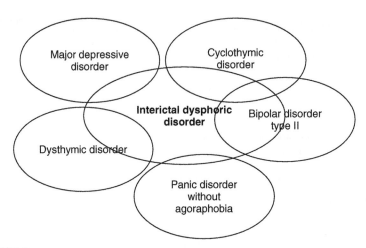

FIGURE 5.1   Overlap between DSM-IV mood and anxiety disorders and the IDD.

brain lesions (with or without an abnormal EEG) (Blumer *et al.*, 1988). However, further studies are needed to clarify whether IDD is really epilepsy-related or relates to the typical phenomenology of affective disorders in chronically ill populations or better, whether IDD is an organic affective disorder syndrome.

THE PROBLEM OF DEPRESSION DIAGNOSIS IN EPILEPSY

To conclude, a definite diagnosis of depression in patients with epilepsy can be difficult because a number of symptoms, which are recognized as diagnostic criteria for a depressive episode by the ICD-10 and DSM-IV, may occur in epilepsy secondary to seizure activity or AED treatment (e.g., loss of energy, insomnia or hypersomnia, increase or decrease in appetite, loss of libido, psychomotor agitation or retardation, diminished ability to think or concentrate). Because these symptoms may be present in patients who are not depressed, physicians need to explore fully the mental status of their patients. Inquiring about anhedonia has been suggested as an excellent predictor of the presence of depression (Kanner, 2006) and the use of self-rating instruments can be revealing. However, one of the most frequent methodological errors in research studies on depression and epilepsy is the sole reliance on screening instruments to diagnose depressive disorders. Firstly, a depressive episode can occur as part of a major depressive disorder, a bipolar disorder or as part of a double depression, which consists of recurrent major depressive episodes during a dysthymic disorder. Secondly, it is established that up to 50% of mood disorders identified in patients with epilepsy present with atypical clinical characteristics that fail to meet any of the DSM Axis I categories (see above) (Kanner, 2006).

In a recent paper, Krishnamoorthy (2006) reviewed instruments available for the evaluation of behavioral disturbances in epilepsy, pointing out that the vast majority of studies use measures or cut-off scores that may not be valid in the epilepsy population. At present, no measures exist that have been developed *de novo* for the assessment of comorbid psychopathology in epilepsy, using modern techniques of questionnaire development.

Krishnamoorthy (2005) suggested an adapted version of the SCID-I, named SCID-E. Mintzer and Lopez (2002) proposed the Epilepsy Addendum for Psychiatric Assessment (EAPA), an instrument expressly designed for use with the MINI. However, the relative benefits of these various instruments, in the assessment of generic psychopathology in community-based studies, are the subject of considerable debate. The psychometric properties of the Beck Depression Inventory (BDI), a well-known self-rating scale, to detect the severity of current (past 2 weeks) depressive symptoms, have been investigated in the epilepsy setting against the SCID-I, showing a good sensitivity (0.93), an acceptable specificity (0.81) and an excellent negative predictive value (0.98) but a very low positive predictive value (0.47) (Jones *et al.*, 2005b). A six-item screening instrument, the Neurological Disorders Depression Inventory for Epilepsy (NDDI-E), was recently validated to screen for major depressive episodes in epilepsy (Gilliam *et al.*, 2006). It has the advantage of being constructed specifically to minimize confounding factors, such as adverse events related to AEDs or cognitive problems associated with epilepsy, and showed an internal consistency of 0.85 and a test-retest reliability of 0.78. A score of 15 or higher has a specificity of 0.90 and a sensitivity of 0.81 for a diagnosis of major depression.

All the instruments discussed so far relate to a diagnosis of a mood disorder characterized by symptoms identical to those of the general population. However, it seems established that affective disorders in many patients with epilepsy are, in fact, different from those of non-epileptic patients and, in some cases such as the IDD, unique. The Seizure Questionnaire (Blumer *et al.*, 2002) contains an inquiry for the eight key symptoms of the IDD. Patient and next of kin answer them jointly, and the examiner for completeness and accuracy then reviews all answers.

We have recently developed a specific questionnaire, named Interictal Dysphoric Disorder Inventory (IDDI) (see page 63). This is a 38-item self-report questionnaire where the eight key symptoms are evaluated in the first 32 items in terms of presence, frequency, severity and global impairment. The time-interval explored is the last 12 months. The six questions of the Appendix concern the time course of the disorder and relations of symptoms with seizures and therapy. As suggested by Blumer *et al.* (2004), a definite diagnosis of IDD is defined by the presence of at least three symptoms of at least "moderate" or "severe" severity that cause "moderate" or "severe" distress. Apart from the diagnosis, it is possible to obtain total and separate scores for key symptoms and the "severeness" of key symptoms, defined by severity, frequency and impairment scores (see page 66). The IDDI represents the first instrument for standardized diagnosis of IDD and its severity. It has been developed from a preliminary version of Krishnamoorthy and Trimble, and studies of validation against DSM-IV criteria for affective disorders are ongoing.

TREATMENT

It is important to state at the outset that there is only one controlled trial of the effects of an intervention for mood disorders in epilepsy, and the evidence for treatment strategies relies heavily on clinical experience.

Psychiatric symptoms temporally related to the occurrence of seizures (preceding or following a seizure or a cluster of seizures; occurring when the patient achieves a sudden and complete seizure control or when seizures worsened) do not need any specific psychotropic treatment and a better control of seizures is often the sole solution.

In the case of a mood disorder characterized by symptoms occurring independently of seizures, psychopharmacotherapy can be required but evidence in favor of a particular drug is lacking. The only published controlled trial involved nomifensine, an antidepressant no longer available (Robertson and Trimble, 1985).

Selective serotonin reuptake inhibitors (SSRIs) have become the first-line of pharmacotherapy for primary major depressive and dysthymic disorders in psychiatric but also in neurological practice. However, studies about efficacy and safety in epilepsy are lacking. During recent years, a number of authors have approached the clinical problem of treating mood disorders in epilepsy from different points of view (Mula *et al.*, 2004; Kanner and Balabanov, 2005; Prueter and Norra, 2005). A few open studies have been published about sertraline (Kanner *et al.*, 2000; Thomè-Souza *et al.*, 2007), citalopram (Hovorka *et al.*, 2000; Kuhn *et al.*, 2003; Specchio *et al.*, 2004), reboxetine (Kuhn *et al.*, 2003), mirtazapine (Kuhn *et al.*, 2003), and fluoxetine (Thomè-Souza *et al.*, 2007). The study by Thomè-Souza *et al.* (2007) is of particular interest because it is the only published paper involving children and adolescents with epilepsy and depression.

In general, all presented studies showed antidepressant drug treatment to be well tolerated, but the reported response rates were highly variable, ranging, for example, with citalopram, between 38% (Khun *et al.*, 2003) and 65% (Hovorka *et al.*, 2000) at eight weeks. It is evident that the reported variability is influenced by the selection of patients, the lack of a rigorous psychiatric assessment for a correct diagnosis (dysthymia, major depression, bipolar depression, IDD), the presence of other comorbid Axis I disorders, the presence of brain damage, cognitive impairment, a family history for mood disorders and so on. All these variables are taken into consideration very rarely in these studies, but they are essential for a correct interpretation of results.

If studies about psychoactive drugs in epilepsy are rare, studies about psychological therapies for mood disorders in epilepsy are really exceptional. We are aware of only two papers, one involving adult patients (Tan and Bruni, 1986) and the other with children (Martinovic *et al.*, 2006). Both showed some utility of cognitive behavioral therapies in the management of mood disorder symptoms in epilepsy.

The issue of psychotropic drug treatment of depression in epilepsy is inter-linked with that of the "proconvulsant" or "anticonvulsant" effects of antide-pressants. Tricyclic antidepressants developed a clinical reputation for convulsant liability soon after their introduction into therapeutics (Dailey and Naritoku, 1996). The concept that antidepressant medications are more likely to produce convulsions in patients with epilepsy than in patients without this disorder is intuitively appealing, and is seemingly compatible with the concept that seizure predisposition is fundamental to the definition of epilepsy. However, it is clear that the biology of seizure predisposition is not a single entity; moreover, it is not clear whether the risk of seizure expression arises from the seizure liability itself or from a more complex predisposition inherent in the mechanisms of comorbidity between affective disorders and the epilepsies (Jobe and Browning, 2005).

A number of authors claimed possible anticonvulsant properties for SSRIs speculating about the antagonistic role of serotonin transmission in epileptogenesis (Favale *et al.*, 1995; Albano *et al.*, 2006). Although some neurobiological explana-tions have been suggested, further studies are needed to clarify whether SSRIs can exhibit direct anticonvulsant properties or are indirectly effective, for example by ameliorating sleep and circadian patterns.

CONCLUSIONS

Although there are a number of psychiatric controversies in epilepsy, that there is a link between depression and epilepsy has never been controversial. At one time the arguments for whether or not the majority of the clinical presentations could be explained by psychosocial factors were prominent, and whatever way one views this, epilepsy is a disorder of lifestyle as such that one would anticipate a close link between depression and epilepsy on those grounds alone. However, the literature as reviewed suggests that the link between depression and epilepsy may not be unidirectional, that there is probably an over-representation of people with complex partial seizures (and possibly temporal lobe epilepsy), as opposed to other forms of epilepsy, that some antiepileptic drugs are more associated with depres-sive syndromes than others (particularly those that are GABAergic), and that sud-den reduction of seizures may precipitate the onset of affective symptoms.

It does seem generally agreed that the clinical picture is not typical in all cases for a DSM-IV categorization, and a number of authors have noted the atypical nature of the clinical picture. The suggestion (with historical reference) is that there is within the overall presentation of patients with depression and epilepsy a sub-group who have an epilepsy-specific affective syndrome which some have referred to as the IDD of epilepsy. In the data presented in this chapter we have suggested that patients presenting with IDD have overlapping features with cyclothymia, and we have suggested that the link between epilepsy and bipolar disorders requires further evaluation.

Finally, we note the scarcity of controlled clinical trials of treatment agents of depression in epilepsy, and hope that this chapter will lead to further work in this area.

REFERENCES

Albano, C., Cupello, A., Mainardi, P., Scarrone, S., Favale, E. (2006). Successful treatment of epilepsy with serotonin reuptake inhibitors: proposed mechanism. *Neurochem Res* **31**, 509–514.

Betts, T. A. (1974). A follow up study of a cohort of patients with epilepsy admitted to psychiatric care in an English city. In *Epilepsy: Proceedings of the Hans Berger Centenary Symposium* (P. Harris, C. Mawdsley, eds), pp. 326–338. Edinburgh: Churchill Livingstone.

Bleuler, E. (1949). *Lehrbuch der Psychiatrie* 8th edition. Berlin: Springer-Verlag.

Blumer, D. (2000). Dysphoric disorders and paroxysmal affects: recognition and treatment of epilepsy-related psychiatric disorders. *Harv Rev Psychiatry* **8**, 8–17.

Blumer, D., Heilbronn, M., Himmelhoch, J. (1988). Indications for carbamazepine in mental illness: atypical psychiatric disorder or temporal lobe syndrome? *Compr Psychiatry* **29**, 108–122.

Blumer, D., Montouris, G., Davies, K., Wyler, A., Phillips, B., Hermann, B. (2002). Suicide in epilepsy: psychopathology, pathogenesis, and prevention. *Epilepsy Behav* **3**, 232–241.

Blumer, D., Montouris, G., Davies, K. (2004). The interictal dysphoric disorder: recognition, pathogenesis, and treatment of the major psychiatric disorder of epilepsy. *Epilepsy Behav* **5**, 826–840.

Boylan, L. S., Flint, L. A., Labovitz, D. L., Jackson, S. C., Starner, K., Devinsky, O. (2004). Depression but not seizure frequency predicts quality of life in treatment-resistant epilepsy. *Neurology* **62**, 258–261.

Bremner, J. D., Narayan, M., Anderson, E. R., Staib, L. H., Miller, H. L., Charney, D. S. (2000). Hippocampal volume reduction in major depression. *Am J Psychiatry* **157**, 115–118.

Bromfield, E., Altschuler, L., Leiderman, D. (1990). Cerebral metabolism and depression in patients with complex partial seizures. *Epilepsia* **31**, 625.

Dailey, J. W., Naritoku, D. K. (1996). Antidepressants and seizures: clinical anecdotes overshadow neuroscience. *Biochem Pharmacol* **52**, 1323–1329.

Edeh, J., Toone, B. (1987). Relationship between interictal psychopathology and type of epilepsy. Results of a survey in general practice. *Br J Psychiatry* **151**, 95–101.

Ettinger, A., Reed, M., Cramer, J. (2004). Epilepsy Impact Project Group. Depression and comorbidity in community-based patients with epilepsy or asthma. *Neurology* **63**, 1008–1014.

Ettinger, A. B., Reed, M. L., Goldberg, J. F., Hirschfeld, R. M. (2005). Prevalence of bipolar symptoms in epilepsy vs other chronic health disorders. *Neurology* **65**, 535–540.

Favale, E., Rubino, V., Mainardi, P., Lunardi, G., Albano, C. (1995). Anticonvulsant effect of fluoxetine in humans. *Neurology* **45**, 1926–1927.

Forsgren, L., Nystrom, L. (1990). An incident case-referent study of epileptic seizures in adults. *Epilepsy Res* **6**, 66–81.

Frodl, T., Meisenzahl, E. M., Zetzsche, T., Born, C., Groll, C., Jager, M., Leinsinger, G., Bottlender, R., Hahn, K., Moller, H. J. (2002). Hippocampal changes in patients with a first episode of major depression. *Am J Psychiatry* **159**, 1112–1118.

Gastaut, H., Morin, G., Lesevre, N. (1955). Etude du comportement des epileptiques psychomoteurs dans l'intervalle de leurs crises: les troubles de l'activitè globale et de la sociabilitè. *Ann Med Psychol (Paris)* **113**, 1–27.

Gilliam, F., Kuzniecky, R., Faught, E., Black, L., Carpenter, G., Schrodt, R. (1997). Patient validated content of epilepsy specific quality of life measurement. *Epilepsia* **38**, 233–236.

Gilliam, F. G., Barry, J. J., Hermann, B. P., Meador, K. J., Vahle, V., Kanner, A. M. (2006). Rapid detection of major depression in epilepsy: a multicentre study. *Lancet Neurol* **5**, 399–405.

Hermann, B. P., Seidenberg, M., Haltiner, A., Wyler, A. R. (1991). Mood state in unilateral temporal lobe epilepsy. *Biol Psychiatry* **30**, 1205–1218.

Hesdorffer, D. C., Hauser, W. A., Annegers, J. F., Cascino, G. (2000). Major depression is a risk factor for seizures in older patients. *Ann Neurol* **47**, 246–249.

Himmelhoch, J. M. (1984). Major mood disorders related to epileptic changes. In *Psychiatric Aspects of Epilepsy* (D. Blumer, ed.), pp. 271–294. Washington, DC: American Psychiatric Press.

Hovorka, J., Herman, E., Nemcova, I. I. (2000). Treatment of interictal depression with citalopram in patients with epilepsy. *Epilepsy Behav* **1**, 444–447.

Jacoby, A., Baker, G. A., Steen, N., Potts, P., Chadwick, D. W. (1996). The clinical course of epilepsy and it psychosocial correlates: findings from a UK community study. *Epilepsia* **37**, 148–161.

Jobe, P. C., Browning, R. S. (2005). The serotonergic and noradrenergic effects of antidepressant drugs are anticonvulsant, not proconvulsant. *Epilepsy Behav* **7**, 602–619.

Jones, J. E., Hermann, B. P., Barry, J. J., Gilliam, F., Kanner, A. M., Meador, K. J. (2005a). Clinical assessment of Axis 1 psychiatric morbidity in chronic epilepsy: a multicentre investigation. *J Neuropsychiatry Clin Neurosci* **17**, 172–179.

Jones, J. E., Hermann, B. P., Woodard, J. L., Barry, J. J., Gilliam, F., Kanner, A. M., Meador, K. J. (2005b). Screening for major depression in epilepsy with common self-report depression inventories. *Epilepsia* **46**, 731–735.

Kanner, A. M. (2003). Depression in epilepsy: a frequently neglected multifaceted disorder. *Epilepsy Behav* **4**, S11–S19.

Kanner, A. M. (2006). Depression and epilepsy: a new perspective on two closely related disorders. *Epilepsy Currents* **6**, 141–146.

Kanner, A. M., Balabanov, A. (2002). Depression and epilepsy: how closely related are they? *Neurology* **58**(8 Suppl 5), S27–39.

Kanner, A. M., Balabanov, A. J. (2005). Pharmacotherapy of mood disorders in epilepsy: the role of newer psychotropic drugs. *Curr Treat Options Neurol* **7**, 281–290.

Kanner, A. M., Kozac, A. M., Frey, M. (2000). The use of sertraline in patients with epilepsy: is it safe? *Epilepsy Behav* **1**, 100–105.

Kanner, A. M., Soto, A., Gross-Kanner, H. (2004). Prevalence and clinical characteristics of postictal psychiatric symptoms in partial epilepsy. *Neurology* **62**, 708–713.

Kraepelin, E. (1923). *Psychiatrie. Band 3*. Leipzig: Johann Ambrosius Barth.

Krishnamoorthy, E.S. (2005). Epidemiology and assessment of psychiatric disorders of epilepsy. PhD Thesis. University College London, London.

Krishnamoorthy, E. S. (2006). The evaluation of behavioral disturbances in epilepsy. *Epilepsia* **47** (Suppl 2), 3–8.

Krishnamoorthy, E. S., Trimble, M. R., Blumer, D. (2007). The classification of neuropsychiatric disorders in epilepsy: a proposal by the ILAE commission on psychobiology of epilepsy. *Epilepsy Behav* **10**, 349–353.

Kuhn, K. U., Quednow, B. B., Thiel, M., Falkai, P., Maier, W., Elger, C. E. (2003). Antidepressive treatment in patients with temporal lobe epilepsy and major depression: a prospective study with three different antidepressants. *Epilepsy Behav* **4**, 674–679.

Martinovic, Z., Simonovic, P., Djokic, R. (2006). Preventing depression in adolescents with epilepsy. *Epilepsy Behav* **9**, 619–624.

Mendez, M. F., Cummings, J. L., Benson, D. F. (1986). Depression in epilepsy. Significance and phenomenology. *Arch Neurol* **43**, 766–770.

Mintzer, S., Lopez, F. (2002). Comorbidity of ictal fear and panic disorder. *Epilepsy Behav* **3**, 330–337.

Mula, M., Monaco, F. (2006). Antiepileptic drug-induced mania in patients with epilepsy: what do we know? *Epilepsy Behav* **9**, 265–267.

Mula, M., Sander, J. W. (2007). Negative effects of antiepileptic drugs on mood in patients with epilepsy. *Drug Safety* **30**, 555–567.

Mula, M., Monaco, F., Trimble, M. R. (2004). Use of psychotropic drugs in patients with epilepsy: interactions and seizure risk. *Expert Rev Neurother* **4**, 953–964.

Mula, M., Trimble, M. R., Sander, J. W. (2007). Are psychiatric adverse events of antiepileptic drugs a unique entity? A study on topiramate and levetiracetam. *Epilepsia* **48**, 2322–2326.

Mula, M., Jauch, R., Cavanna, A., Collimedaglia, L., Barbagli, D., Gaus, V., Kretz, R., Viana, M., Tota, G., Israel, H., Reuter, U., Martus, P., Cantello, R., Monaco, F., Schmitz, B. (2008). Clinical and psychopathological definition of the interictal dysphoric disorder of epilepsy. *Epilepsia* **49**, 650–656.

Mulder, D. W., Daly, D. (1952). Psychiatric symptoms associated with lesions of the temporal lobe. *JAMA* **141**, 173–176.

Nishida, T., Kudo, T., Inoue, Y., Nakamura, F., Yoshimura, M., Matsuda, K., Yagi, K., Fujiwara, T. (2006). Postictal mania versus postictal psychosis: differences in clinical features, epileptogenic zone, and brain functional changes during postictal period. *Epilepsia* **47**, 2104–2114.

O'Donoghue, M. F., Goodridge, D. M., Redhead, K., Sander, J. W., Duncan, J. S. (1999). Assessing the psychosocial consequences of epilepsy: a community-based study. *Br J Gen Pract* **49**, 211–214.

Perrine, K., Hermann, B. P., Meador, K. J., Vickrey, B. G., Cramer, J. A., Hays, R. D., Devinsky, O. (1995). The relationship of neuropsychological functioning to quality of life in epilepsy. *Arch Neurol* **52**, 997–1003.

Prueter, C., Norra, C. (2005). Mood disorders and their treatment in patients with epilepsy. *J Neuropsychiatry Clin Neurosci* **17**, 20–28.

Quiske, A., Helmstaedter, C., Lux, S., Elger, C. E. (2000). Depression in patients with temporal lobe epilepsy is related to mesial temporal sclerosis. *Epilepsy Res* **39**, 121–125.

Ring, H. A., Crellin, R., Kirker, S., Reynolds, E. H. (1993). Vigabatrin and depression. *J Neurol Neurosurg Psychiatry* **56**, 925–928.

Ring, H. A., Moriarty, J., Trimble, M. R. (1998). A prospective study of the early post surgical psychiatric associations of epilepsy surgery. *J Neurol Neurosurg Psychiatry* **64**, 601–604.

Robertson, M. M. (1998). Forced normalisation and the aetiology of depression in epilepsy. In *Forced Normalisation and Alternative Psychoses of Epilepsy* (M. R. Trimble, B. Schmitz, eds), pp. 143–167. Petersfield: Wrightson Biomedical Publishing.

Robertson, M. M., Trimble, M. R. (1985). The treatment of depression in patients with epilepsy. A double-blind trial. *J Affect Disord* **9**, 127–136.

Robertson, M. M., Trimble, M. R., Townsend, H. R. (1987). Phenomenology of depression in epilepsy. *Epilepsia* **28**, 364–372.

Schmitz, E. B., Moriarty, J., Costa, D. C., Ring, H. A., Ell, P. J., Trimble, M. R. (1997). Psychiatric profiles and patterns of cerebral blood flow in focal epilepsy: interactions between depression, obsessionality, and perfusion related to the laterality of the epilepsy. *J Neurol Neurosurg Psychiatry* **62**, 458–463.

Specchio, L. M., Iudice, A., Specchio, N., La Neve, A., Spinelli, A., Galli, R., Rocchi, R., Ulivelli, M., de Tommaso, M., Pizzanelli, C., Murri, L. (2004). Citalopram as treatment of depression in patients with epilepsy. *Clin Neuropharmacol* **27**, 133–136.

Tan, S. Y., Bruni, J. (1986). Cognitive-behavior therapy with adult patients with epilepsy: a controlled outcome study. *Epilepsia* **27**, 225–233.

Tellez-Zenteno, J. F., Patten, S. B., Wiebe, S. (2005). Psychiatric comorbidity in epilepsy: a population-based analysis. *Epilepsia* **46**(Suppl 8), 264.

Thomas, L., Trimble, M. R., Schmitz, B., Ring, H. (1996). Vigabatrin and behaviour disorders – a retrospective survey. *Epilepsy Res* **25**, 21–27.

Thomè-Souza, M. S., Kuczynski, E., Valente, K. D. (2007). Sertraline and fluoxetine: safe treatments for children and adolescents with epilepsy and depression. *Epilepsy Behav* **10**, 417–425.

Trimble, M. R. (1996). *Biological Psychiatry* 2nd edition. Chichester: J Wylie & Sons.

Trimble, M. R. (1998). Forced normalisation and the role of anticonvulsants. In *Forced Normalisation and Alternative Psychoses of Epilepsy* (M. R. Trimble, B. Schmitz, eds), pp. 169–178. Petersfield: Wrighton Biomedical Publishing.

Victoroff, J. I., Benson, F., Grafton, S. T., Engel, J., Jr., Mazziotta, J. C. (1994). Depression in complex partial seizures. Electroencephalography and cerebral metabolic correlates. *Arch Neurol* **51**, 155–163.

Wolf, P. (1982). Manic episodes in epilepsy. In *Advances in Epileptology: XIIIth Epilepsy International Symposium* (H. Akimoto, H. Kazamatsuri, M. Seino, A. A. Ward, Jr., eds), pp. 237–240. New York: Raven Press.

INTERICTAL DYSPHORIC DISORDER INVENTORY (IDDI)
Mula and Schmitz 2005
(modified from Krishnamoorthy and Trimble)

Patient code: _____   Date of evaluation: _____

Some people with epilepsy experience changes in their mood, emotions and feelings from time to time. We would like to ask you about any such changes you experienced **in the last 12 months**. Please put a cross to the right answers.

1. **Anergia:**

1.1 Do you feel you lack in energy from time to time?	No (0) Yes (1)

Further questions about anergia:

1.2 How often does this lack in energy occur?	Never (0) Rarely (1) Sometimes (2) Often (3)
1.3 How severe is this lack in energy usually?	Not present (0) Mild (1) Moderate (2) Severe (3)
1.4 How much do you feel impaired by this lack in energy when it occurs?	Not at all (0) Mildly (1) Moderately (2) Severely (3)

2. **Pain:**

2.1 Do you suffer from many aches and pain from time to time (e.g. headaches, stomachaches, abdominal pain, back pain)?	No (0) Yes (1)

Further questions about pain:

2.2 How often does pain occur?	Never (0) Rarely (1) Sometimes (2) Often (3)
2.3 How severe is this pain usually?	Not present (0) Mild (1) Moderate (2) Severe (3)
2.4 How much do you feel impaired by this pain, when it occurs?	Not at all (0) Mildly (1) Moderately (2) Severely (3)

3. **Insomnia:**

3.1 Do you have trouble with your sleep from time to time?	No (0) Yes (1)

Further questions about insomnia:

3.2 How often do these problems occur?	Never (0) Rarely (1) Sometimes (2) Often (3)
3.3 How severe are these problems usually?	Not present (0) Mild (1) Moderate (2) Severe (3)
3.4 How much do you feel impaired by these problems?	Not at all (0) Mildly (1) Moderately (2) Severely (3)

4. Fear/panic:

4.1 Do you experience feelings of fear or feel panicky from time to time?	No (0) Yes (1)

Further questions about fear or panic:

4.2 How often do these feelings of fear or panic occur?	Never (0) Rarely (1) Sometimes (2) Often (3)
4.3 How severe are these feelings of fear or panic usually?	Not present (0) Mild (1) Moderate (2) Severe (3)
4.4 How much do you feel impaired by these feelings of fear or panic, when they occur?	Not at all (0) Mildly (1) Moderately (2) Severely (3)

5. Anxiety:

5.1 Do you have frequent worries, feelings of oppression, agitation or anxiety from time to time?	No (0) Yes (1)

Further questions about anxiety:

5.2 How often does anxiety occur?	Never (0) Rarely (1) Sometimes (2) Often (3)
5.3 How severe is this anxiety usually?	Not present (0) Mild (1) Moderate (2) Severe (3)
5.4 How much do you feel impaired by this anxiety, when it occurs?	Not at all (0) Mildly (1) Moderately (2) Severely (3)

6. Depression:

6.1 Do you feel in low spirits, depressed or find difficult to take pleasure in most activities from time to time?	No (0) Yes (1)

Further questions about depressed mood:

6.2 How often does this occur?	Never (0) Rarely (1) Sometimes (2) Often (3)
6.3 How severe is this usually?	Not present (0) Mild (1) Moderate (2) Severe (3)
6.4 How much do you feel impaired by this, when it occurs?	Not at all (0) Mildly (1) Moderately (2) Severely (3)

7. Euphoria:

7.1 Do you feel cheerful, very happy, full of energy without good reasons from time to time?	No (0) Yes (1)

Further questions about euphoric moods:

7.2 How often does this occur?	Never (0) Rarely (1) Sometimes (2) Often (3)
7.3 How severe is this usually?	Not present (0) Mild (1) Moderate (2) Severe (3)
7.4 How much do you feel impaired by this?	Not at all (0) Mildly (1) Moderately (2) Severely (3)

8. Irritability:

8.1 Do you feel irritable, experience bad temper or fly off the handle easily over little things from time to time?	No (0) Yes (1)

Further questions about irritability moods:

8.2 How often does this irritability occur?	Never (0) Rarely (1) Sometimes (2) Often (3)
8.3 How severe is this irritability usually?	Not present (0) Mild (1) Moderate (2) Severe (3)
8.4 How much do you feel impaired by this irritability?	Not at all (0) Mildly (1) Moderately (2) Severely (3)

APPENDIX: QUESTIONS CONCERNING THE TEMPORAL RELATIONS OF THE ABOVE MENTIONED COMPLAINTS:

A. Do the above symptoms occur temporally independently from each other?	No ( ) Yes ( )
B. How often do these symptoms occur?	____times a day ____times a week ____times a month
C. How long do these symptoms last?	A few hours ( ) One day ( ) A few days/less than a week ( ) A week or more ( ) Time periods differ ( ) Chronic ( )
D. Is the occurrence of these symptoms related to seizures in any way?	No ( ) Yes ( )
E. If Yes. In what temporal relation to your seizures do these symptoms occur?	Before seizure ( ) After seizure ( ) During seizure ( ) During periods when you are free of seizures ( )
F. Is the occurrence of these symptoms more noticeable when your therapy is changed?	No ( ) Yes ( )

IDDI Scoring

IDD definite diagnosis: at least three symptoms of at least "moderate" or "severe" severity and causing "moderate" to "severe" distress.

Symptoms-scoring:

IDDI total score: (Yes answers 1.1 + 2.1 + 3.1 + 4.1 + 5.1 + 6.1 + 7.1 + 8.1)/8

IDDI labile depressive symptoms score: (Yes answers 1.1 + 2.1 + 3.1 + 6.1)/4

IDDI labile affective symptoms score: (Yes answers 4.1 + 5.1)/2

IDDI specific symptoms score: (Yes answers 7.1 + 8.1)/2

Severeness scoring:

Total severeness: total sum of frequency (X.2 for each item), severity (X.3 for each item) and impairment (X.4 for each item) scores

Labile depressive symptoms severeness: sum of frequency, severity and impairment scores for 1, 2, 3 and 6

Labile affective symptoms severeness: sum of frequency, severity and impairment scores for 4 and 5

Specific symptoms severeness: sum of frequency, severity and impairment scores for 7 and 8

CHAPTER 6

Is Depression in Epilepsy the Expression of a Neurological Disorder?

Sean T. Hwang and Frank G. Gilliam

INTRODUCTION

Depression is the most common psychiatric comorbidity associated with epilepsy. Though estimates of the prevalence of major depression in patients with epilepsy vary due to methodology and patient selection, the prevalence ranges between 20% and 50% in patients with recurrent seizures, and 3% and 9% in patients with well-controlled seizures (Mendez *et al.*, 1986; Edeh and Toone, 1987; Jacoby *et al.*, 1996; O'Donoghue *et al.*, 1999; Wiegartz *et al.*, 1999; Hermann *et al.*, 2000; Beghi *et al.*, 2002; Kanner, 2003b; Boylan *et al.*, 2004; Gaitatzis *et al.*, 2004). This contrasts with an estimated baseline 13–16% lifetime prevalence of major depression in the general population (Kessler *et al.*, 2003; Hasin *et al.*, 2005). Unfortunately, depression in epilepsy also commonly occurs in children (Ettinger *et al.*, 1998). Despite the frequency with which depression occurs in the setting of epilepsy, it remains underrecognized and often neglected (Mendez *et al.*, 1986; Wiegartz *et al.*, 1999; Kanner *et al.*, 2000; Gilliam, 2002).

Differences in the characterization of depression in the setting of epilepsy vs. primary mood disorders have been described at least since the time of Kraepelin (Kraepelin, 1923). While patients with epilepsy may present with syndromes meeting DSM-IV criteria for major depression, a proportion of patients may alternatively present with a more atypical chronic intermittent dysthymic state. Peri-ictal depressive symptoms may also occur, as well as iatrogenic mood fluctuations in the setting of antiepileptic drug (AED) therapy or surgical treatment, adding to the pleomorphic presentation of depression in epilepsy. Variation in mood has been observed to occur as a prodrome up to 3 days prior to a seizure (Blanchet and Frommer, 1986), and lasting hours to days postictally (Kanner *et al.*, 2004).

An estimated 25% of auras have an emotional component, with affective changes apparent in 15% (Kanner and Balabanov, 2002). The qualitatively different, more pleomorphic, less severe, waxing and waning, chronic state of interictal depression has been described more recently by Blumer as the "interictal dysphoric disorder" (Blumer *et al.*, 2004) or by Kanner as the "dysthymic-like disorder of epilepsy" (Kanner *et al.*, 2000). Notable also is the lack of a substantial difference in prevalence based on sex among patients with comorbid depression and epilepsy, which differs in comparison to the greater female predominance observed with idiopathic major depressive illness (Barry, 2003). It is important to appreciate that both major depression and atypical depression in the setting of epilepsy carry significant morbidity and warrant appropriate therapy.

Mood has been consistently shown to be an essential factor affecting health-related quality of life in patients with epilepsy (Figure 6.1) (Perrine *et al.*, 1995; Gilliam *et al.*, 1997; Lehrner *et al.*, 1999; Gilliam, 2002; Cramer *et al.*, 2003; Boylan *et al.*, 2004). In one analysis, mood was found to be the most powerful predictor of subjective health, independent of seizures, driving status, medication requirement, or employment (Gilliam *et al.*, 1999). In people with epilepsy, depression has also been shown to be associated with increased utilization of medical services and health care costs (Cramer *et al.*, 2004).

Suicide is a major cause of mortality in patients with epilepsy. A review by Robertson found the prevalence rate of suicide and suicide attempt to be between

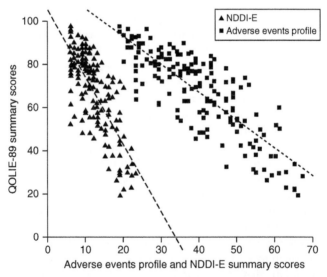

FIGURE 6.1   Scatter plot of the correlation of quality of life in epilepsy inventory-89 (QOLIE-89) summary scores with the neurological disorders depression inventory for epilepsy (NDDI-E) (partial r = −0·39, $p < 0.0001$) and the adverse events profile (partial r = −0.60, $p < 0.0001$). Adjusted R2 was 0.72 ($p < 0.0001$) for regression model with the QOLIE-89 as the dependent variable. *Source*: Gilliam *et al.* (2006).

5% and 14% in people with epilepsy, up to 10 times more common than among the general population (Robertson, 1997). A separate meta-analysis by Harris and Barraclough found the overall lifetime prevalence suicide rate in people with epilepsy to be five times higher than in the general population. For people with temporal lobe epilepsy (TLE), the rate was 6–25 times higher than in the general population (Harris and Barraclough, 1997). Rafnsson reported a standardized mortality ratio (SMR) for suicide of 5.8 for patients with epilepsy, the highest for all causes of death in a population-based incidence cohort study from Iceland (Rafnsson et al., 2001). Similarly, in a Swedish study of cause-specific mortality in 9,061 previously hospitalized epilepsy patients, the SMR for suicide was 3.8 (Nilsson et al., 2003). In a review of the literature by Jones et al. (2003), the collective data yielded the lifetime average suicide rate of approximately 11.5% in people with epilepsy, compared with that of 1.1–1.2% in the general population. The coexistence of an anxiety disorder to those patients with epilepsy and depression has been cited to be as high as 73% (Jones et al., 2005), which may further compound the risk for suicide (Sareen et al., 2005).

Evidence has emerged to suggest that depressive disorders may also detrimentally affect seizure control in a reciprocal manner. In a study of 890 patients with new onset epilepsy, Mohanraj and Brodie (2003) reported that those with a history of psychiatric illness were more than three times less likely to be seizure-free on AEDs than patients without such history (median follow-up of 79 months). In a different study by Anhoury et al. (2000) of 121 patients with epilepsy who had undergone a temporal lobectomy, there were significantly worse postsurgical seizure outcomes for patients with a psychiatric history compared with those without. Kanner et al. evaluated 90 patients who underwent a temporal lobectomy for factors that were predictive of failure to achieve seizure freedom (mean follow-up of 6.5 years). Using a multivariate logistic regression model, they examined the covariates of a prior history of depression, cause of temporal lobe epilepsy (TLE), duration of epilepsy, and occurrence of generalized tonic–clonic seizures. A lifetime history of depression was the only predictor of persistent auras in the absence of disabling seizures, while a lifetime history of depression and the underlying cause of the TLE were both predictive of persistent disabling seizures (Kanner et al., 2003). An increase in compliance with antiepileptic medications may occur in association with improvements in mood (Helgeson et al., 1990).

While epilepsy is a well-recognized risk factor for depression, of interest is the increasingly apparent bi-directional relationship between depression and epilepsy, which suggests common pathogenic mechanisms for the two disorders (Kanner, 2006). Data from three large population-based studies indicate that there is a 1.7–7 fold risk of developing epilepsy in patients with a history of depression (Forsgren and Nystrom, 1990; Hesdorffer et al., 2000, 2006). History of suicide attempt, as an independent risk factor, increased the risk of developing unprovoked seizures by more than five fold (Hesdorffer et al., 2006).

The prevalent belief that depression in epilepsy simply occurs as a reactive phenomenon in the setting of a chronic medical illness may serve to compound the problematic tendency of physicians and patients to fail to recognize or appropriately treat this common and disabling entity (Barry, 2003). Psychosocial factors may contribute to the pathogenesis of depressive symptoms in epilepsy. Factors that have been ascribed to the development of depression in epilepsy include the perceived social stigma of the disorder, social isolation, poor vocational adjustment, and attributional style (Hermann *et al.*, 1996). However, in a retrospective comparison, Schmitz *et al.* identified no significant differences in psychosocial variables between 25 patients with epilepsy and comorbid depression, and 50 non-psychiatric patients with epilepsy. The authors therefore challenged the hypothesis that social and vocational issues could predict patients with epilepsy who develop depression (Schmitz *et al.*, 1999).

Recent research has offered to further elucidate some of the common pathogenic neurobiological mechanisms underlying both depression and epilepsy (Gilliam *et al.*, 2004b; Hecimovic *et al.*, 2003; Jobe, 2003; Kanner, 2005), the focus of which will be the remainder of this chapter. Substantial data exist in support of monoamine neurotransmitter abnormalities in the pathogenesis of depression, and evidence continues to grow in support of their role in affecting the seizure threshold of specific brain regions. Histopathologic and neuroimaging studies continue to provide insight into structural and functional similarities in the two disorders, and may facilitate a better understanding of the affected networks and interrelated pathways.

SHARED NEUROTRANSMITTER ABNORMALITIES IN DEPRESSION AND EPILEPSY

Both depression and epilepsy have been linked to the abnormal activity of several neurotransmitters, including serotonin (5-hydroxytryptamine, 5-HT), norepinephrine (NE), dopamine (DA), γ-aminobutyric acid (GABA), and glutamate (Ressler and Nemeroff, 2000; Nemeroff and Owens, 2002; Nestler *et al.*, 2002; Jobe, 2003). Neurotransmitter abnormalities in depression form the theoretical basis for psychopharmacologic treatment with monoamine oxidase inhibitors (MAOIs), tricyclic antidepressants (TCAs), selective serotonin reuptake inhibitors (SSRIs), and serotonin-norepinephrine reuptake inhibitors (SNRIs); as well as for the use of psychoactive AEDs such as valproic acid (VPA), carbamazepine (CBZ), oxcarbazepine (OXC), and lamotrigine (LTG) in the treatment of mood disorders.

Abnormalities in 5-HT and NE networks, in particular, highlight the biological link between the affective disorders and epilepsy. A review by Jobe (2003) offers an extensive overview of several studies that support this association. In addition to being linked to the pathophysiologic mechanisms of depression, decreased 5-HT and NE activity has been demonstrated to facilitate kindling, exacerbate seizure severity, and lower seizure threshold in several animal models of epilepsy (Jobe, 2003).

The genetically epilepsy-prone rat (GEPR), which harbors inborn defects in pre- and postsynaptic transmission of both 5-HT and NE, serves as a valuable animal model to explore the common pathogenic mechanisms in depression and epilepsy. Two strains, GEPR-3 and GEPR-9, are predisposed to sound-induced generalized tonic–clonic seizures, and an accelerated rate of seizure kindling. They have also been found to experience behavioral analogs of depression (Jobe, 2003). Deficiencies in NE in GEPRs result from deficient arborization of neurons arising from the locus coeruleus, with excessive presynaptic suppression of NE release, and lack of compensatory postsynaptic upregulation (Yan et al., 1993; Clough et al., 1998; Ryu et al., 1999). Abnormal serotonergic arborization, a decreased density of postsynaptic 5-HT1A receptors, and excessive feedback inhibition via the 5-HT1b autoreceptor may also contribute to 5-HT abnormalities in the brains of GEPRs (Dailey et al., 1992; Statnick et al., 1996a).

Selective destruction of noradrenergic and serotonergic neurons significantly reduced the anticonvulsant effect of vagal nerve stimulation in rats (Browning et al., 1997). Deletion of 5-HT or NE neurons via neurotoxic exposure has also been shown to result in seizure exacerbation in GEPRs (Wang et al., 1994; Statnick et al., 1996b).

Substances that interfere with the synthesis or release of NE or 5-HT have been shown to provoke seizures in GEPRs. These include reserpine and tetrabenazine, which inactivate NE storage vesicles; α-methyl-*m*-tyrosine, a false NE transmitter; α-methyl-*p*-tyrosine, a NE synthesis inhibitor; and *p*-chlorophenylalanine, a 5-HT synthesis inhibitor (Jobe, 2003). 5-HT depleting drugs may also block the anticonvulsant effect of CBZ in GEPRs (Dailey et al., 1997).

Conversely, substances that increase 5-HT and NE transmission may have an anticonvulsant effect in experimental animal models (Jobe, 2003). As reviewed in an article by Kanner highlighting the work of several different groups, certain SSRIs and MAOIs have been shown to be anticonvulsants in GEPRs, and baboons, as well as non-genetically prone cats, rabbits, and rhesus monkeys (Kanner, 2005). The 5-HT precursor, 5-hydroxy-l-tryptophan (5-HTP) in combination with an SSRI or with an MAOI has been shown to have an anticonvulsant effect in GEPRs (Yan et al., 1995; Jobe, 2003). The AEDs VPA, CBZ, and LTG have also been shown to cause an increase in 5-HT (Yan et al., 1992; Dailey et al., 1997; Southam et al., 1998).

Areas of the brain involved in common epilepsy syndromes, such as the hippocampus and prefrontal regions, normally contain a high relative density of 5-HT1A receptors (Theodore, 2003). 5-HT1A receptors are expressed pre- and postsynaptically. The presynaptic 5-HT1A autoreceptors are located on serotonergic neurons of the raphe nuclei. Activation of presynaptic receptors leads to a reduced firing of 5-HT cell bodies in the raphe nuclei and reduced serotonin release. Postsynaptic 5-HT1A receptors are localized to the axon hillock of pyramidal neurons, GABAergic interneurons, and glia cells in the neocortex and limbic structures. Stimulation of these populations of receptors enhances serotonergic transmission, resulting in a potassium mediated membrane hyperpolarizing response, and tonic inhibition of pyramidal neurons (Savic et al., 2004). Postsynaptic 5-HT1A receptors have been demonstrated to exert their antiepileptic effect by this mechanism in hippocampal-kindled seizures in cats and in intrahippocampal kainic acid induced seizures in rats (Beck and Choi, 1991; Okuhara and Beck, 1994).

Less experimental data is available on the impact of pharmacological modulation of 5-HT and NE activity in humans (Kanner, 2005). Reserpine, which causes the depletion of monoamines, is associated with an increase in seizure frequency and severity in patients with epilepsy. A reduction in the seizure threshold and a worsened severity of seizures has also been observed in schizophrenic patients using reserpine while undergoing electroshock therapy. In the only double-blind, placebo-controlled study to date, imipramine, a TCA with NE and 5-HT reuptake inhibitory effects, was reported to suppress absence and myoclonic seizures (Fromm et al., 1978). There have been reports of improved seizure frequency in epilepsy from open trials with the use of the TCA doxepin (Ojemann et al., 1983), and the SSRIs fluoxetine (Favale et al., 1995) and citalopram (Hovorka et al., 2000; Specchio et al., 2004; Albano et al., 2006). To date, no controlled trials studying the effects of SSRI/SNRIs on seizures have been completed.

SHARED STRUCTURAL AND NEUROPATHOLOGICAL ABNORMALITIES IN DEPRESSION AND EPILEPSY

Common structural and functional abnormalities of overlapping neuroanatomic regions have been observed in primary depression and in epileptic disorders associated with comorbid depression (Kanner and Balabanov, 2002). Structural changes have been identified in the amygdala, hippocampus, and entorhinal cortex; temporal lateral neocortex; prefrontal, orbitofrontal, and mesial frontal cortex; and to a lesser degree, the thalamic nuclei and basal ganglia.

In review of the data relevant to volumetric structural changes that have been observed in affective disorders, some authors have asserted that the abnormalities plot along the course of a "limbic–cortical–striatal–pallidal–thalamic tract" (Sheline, 2003). As summarized extensively by Sheline (2003), in patients with depression, volumetric magnetic resonance imaging (MRI) studies have identified diffuse cortical and subcortical atrophy and ventricular enlargement; as well as variable focal structural changes in the amygdala, hippocampal formation, entorhinal cortex, cingulate gyrus, orbitofrontal cortex, and prefrontal cortex. Variability in the results between studies may occur as a consequence of both clinical and methodological factors. Patient characteristics, such as age, comorbidities, disease severity, and duration of disease may differ. Equipment used, methods of measurement, technique, and resolution may also vary between studies.

Reduction in frontal cortical volume ranges from 7% overall in frontal lobe volume loss in major depressive disorder (MDD) (Coffey et al., 1993) to up to 48% in the subgenual prefrontal cortex (Drevets et al., 1997). A study by Kumar et al. (1998) found decreased prefrontal lobe volumes displaying a significant linear trend with severity of depressive illness, after controlling for age and gender.

Reductions in bilateral hippocampal volumes in otherwise healthy patients with a history of primary MDD in remission in comparison to age, sex, and height-matched controls has been highlighted in at least two studies (Sheline *et al.*, 1996, 1999). In these studies, reduction in left hippocampal volume was significantly correlated with total lifetime duration of depression. A different study by Sheline *et al.* (2003) of 38 females with a history of MDD in remission found a significant correlation between hippocampal volume loss and the duration of medically untreated depression; an association was not retained for patients who were or had been taking antidepressant medication.

The neuropathological data has been supportive. In a histopathologic study of brains from depressed patients, Rajkowska *et al.* reported a decrease in cortical thickness, neuronal size, and neuronal and glial densities in layers II, III, and IV of the rostral orbitofrontal region. Reductions in glial density, with small associated decreases in neuronal size, were also identified in layers V and VI of the caudal orbitofrontal cortex. A decrease in the neuronal size and glial density in all cortical layers was observed in the dorsolateral prefrontal cortex (Rajkowska *et al.*, 1999).

Few neuropathologic studies of the hippocampal formation in patients with primary MDD are available. Lucassen *et al.* compared 15 hippocampi of patients with a history of MDD with 16 matched controls and nine steroid-treated patients. Rare but convincing apoptosis, probably neuronal, was identified in the entorhinal cortex, subiculum, dentate gyrus, CA1 and CA4 in 11 of 15 patients with depression, three steroid-treated patients, and one control (Lucassen *et al.*, 2001). In a different postmortem study, glial density was substantially reduced in the amygdala and to a lesser degree in the entorhinal cortex in MDD cases (Bowley *et al.*, 2002).

The structural changes in primary MDD have been attributed to several different, possibly concomitant, pathogenic mechanisms. Elevated glucocorticoid (GC) levels, a stress-induced reduction in neurotrophic factors, stress-induced impaired neurogenesis, and glial loss have been offered as possible explanations (Sheline, 2003). Excessive activation of the hypothalamic–pituitary–adrenal (HPA) axis in depressed patients, potentially through chronic stress, may result in chronic hypercortisolemia. Nearly half of all individuals with depression display impaired dexamethasone suppression of adrenocorticotropic hormone (ACTH) and cortisol, changes which are reversible on antidepressant treatment (Holsboer, 2001). In rats, various acute stressors have been associated with increases of corticotropin-releasing hormone (CRH) in the amygdala and hypothalamus (Hatalski *et al.*, 1998; Hand *et al.*, 2002). A downregulation in CRH has been observed in the frontal cortex of depressed humans compared with controls (Nemeroff and Owens, 2002).

Studies in animals have shown prolonged GC exposure to be associated with damage to hippocampal neurons, impaired granule cell neuron development in the adult hippocampal dentate gyrus, reversible shortening of apical dendrites in CA3, and potentially cell death in extreme and prolonged conditions (Watanabe *et al.*, 1992; Fuchs and Gould, 2000; Sapolsky, 2000; Reul and Holsboer, 2002; Holsboer, 2003). Lupien *et al.* (1998) reported significantly reduced hippocampal volume and deficits in hippocampus-dependent memory tasks in aged humans with prolonged cortisol elevations in comparison to controls. In the setting of

Cushing's disease, cerebral atrophy was at least partially reversible following normalization of GC levels (Bourdeau *et al.*, 2002).

An investigation of 17 patients on long-term corticosteroid therapy found abnormal creatine (Cr)/N-acetylaspartate (NAA) ratios in the mesial temporal regions, and higher depression rating scores vs. controls. The findings suggest that chronic exposure to elevated corticosteroid levels can induce limbic dysfunction and subsequent depression (Brown *et al.*, 2004). The chronic administration of CRH to normal volunteers resulted in alterations to the HPA axis identical to those found in patients with MDD (Holsboer, 2003).

A decrease in brain-derived neurotrophic factor (BDNF) levels in the dentate gyrus and pyramidal cell layer of the hippocampus, amygdala, and neocortical regions may be modulated by GCs and serotonergic transmission, and potentially reversed with antidepressant therapy (Nibuya *et al.*, 1995; Smith *et al.*, 1995). In animal studies, acute and chronic stress has been demonstrated to decrease levels of BDNF expression in the dentate gyrus and pyramidal layer of the hippocampus (Smith *et al.*, 1995). Antidepressant drugs, alternatively, have been shown to increase hippocampal BDNF levels in humans (Chen *et al.*, 2001). Hypothetically, the upregulation of BDNF may be neuroprotective, and may also repair existing damage to hippocampal neurons (Hecimovic *et al.*, 2003).

In epilepsy, areas of abnormalities most relevant to depression include the mesial and orbitofrontal regions, mesial temporal lobe, and subcortical structures. Prevalence rates of depression range from 19% to 65% among patients with epilepsy of mesial temporal or frontal lobe origin (Kanner and Balabanov, 2002).

Neuroimaging studies in patients with epilepsy and comorbid depression have identified an association between the severity of depression and presence of mesial temporal structural abnormalities, as demonstrated by MRI. Quiske *et al.* (2000) evaluated a surgical sample of 60 patients with TLE by MRI and Beck Depression Inventory (BDI), revealing significantly higher depression scores for patients with mesial temporal sclerosis (MTS), regardless of lateralization.

The magnitude of hippocampal volume loss in TLE is significantly greater than what is typically observed in MDD. In MTS, the neuropathologic findings also differ from those of primary MDD with substantial neuronal loss and astrocytosis in the hippocampus, amygdala, entorhinal cortex, and parahippocampal gyrus. Neuronal cell loss is most prominent in areas CA1 and CA4, the dentate gyrus, and subiculum of the hippocampus (Mathern *et al.*, 1997).

SHARED FUNCTIONAL NEUROIMAGING ABNORMALITIES IN DEPRESSION AND EPILEPSY

Utilizing various imaging techniques, functional abnormalities in both epilepsy and depression have been identified in common subcortical structures, as well as similar regions of the frontal and temporal lobes.

Frontal lobe involvement in primary depression has been demonstrated with functional neuroimaging and neuropsychological studies (Baxter et al., 1989). Frontal lobe associated executive dysfunction has been consistently shown in studies on depressive illness. Abnormalities on neuropsychological testing have been correlated to reduced blood flow in the mesial prefrontal cortex (Bench et al., 1993; Dolan et al., 1994).

Frontal lobe mediated executive dysfunction has also been recognized in TLE (Hermann et al., 1988; Horner et al., 1996; Martin et al., 2000), prominently among patients with comorbid depression, correlating with bilateral inferior frontal hypometabolism by ^{18}F-flourodeoxyglucose positron emission tomography (FDG-PET) (Bromfield et al., 1992; Jokeit et al., 1997). In a study by Victoroff et al. (1994) of 53 patients with epilepsy, evaluated by FDG-PET and psychiatric interview, severity of temporal hypometabolism was also found to have significant correlation with a history of MDD, more frequently with left temporal lateralization (Victoroff et al., 1994).

Gilliam et al. performed a pilot study of the association between depressive symptoms, clinical variables, and FDG-PET in 62 consecutive patients with refractory localization-related epilepsy. Age, seizure rate, number of current medications, and scores on the adverse events profile were similar between 55 patients with an abnormal PET compared to the 7 patients with a normal PET. BDI scores were, in contrast, significantly higher in the group of patients with an abnormal PET scan, the majority of which showed abnormalities in the temporal lobes (Gilliam et al., 2004a).

Using (99 m)Tc-HMPAO single-photon emission-computed tomography (SPECT) in a study of 31 patients with localization-related epilepsy, higher scores on the BDI were associated with decreased contralateral temporal and bilateral frontal perfusion in patients with left hemispheric epilepsy (Schmitz et al., 1997).

As discussed previously, deficits in serotonin may predispose to the development of major depression. Activation of 5-HT1A receptors located on serotonin cell bodies in the midbrain dorsal raphe nucleus inhibits the firing of serotonergic neurons and diminishes the release of 5-HT in the prefrontal cortex. As reviewed by Kanner, decreased serotonergic innervation in terminal tissues has been suggested by studies of brain, plasma, and platelets of depressed patients (Kanner, 2005).

Studies have identified serotonergic dysfunction in the ventral prefrontal cortex to be associated with patients with suicidal behavior (Oquendo et al., 2003). A lower density of 5-HT1A receptors in the hippocampus and a lower affinity of 5-HT1 binding sites in the amygdala have been observed in pathologic specimens of untreated depressed patients who committed suicide (Cheetham et al., 1990). The binding of the radiolabel [3H]8-hydroxy-2-(di-n-propyl)aminotetralin ([3H]8-OH-DPAT), an agonist at inhibitory 5-HT1A autoreceptors, was excessively increased in the midbrain dorsal raphe nuclei of suicide victims with MDD in comparison with normal controls, offering further evidence of diminished activity of serotonergic neurons (Stockmeier et al., 1998).

Using the 5-HT1A receptor antagonist [Carbonyl-11C]WAY-100635, a PET study of 25 medicated and medically untreated patients with depression revealed decreased serotonergic binding potential in the frontal, temporal, and limbic cortex as compared to healthy controls (Sargent et al., 2000). Another PET investigation using [Carbonyl-11C]WAY-100635, similarly measured decreased 5-HT1A binding in the mesial temporal cortex and the raphe nuclei in 12 patients with familial recurrent major depressive episodes when compared to controls (Drevets et al., 1999).

The decreased availability of 5-HT1A receptors may reflect receptor loss, decreased expression of receptors, altered receptor function, decreased receptor affinity, or at least theoretically, increased receptor occupancy. Downregulation of autoreceptors in the raphe may result from reciprocal inter-connections from regions of impaired serotonergic transmission in the amygdala, hippocampus, insular and cingulate cortex. In the case of epileptic patients with associated structural abnormalities, such as hippocampal sclerosis, associated neuronal loss may contribute to a low 5-HT1A receptor binding potential. Potentially, downregulation of postsynaptic 5-HT1A receptors may occur as a result of repetitive seizures (Savic et al., 2004).

A PET study of serotonergic activity using [Carbonyl-11C]WAY-100635 by Savic et al. (2004) demonstrated significantly reduced 5-HT1A receptor binding potential in the hippocampus, amygdala, anterior cingulate, and lateral temporal neocortex ipsilateral to the seizure focus, as well as the insula, midbrain raphe nuclei, and to a lesser extent, the contralateral hippocampus of 14 patients with TLE. Six of the 14 patients met criteria for the interictal dysphoric syndrome. However, of these structures, only the anterior cingulate region was significantly correlated with depression rating scores. The study was not powered to detect small potential associations. As extratemporal limbic changes were observed only ipsilaterally to the side of seizure activity, the authors favored the hypothesis that they were related to the epileptogenic process rather than being secondary to the depressive state. They further hypothesized that the interictal affective disorder in mesial TLE may result from an increased vulnerability due to impaired serotonergic transmission in limbic structures of the seizure-generating hemisphere.

Another group, using the radioligand [^{18}F]trans-4-fluoro-N-2-[4-(2-methoxy-phenyl)piperazin-1-yl]ethyl]-N-(2-pyridinyl)cyclohexanecarboxamide or ([^{18}F] FC-WAY), a selective 5-HT1A receptor antagonist, also found lower serotonergic binding in patients with TLE ipsilateral to the epileptic focus in the inferior medial and lateral temporal regions, insula, and raphe by PET. After partial volume correction, decreased serotonergic activity in the mesial, but not lateral temporal structures and insula remained highly significant. A significant inverse relationship between BDI scores and 5-HT1A binding potential in the hippocampus ipsilateral to the seizure focus was reported (Giovacchini et al., 2005). Differences between the results of the studies may have been accounted for in part by partial volume effects, patient sample sizes, the use of different radioligands, and variations in analytic methods.

Theodore *et al.* (2007) conducted a similar PET study using the radioligand [^{18}F]FCWAY. Depressive symptoms measured with the BDI were examined in relation to 5-HT1A receptor binding as assessed by PET, MTS on MRI, and epileptic focus laterality in 45 patients with TLE. Corrected for partial volume effects, 5-HT1A binding was significantly lower in the hippocampus ipsilateral to the epileptic focus, however bilateral hippocampal reductions were also significantly present. Again, there was a significant inverse relation between ipsilateral 5-HT1A binding potential and the BDI. Side of focus, the presence of MTS, and patient gender did not affect the BDI significantly.

Using 4,2-(methoxyphenyl)-1-[2-(*N*-2-pyridinyl)-ρ-fluorobenzamido]ethylpiperazine (^{18}F-MPPF), a different 5-HT1A tracer, investigators were able to demonstrate that a decrease in binding potential of 5-HT1A in patients with TLE was significantly more prominent in regions of seizure onset and propagation, as defined by intracranial recordings with stereoelectroencephalography (SEEG). Again, 5-HT1A binding remained decreased even in the setting of a normal quantitative and qualitative MRI (Merlet *et al.*, 2004).

Of interest, in a PET study of 11 patients with juvenile myoclonic epilepsy, using [Carbonyl-11C]WAY-100635, a decreased serotonergic binding potential was also observed in the dorsolateral prefrontal cortex, raphe nuclei, and hippocampi when compared to 11 control subjects (Meschaks *et al.*, 2005).

In a recent ^{1}H magnetic resonance spectroscopy (^{1}H-MRSI) study of 31 patients with TLE, Gilliam *et al.* (2007) found a significant correlation of severity of depression symptoms with the extent of voxels containing an abnormal hippocampal Cr/NAA metabolite ratio. All subjects had at least one complex partial or generalized tonic–clonic seizure within the last 3 months, and all patients were on at least one antiepileptic medication. Linear regression analysis confirmed an independent association of ^{1}H-MRSI abnormalities with the depression scale of the Profile of Mood States (POMS), and the absence of association with other measured seizure or self-perceived disability variables. Overall, the model explained 57% of the variance in depressive symptoms. The study is unique in demonstrating the degree of depression correlating with spatial involvement of a biomarker of neuronal dysfunction within the limbic region. Given the association of areas of decreased NAA concentration with cerebral regions of interictal spiking and seizure onset (Garcia *et al.*, 1997; Shih *et al.*, 2004), the authors proposed that depression symptoms in TLE may be due to hyperexcitable neurons in the limbic network (Figures 6.2 and 6.3).

Aberrancies in NAA levels and glucose uptake may represent differing mechanisms of cellular metabolic dysfunction. While glucose metabolism may be more dependent on pyramidal cell activation and regional neuron/glia ratios, NAA synthesis is dependent on mitochondrial enzymes, which may be potentially more sensitive to hippocampal functional disruption. A comparison of ^{1}H-MRSI with FDG-PET in TLE, demonstrated that hippocampal Cr/NAA measures did not correlate with glucose metabolism (Knowlton *et al.*, 2002). These differences may

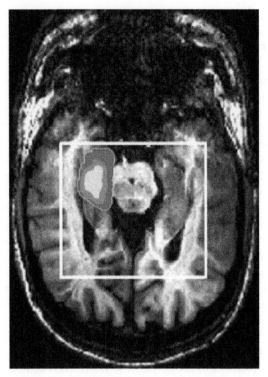

FIGURE 6.2   An example of a ^{1}H-MRSI ratio map in a patient with mesial TLE. The map of the region of abnormality was determined by inclusion of all voxels within the hippocampi that had an abnormal creatine/N-acetylaspartate (Cr/NAA) ratio defined as more than two standard deviations beyond normal. The degree of elevation of Cr/NAA is color coded and corresponds to the abnormal ratios; values of 0.9–1.6 by increments of 0.1. *Source*: Gilliam *et al.* (2007).

account in part for discrepancies in the results of studies of depression in TLE utilizing ^{1}H-MRSI and FDG-PET.

Potential confounders of the results of any functional imaging studies in patients with epilepsy to be considered include the postictal effects of seizures, and chronic effects of antiepileptic medications.

THE TREATMENT OF PATIENTS WITH EPILEPSY AND DEPRESSION

Several factors may contribute to the underrecognition of depression in epilepsy and lack of adequate therapeutic intervention. These may include the failure to appreciate the impact of depression on quality of life (Wiegartz *et al.*, 1999; Gilliam, 2002), concerns over the epileptogenic potential of the antidepressants

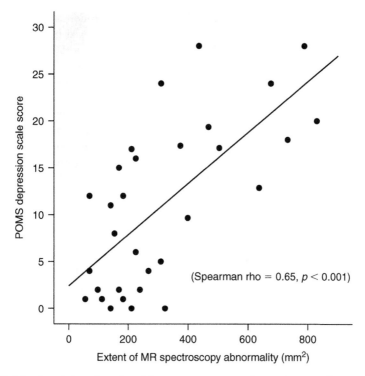

FIGURE 6.3   Correlation of severity of depression symptoms with spatial extent of hippocampal dysfunction defined by ^{1}H-MR spectroscopic creatine/N-acetylaspartate ratio maps ($n = 31$). *Source:* Gilliam *et al.* (2007).

(Pisani *et al.*, 1999), and uncertainty in regards to the efficacy of standard antidepressant therapy in patients with epilepsy (Kanner, 2003a; Gilliam *et al.*, 2004b).

Widespread concern in regards to the potential for antidepressants to worsen seizures appears to be mostly unsubstantiated. There are no reports in the literature of TCA-induced seizures at therapeutic serum levels. Patients who suffered seizures at therapeutic doses were discovered to be slow metabolizers of the drug. Rapid increases in medication, an abnormal EEG, the presence of CNS pathology, and personal and family history of epilepsy were other risk factors for developing seizures in non-epileptic patients (Kanner, 2003a).

The actual risk of causing seizures to worsen in patients with epilepsy by the use of antidepressant medications appears to be low. In a prospective study by Kanner *et al.* of 100 patients with epilepsy treated with sertraline, an SSRI, only one patient was found to have definite worsening of seizure rate and severity, and five others with probable but not definite worsening. Four of these patients remained on sertraline, and none of the patients suffered further seizure exacerbation after adjustment of their AEDs. Resolution of depressive symptoms was achieved in 54% of patients (Kanner *et al.*, 2000). In a retrospective 2-month study of the TCA doxepin in 19 patients with epilepsy, Ojemann *et al.* (1983) reported two patients with increased seizures, and 15 experiencing a reduction in seizures (Ojemann *et al.*, 1983). As discussed earlier, some antidepressant medications may also exhibit anticonvulsant effects, particularly in low doses, in experimental models of epilepsy and in humans.

The actual treatment of depression in persons with epilepsy is largely based on empiric evidence and anecdotal experience. A practical approach has been proposed by Kanner: Attempts should be made to exclude iatrogenic causes of depressive symptoms, such as the withdrawal of mood stabilizing AEDs such as CBZ, VPA, or LTG, or initiation or titration of an AED with known negative psychotropic properties such as phenobarbital, primidone, tiagabine, topiramate, felbamate, and vigabatrin. In the setting of an AED with negative psychotropic properties, the medication may need to be decreased in dose, discontinued, or replaced with an alternative medication. In patients for whom these options are not available, the initiation of an SSRI may be considered. The patient should be started on a small dose of medication, with small incremental increases until desired clinical effect, in order to minimize potential exacerbation of seizures (Kanner, 2003a).

As first line pharmacologic therapy, SSRIs with minimal effects on CYP450 isoenzymes may be preferred, such as citalopram and sertraline. Open trials of the SSRIs sertraline, paroxetine, and fluoxetine and the SNRI venlafaxine suggest that many patients with epilepsy may experience remission of their symptoms of depression (Kanner *et al.*, 2000).

The only prospective randomized, double-blind, placebo-controlled trial of antidepressants in epilepsy was reported in 1985 by Robertson and Trimble (1985). The efficacy of amitriptyline, nomifensine, and placebo were compared in 42 patients with epilepsy and depression. Despite a 50% improvement in the mean BDI and Hamilton Depression Rating Scale (HAMD) scores after treatment, similar degrees of improvement in the placebo arm resulted in no significant difference between groups at 6 weeks. Small patient sample sizes in each treatment group limited the interpretation of the trial results. Of note, no difference was observed in seizure rates between groups placed on TCA or placebo.

Blumer conducted an open label, uncontrolled study involving the addition of an SSRI to patients with interictal affective disorder previously refractory to a TCA, followed over a 20-month period (Blumer, 1997). Within the limitations of the study, treatment was felt to be highly satisfactory with 15 of 22 patients becoming good to excellent responders.

An uncontrolled study by Hovorka *et al.* (2000) sought to assess the efficacy and safety of the SSRI citalopram in 43 depressed patients with epilepsy. A significant decrease in HAMD scores was observed after 4 weeks and 8 weeks of treatment. There were 28 (65.1%) responders after 8 weeks, all with a decrease on the HAMD of greater than 50%. No seizure worsening was observed in any patient.

Kuhn *et al.* (2003) conducted a *post hoc* analysis of 75 adult patients with TLE and MDD treated prospectively with antidepressant medications. Of these, 33 patients were treated with citalopram, 27 with mirtazapine, and 15 with reboxetine. Depressive symptoms were rated at baseline, at 4 weeks, and between weeks 20 and 30. The rates of response to treatment, defined as a reduction in the HAMD score of greater than 50% between baseline and weeks 20–30, were not significantly different at 51.9% for mirtazapine, 36.4% for citalopram, and 53.3% for reboxetine. Dropout rates were notably high at 74.1% for mirtazapine, 48.5% for citalopram and 40.0% for reboxetine by weeks 20–30, and significantly different only for mirtazapine. There was no increase in frequency or severity of seizures as assessed by clinical judgment.

In an open, multicentered, uncontrolled study by Specchio *et al.* (2004), patients with epilepsy and depression underwent treatment with citalopram for 4 consecutive months (Specchio *et al.*, 2004). Clinical assessments were performed at baseline, and at 2 and 4 months of therapy. Of 45 enrolled patients, none had a deterioration of seizure frequency. An overall improvement in seizure frequency was observed in the 39 patients who completed the study. Most patients (67%) experienced a marked or moderate improvement of depressive symptoms during the treatment period.

Thome-Souza *et al.* (2007) sought to evaluate the influence of SSRIs on the severity and frequency of seizures in children with epilepsy and MDD, in addition to the efficacy of SSRIs in the treatment of depressive symptoms. In their uncontrolled study of 36 children with epilepsy and depression treated with fluoxetine or sertraline, seizures worsened in only two patients within 3 months after beginning an SSRI. All patients remaining on an SSRI had clinical improvement in their depressive symptoms. The authors concluded that SSRIs were a good therapeutic option in children with epilepsy, efficacious in the remission of depressive symptoms, generally associated with sustained maintenance of seizure control, and had few adverse effects.

CONCLUSIONS

Comorbid depressive disorders and suicidal behavior are more likely in patients with epilepsy than in healthy individuals. Likewise, epileptic disorders may be more common in patients with depression. Evidence exists for a close, potentially bi-directional relationship between the two disease states, with common underlying anatomic, pathologic, and functional abnormalities.

Unresolved issues include establishing a direct causal link between depression and epilepsy, determining optimal treatment strategies for patients with depressive disorders in the setting of epilepsy, and whether the pathophysiologic characteristics of this subgroup of patients and their response to treatment may substantially differ from patients with primary depression. Preliminary data with standard antidepressant regimens, and psychiatric referral when appropriate, appear to be effective initial interventions. It is unknown at this time whether early treatment of a depressive disorder may lower the risk of developing epileptic seizures (Kanner, 2006). Additional study is required.

Further awareness and research promises to offer an improved quality of living for a substantial number of patients with epilepsy and depression, a segment of the neurological community long neglected.

REFERENCES

Albano, C., Cupello, A., Mainardi, P., *et al.* (2006). Successful treatment of epilepsy with serotonin reuptake inhibitors: proposed mechanism. *Neurochem Res* **31**, 509–514.

Anhoury, S., Brown, R. J., Krishnamoorthy, E. S., *et al.* (2000). Psychiatric outcome after temporal lobectomy: a predictive study. *Epilepsia* **41**, 1608–1615.

Barry, J. J. (2003). The recognition and management of mood disorders as a comorbidity of epilepsy. *Epilepsia* **44**(Suppl 4), 30–40.

Baxter, L. R., Jr., Schwartz, J. M., Phelps, M. E., *et al.* (1989). Reduction of prefrontal cortex glucose metabolism common to three types of depression. *Arch Gen Psychiatry* **46**, 243–250.

Beck, S. G., Choi, K. C. (1991). 5-Hydroxytryptamine hyperpolarizes CA3 hippocampal pyramidal cells through an increase in potassium conductance. *Neurosci Lett* **133**, 93–96.

Beghi, E., Spagnoli, P., Airoldi, L., *et al.* (2002). Emotional and affective disturbances in patients with epilepsy. *Epilepsy Behav* **3**, 255–261.

Bench, C. J., Friston, K. J., Brown, R. G., *et al.* (1993). Regional cerebral blood flow in depression measured by positron emission tomography: the relationship with clinical dimensions. *Psychol Med* **23**, 579–590.

Blanchet, P., Frommer, G. P. (1986). Mood change preceding epileptic seizures. *J Nerv Ment Dis* **174**, 471–476.

Blumer, D. (1997). Antidepressant and double antidepressant treatment for the affective disorder of epilepsy. *J Clin Psychiatry* **58**, 3–11.

Blumer, D., Montouris, G., Davies, K. (2004). The interictal dysphoric disorder: recognition, pathogenesis, and treatment of the major psychiatric disorder of epilepsy. *Epilepsy Behav* **5**, 826–840.

Bourdeau, I., Bard, C., Noel, B., *et al.* (2002). Loss of brain volume in endogenous Cushing's syndrome and its reversibility after correction of hypercortisolism. *J Clin Endocrinol Metab* **87**, 1949–1954.

Bowley, M. P., Drevets, W. C., Ongur, D., *et al.* (2002). Low glial numbers in the amygdala in major depressive disorder. *Biol Psychiatry* **52**, 404–412.

Boylan, L. S., Flint, L. A., Labovitz, D. L., *et al.* (2004). Depression but not seizure frequency predicts quality of life in treatment-resistant epilepsy. *Neurology* **62**, 258–261.

Bromfield, E. B., Altshuler, L., Leiderman, D. B., *et al.* (1992). Cerebral metabolism and depression in patients with complex partial seizures. *Arch Neurol* **49**, 617–623.

Brown, E. S., Woolston, D. J., Frol, A., *et al.* (2004). Hippocampal volume, spectroscopy, cognition, and mood in patients receiving corticosteroid therapy. *Biol Psychiatry* **55**, 538–545.

Browning, R. A., Clark, K. B., Naritoku, D. K., Smith, D. C., Jensen, R. A. (1997). Loss of anticonvulsant effect of vagus nerve stimulation in the pentylenetetrazol seizure model following treatment with 6-hydroxydopamine or 5,7-dihydroxy-tryptamine [abstract]. *Proc Soc Neurosci* **2**, 23–35.

Cheetham, S. C., Crompton, M. R., Katona, C. L., *et al.* (1990). Brain 5-HT1 binding sites in depressed suicides. *Psychopharmacology (Berlin)* **102**, 544–548.

Chen, B., Dowlatshahi, D., MacQueen, G. M., *et al.* (2001). Increased hippocampal BDNF immunoreactivity in subjects treated with antidepressant medication. *Biol Psychiatry* **50**, 260–265.

Clough, R. W., Peterson, B. R., Steenbergen, J. L., *et al.* (1998). Neurite extension of developing noradrenergic neurons is impaired in genetically epilepsy-prone rats (GEPR-3s): an *in vitro* study on the locus coeruleus. *Epilepsy Res* **29**, 135–146.

Coffey, C. E., Wilkinson, W. E., Weiner, R. D., et al. (1993). Quantitative cerebral anatomy in depression. A controlled magnetic resonance imaging study. *Arch Gen Psychiatry* **50**, 7–16.

Cramer, J. A., Blum, D., Reed, M., et al. (2003). The influence of comorbid depression on quality of life for people with epilepsy. *Epilepsy Behav* **4**, 515–521.

Cramer, J. A., Blum, D., Fanning, K., et al. (2004). The impact of comorbid depression on health resource utilization in a community sample of people with epilepsy. *Epilepsy Behav* **5**, 337–342.

Dailey, J. W., Mishra, P. K., Ko, K. H., et al. (1992). Serotonergic abnormalities in the central nervous system of seizure-naive genetically epilepsy-prone rats. *Life Sci* **50**, 319–326.

Dailey, J. W., Reith, M. E., Yan, Q. S., et al. (1997). Carbamazepine increases extracellular serotonin concentration: lack of antagonism by tetrodotoxin or zero Ca^{2+}. *Eur J Pharmacol* **328**, 153–162.

Dolan, R. J., Bench, C. J., Brown, R. G., et al. (1994). Neuropsychological dysfunction in depression: the relationship to regional cerebral blood flow. *Psychol Med* **24**, 849–857.

Drevets, W. C., Price, J. L., Simpson, J. R., Jr., et al. (1997). Subgenual prefrontal cortex abnormalities in mood disorders. *Nature* **386**, 824–827.

Drevets, W. C., Frank, E., Price, J. C., et al. (1999). PET imaging of serotonin 1A receptor binding in depression. *Biol Psychiatry* **46**, 1375–1387.

Edeh, J., Toone, B. (1987). Relationship between interictal psychopathology and the type of epilepsy. Results of a survey in general practice. *Br J Psychiatry* **151**, 95–101.

Ettinger, A. B., Weisbrot, D. M., Nolan, E. E., et al. (1998). Symptoms of depression and anxiety in pediatric epilepsy patients. *Epilepsia* **39**, 595–599.

Favale, E., Rubino, V., Mainardi, P., et al. (1995). Anticonvulsant effect of fluoxetine in humans. *Neurology* **45**, 1926–1927.

Forsgren, L., Nystrom, L. (1990). An incident case-referent study of epileptic seizures in adults. *Epilepsy Res* **6**, 66–81.

Fromm, G. H., Wessel, H. B., Glass, J. D., et al. (1978). Imipramine in absence and myoclonic-astatic seizures. *Neurology* **28**, 953–957.

Fuchs, E., Gould, E. (2000). Mini-review: *in vivo* neurogenesis in the adult brain: regulation and functional implications. *Eur J Neurosci* **12**, 2211–2214.

Gaitatzis, A., Carroll, K., Majeed, A., et al. (2004). The epidemiology of the comorbidity of epilepsy in the general population. *Epilepsia* **45**, 1613–1622.

Garcia, P. A., Laxer, K. D., van der Grond, J., et al. (1997). Correlation of seizure frequency with *N*-acetyl-aspartate levels determined by 1H magnetic resonance spectroscopic imaging. *Magn Reson Imaging* **15**, 475–478.

Gilliam, F. (2002). Optimizing health outcomes in active epilepsy. *Neurology* **58**, S9–S20.

Gilliam, F., Kuzniecky, R., Faught, E., et al. (1997). Patient-validated content of epilepsy-specific quality-of-life measurement. *Epilepsia* **38**, 233–236.

Gilliam, F., Kuzniecky, R., Meador, K., et al. (1999). Patient-oriented outcome assessment after temporal lobectomy for refractory epilepsy. *Neurology* **53**, 687–694.

Gilliam, F. G., Barry, J. J., Hermann, B. P., et al. (2006). Rapid detection of major depression in epilepsy: a multicentre study. *Lancet Neurol* **5**, 399–405.

Gilliam, F. G., Fessler, A. J., Baker, G., et al. (2004a). Systematic screening allows reduction of adverse antiepileptic drug effects: a randomized trial. *Neurology* **62**, 23–27.

Gilliam, F. G., Santos, J., Vahle, V., et al. (2004b). Depression in epilepsy: ignoring clinical expression of neuronal network dysfunction? *Epilepsia* **45**(Suppl 2), 28–33.

Gilliam, F. G., Maton, B. M., Martin, R. C., et al. (2007). Hippocampal 1H-MRSI correlates with severity of depression symptoms in temporal lobe epilepsy. *Neurology* **68**, 364–368.

Giovacchini, G., Toczek, M. T., Bonwetsch, R., et al. (2005). 5-HT 1A receptors are reduced in temporal lobe epilepsy after partial-volume correction. *J Nucl Med* **46**, 1128–1135.

Hand, G. A., Hewitt, C. B., Fulk, L. J., et al. (2002). Differential release of corticotropin-releasing hormone (CRH) in the amygdala during different types of stressors. *Brain Res* **949**, 122–130.

Harris, E. C., Barraclough, B. (1997). Suicide as an outcome for mental disorders. A meta-analysis. *Br J Psychiatry* **170**, 205–228.

Hasin, D. S., Goodwin, R. D., Stinson, F. S., et al. (2005). Epidemiology of major depressive disorder: results from the National Epidemiologic Survey on Alcoholism and Related Conditions. *Arch Gen Psychiatry* **62**, 1097–1106.

Hatalski, C. G., Guirguis, C., Baram, T. Z. (1998). Corticotropin releasing factor mRNA expression in the hypothalamic paraventricular nucleus and the central nucleus of the amygdala is modulated by repeated acute stress in the immature rat. *J Neuroendocrinol* **10**, 663–669.

Hecimovic, H., Goldstein, J. D., Sheline, Y. I., et al. (2003). Mechanisms of depression in epilepsy from a clinical perspective. *Epilepsy Behav* **4**(Suppl 3), S25–S30.

Helgeson, D. C., Mittan, R., Tan, S. Y., et al. (1990). Sepulveda Epilepsy Education: the efficacy of a psychoeducational treatment program in treating medical and psychosocial aspects of epilepsy. *Epilepsia* **31**, 75–82.

Hermann, B. P., Wyler, A. R., Richey, E. T. (1988). Wisconsin Card Sorting Test performance in patients with complex partial seizures of temporal-lobe origin. *J Clin Exp Neuropsychol* **10**, 467–476.

Hermann, B. P., Trenerry, M. R., Colligan, R. C. (1996). Learned helplessness, attributional style, and depression in epilepsy. Bozeman Epilepsy Surgery Consortium. *Epilepsia* **37**, 680–686.

Hermann, B. P., Seidenberg, M., Bell, B. (2000). Psychiatric comorbidity in chronic epilepsy: identification, consequences, and treatment of major depression. *Epilepsia* **41**(Suppl 2), S31–S41.

Hesdorffer, D. C., Hauser, W. A., Annegers, J. F., et al. (2000). Major depression is a risk factor for seizures in older adults. *Ann Neurol* **47**, 246–249.

Hesdorffer, D. C., Hauser, W. A., Olafsson, E., et al. (2006). Depression and suicide attempt as risk factors for incident unprovoked seizures. *Ann Neurol* **59**, 35–41.

Holsboer, F. (2001). Stress, hypercortisolism and corticosteroid receptors in depression: implications for therapy. *J Affect Disord* **62**, 77–91.

Holsboer, F. (2003). Corticotropin-releasing hormone modulators and depression. *Curr Opin Investig Drugs* **4**, 46–50.

Horner, M. D., Flashman, L. A., Freides, D., et al. (1996). Temporal lobe epilepsy and performance on the Wisconsin Card Sorting Test. *J Clin Exp Neuropsychol* **18**, 310–313.

Hovorka, J., Herman, E., Nemcova, I. I. (2000). Treatment of interictal depression with citalopram in patients with epilepsy. *Epilepsy Behav* **1**, 444–447.

Jacoby, A., Baker, G. A., Steen, N., et al. (1996). The clinical course of epilepsy and its psychosocial correlates: findings from a UK community study. *Epilepsia* **37**, 148–161.

Jobe, P. C. (2003). Common pathogenic mechanisms between depression and epilepsy: an experimental perspective. *Epilepsy Behav* **4**(Suppl 3), S14–S24.

Jokeit, H., Seitz, R. J., Markowitsch, H. J., et al. (1997). Prefrontal asymmetric interictal glucose hypometabolism and cognitive impairment in patients with temporal lobe epilepsy. *Brain* **120**(Pt 12), 2283–2294.

Jones, J. E., Hermann, B. P., Barry, J. J., et al. (2003). Rates and risk factors for suicide, suicidal ideation, and suicide attempts in chronic epilepsy. *Epilepsy Behav* **4**(Suppl 3), S31–S38.

Jones, J. E., Hermann, B. P., Barry, J. J., et al. (2005). Clinical assessment of Axis I psychiatric morbidity in chronic epilepsy: a multicenter investigation. *J Neuropsychiatry Clin Neurosci* **17**, 172–179.

Kanner, A. M. (2003a). Depression in epilepsy: a frequently neglected multifaceted disorder. *Epilepsy Behav* **4**(Suppl 4), 11–19.

Kanner, A. M. (2003b). Depression in epilepsy: prevalence, clinical semiology, pathogenic mechanisms, and treatment. *Biol Psychiatry* **54**, 388–398.

Kanner, A. M. (2005). Depression in epilepsy: a neurobiologic perspective. *Epilepsy Curr* **5**, 21–27.

Kanner, A. M. (2006). Epilepsy, suicidal behaviour, and depression: do they share common pathogenic mechanisms? *Lancet Neurol* **5**, 107–108.

Kanner, A. M., Balabanov, A. (2002). Depression and epilepsy: how closely related are they? *Neurology* **58**, S27–S39.

Kanner, A. M., Kozak, A. M., Frey, M. (2000). The use of sertraline in patients with epilepsy: is it safe? *Epilepsy Behav* **1**, 100–105.

Kanner, A. M., Tilwalli, S., Smith, M. C., *et al*. (2003). A presurgical history of depression is associated with a worse postsurgical seizure outcome following a temporal lobectomy. *Neurology* **62**, A389.

Kanner, A. M., Soto, A., Gross-Kanner, H. (2004). Prevalence and clinical characteristics of postictal psychiatric symptoms in partial epilepsy. *Neurology* **62**, 708–713.

Kessler, R. C., Berglund, P., Demler, O., *et al*. (2003). The epidemiology of major depressive disorder: results from the National Comorbidity Survey Replication (NCS-R). *JAMA* **289**, 3095–3105.

Knowlton, R. C., Abou-Khalil, B., Sawrie, S. M., *et al*. (2002). *In vivo* hippocampal metabolic dysfunction in human temporal lobe epilepsy. *Arch Neurol* **59**, 1882–1886.

Kraepelin, E. (1923). *Psychiatrie*. Leipzig: Johann Ambrosius Barth.

Kuhn, K. U., Quednow, B. B., Thiel, M., *et al*. (2003). Antidepressive treatment in patients with temporal lobe epilepsy and major depression: a prospective study with three different antidepressants. *Epilepsy Behav* **4**, 674–679.

Kumar, A., Jin, Z., Bilker, W., *et al*. (1998). Late-onset minor and major depression: early evidence for common neuroanatomical substrates detected by using MRI. *Proc Natl Acad Sci USA* **95**, 7654–7658.

Lehrner, J., Kalchmayr, R., Serles, W., *et al*. (1999). Health-related quality of life (HRQOL), activity of daily living (ADL) and depressive mood disorder in temporal lobe epilepsy patients. *Seizure* **8**, 88–92.

Lucassen, P. J., Muller, M. B., Holsboer, F., *et al*. (2001). Hippocampal apoptosis in major depression is a minor event and absent from subareas at risk for glucocorticoid overexposure. *Am J Pathol* **158**, 453–468.

Lupien, S. J., de Leon, M., de Santi, S., *et al*. (1998). Cortisol levels during human aging predict hippocampal atrophy and memory deficits. *Nat Neurosci* **1**, 69–73.

Martin, R. C., Sawrie, S. M., Gilliam, F. G., *et al*. (2000). Wisconsin Card Sorting performance in patients with temporal lobe epilepsy: clinical and neuroanatomical correlates. *Epilepsia* **41**, 1626–1632.

Mathern, G. W., Babb, T. L., Armstrong, D. L. (1997). Hippocampal sclerosis. In *Epilepsy: A Comprehensive Textbook* (J. Engel, T. A. Pedley, eds), pp. 133–155. Philadelphia, PA: Lippincott-Raven.

Mendez, M. F., Cummings, J. L., Benson, D. F. (1986). Depression in epilepsy. Significance and phenomenology. *Arch Neurol* **43**, 766–770.

Merlet, I., Ostrowsky, K., Costes, N., *et al*. (2004). 5-HT1A receptor binding and intracerebral activity in temporal lobe epilepsy: an [18F]MPPF-PET study. *Brain* **127**, 900–913.

Meschaks, A., Lindstrom, P., Halldin, C., *et al*. (2005). Regional reductions in serotonin 1A receptor binding in juvenile myoclonic epilepsy. *Arch Neurol* **62**, 946–950.

Mohanraj, R., Brodie, M. J. (2003). Predicting outcomes in newly diagnosed epilepsy. *Epilepsia* **44**(suppl 9).

Nemeroff, C. B., Owens, M. J. (2002). Treatment of mood disorders. *Nat Neurosci* **5**(Suppl), 1068–1070.

Nestler, E. J., Barrot, M., DiLeone, R. J., *et al*. (2002). Neurobiology of depression. *Neuron* **34**, 13–25.

Nibuya, M., Morinobu, S., Duman, R. S. (1995). Regulation of BDNF and trkB mRNA in rat brain by chronic electroconvulsive seizure and antidepressant drug treatments. *J Neurosci* **15**, 7539–7547.

Nilsson, L., Ahlbom, A., Farahmand, B. Y., *et al*. (2003). Mortality in a population-based cohort of epilepsy surgery patients. *Epilepsia* **44**, 575–581.

O'Donoghue, M. F., Goodridge, D. M., Redhead, K., *et al*. (1999). Assessing the psychosocial consequences of epilepsy: a community-based study. *Br J Gen Pract* **49**, 211–214.

Ojemann, L. M., Friel, P. N., Trejo, W. J., *et al*. (1983). Effect of doxepin on seizure frequency in depressed epileptic patients. *Neurology* **33**, 646–648.

Okuhara, D. Y., Beck, S. G. (1994). 5-HT1A receptor linked to inward-rectifying potassium current in hippocampal CA3 pyramidal cells. *J Neurophysiol* **71**, 2161–2167.

Oquendo, M. A., Placidi, G. P., Malone, K. M., *et al*. (2003). Positron emission tomography of regional brain metabolic responses to a serotonergic challenge and lethality of suicide attempts in major depression. *Arch Gen Psychiatry* **60**, 14–22.

Perrine, K., Hermann, B. P., Meador, K. J., *et al.* (1995). The relationship of neuropsychological functioning to quality of life in epilepsy. *Arch Neurol* **52**, 997–1003.

Pisani, F., Spina, E., Oteri, G. (1999). Antidepressant drugs and seizure susceptibility: from *in vitro* data to clinical practice. *Epilepsia* **40**(Suppl 10), S48–S56.

Quiske, A., Helmstaedter, C., Lux, S., *et al.* (2000). Depression in patients with temporal lobe epilepsy is related to mesial temporal sclerosis. *Epilepsy Res* **39**, 121–125.

Rafnsson, V., Olafsson, E., Hauser, W. A., *et al.* (2001). Cause-specific mortality in adults with unprovoked seizures. A population-based incidence cohort study. *Neuroepidemiology* **20**, 232–236.

Rajkowska, G., Miguel-Hidalgo, J. J., Wei, J., *et al.* (1999). Morphometric evidence for neuronal and glial prefrontal cell pathology in major depression. *Biol Psychiatry* **45**, 1085–1098.

Ressler, K. J., Nemeroff, C. B. (2000). Role of serotonergic and noradrenergic systems in the pathophysiology of depression and anxiety disorders. *Depress Anxiety* **12**(Suppl 1), 2–19.

Reul, J. M., Holsboer, F. (2002). Corticotropin-releasing factor receptors 1 and 2 in anxiety and depression. *Curr Opin Pharmacol* **2**, 23–33.

Robertson, M. M. (1997). Suicide, parasuicide, and epilepsy. In *Epilepsy: A Comprehensive Textbook* (J. Engle, T. A. Pedley, eds). Philadelphia, PA: Lippincott-Raven. 2141–2151.

Robertson, M. M., Trimble, M. R. (1985). The treatment of depression in patients with epilepsy. A double-blind trial. *J Affect Disord* **9**, 127–136.

Ryu, J. R., Jobe, P. C., Milbrandt, J. C., *et al.* (1999). Morphological deficits in noradrenergic neurons in GEPR-9s stem from abnormalities in both the locus coeruleus and its target tissues. *Exp Neurol* **156**, 84–91.

Sapolsky, R. M. (2000). Glucocorticoids and hippocampal atrophy in neuropsychiatric disorders. *Arch Gen Psychiatry* **57**, 925–935.

Sareen, J., Cox, B. J., Afifi, T. O., *et al.* (2005). Anxiety disorders and risk for suicidal ideation and suicide attempts: a population-based longitudinal study of adults. *Arch Gen Psychiatry* **62**, 1249–1257.

Sargent, P. A., Kjaer, K. H., Bench, C. J., *et al.* (2000). Brain serotonin1A receptor binding measured by positron emission tomography with [11C]WAY-100635: effects of depression and antidepressant treatment. *Arch Gen Psychiatry* **57**, 174–180.

Savic, I., Lindstrom, P., Gulyas, B., *et al.* (2004). Limbic reductions of 5-HT1A receptor binding in human temporal lobe epilepsy. *Neurology* **62**, 1343–1351.

Schmitz, E. B., Moriarty, J., Costa, D. C., *et al.* (1997). Psychiatric profiles and patterns of cerebral blood flow in focal epilepsy: interactions between depression, obsessionality, and perfusion related to the laterality of the epilepsy. *J Neurol Neurosurg Psychiatry* **62**, 458–463.

Schmitz, E. B., Robertson, M. M., Trimble, M. R. (1999). Depression and schizophrenia in epilepsy: social and biological risk factors. *Epilepsy Res* **35**, 59–68.

Sheline, Y. I. (2003). Neuroimaging studies of mood disorder effects on the brain. *Biol Psychiatry* **54**, 338–352.

Sheline, Y. I., Wang, P. W., Gado, M. H., *et al.* (1996). Hippocampal atrophy in recurrent major depression. *Proc Natl Acad Sci USA* **93**, 3908–3913.

Sheline, Y. I., Sanghavi, M., Mintun, M. A., *et al.* (1999). Depression duration but not age predicts hippocampal volume loss in medically healthy women with recurrent major depression. *J Neurosci* **19**, 5034–5043.

Sheline, Y. I., Gado, M. H., Kraemer, H. C. (2003). Untreated depression and hippocampal volume loss. *Am J Psychiatry* **160**, 1516–1518.

Shih, J. J., Weisend, M. P., Lewine, J., *et al.* (2004). Areas of interictal spiking are associated with metabolic dysfunction in MRI-negative temporal lobe epilepsy. *Epilepsia* **45**, 223–229.

Smith, M. A., Makino, S., Kvetnansky, R., *et al.* (1995). Effects of stress on neurotrophic factor expression in the rat brain. *Ann NY Acad Sci* **771**, 234–239.

Southam, E., Kirkby, D., Higgins, G. A., *et al.* (1998). Lamotrigine inhibits monoamine uptake *in vitro* and modulates 5-hydroxytryptamine uptake in rats. *Eur J Pharmacol* **358**, 19–24.

Specchio, L. M., Iudice, A., Specchio, N., *et al.* (2004). Citalopram as treatment of depression in patients with epilepsy. *Clin Neuropharmacol* **27**, 133–136.

Statnick, M. A., Dailey, J. W., Jobe, P. C., et al. (1996a). Abnormalities in 5-HT1A and 5-HT1B receptor binding in severe-seizure genetically epilepsy-prone rats (GEPR-9s). *Neuropharmacology* **35**, 111–118.

Statnick, M. A., Maring-Smith, M. L., Clough, R. W., et al. (1996b). Effect of 5,7-dihydroxytryptamine on audiogenic seizures in genetically epilepsy-prone rats. *Life Sci* **59**, 1763–1771.

Stockmeier, C.A., Shapiro, L.A., Dilley, G.E. et al. (1998). Increase in serotonin-1A autoreceptors in the midbrain of suicide victims with major depression – postmortem evidence for decreased serotonin activity. *J Neuroscience* **18**, 7394–7401.

Theodore, W. H. (2003). Does serotonin play a role in epilepsy? *Epilepsy Curr* **3**, 173–177.

Theodore, W. H., Hasler, G., Giovacchini, G., et al. (2007). Reduced hippocampal 5HT1A PET receptor binding and depression in temporal lobe epilepsy. *Epilepsia* **48**, 1526–1530.

Thome-Souza, M.S., Kuczynski, E., Valente, K.D. (2007). Sertraline and fluoxetine: safe treatments for children and adolescents with epilepsy and depression. *Epilepsy Behav* **10**, 417–425.

Victoroff, J. I., Benson, F., Grafton, S. T., et al. (1994). Depression in complex partial seizures. Electroencephalography and cerebral metabolic correlates. *Arch Neurol* **51**, 155–163.

Wang, C., Mishra, P. K., Dailey, J. W., et al. (1994). Noradrenergic terminal fields as determinants of seizure predisposition in GEPR-3s: a neuroanatomic assessment with intracerebral microinjections of 6-hydroxydopamine. *Epilepsy Res* **18**, 1–9.

Watanabe, Y., Gould, E., Cameron, H. A., et al. (1992). Phenytoin prevents stress- and corticosterone-induced atrophy of CA3 pyramidal neurons. *Hippocampus* **2**, 431–435.

Wiegartz, P., Seidenberg, M., Woodard, A., et al. (1999). Co-morbid psychiatric disorder in chronic epilepsy: recognition and etiology of depression. *Neurology* **53**, S3–S8.

Yan, Q. S., Mishra, P. K., Burger, R. L., et al. (1992). Evidence that carbamazepine and antiepilepsirine may produce a component of their anticonvulsant effects by activating serotonergic neurons in genetically epilepsy-prone rats. *J Pharmacol Exp Ther* **261**, 652–659.

Yan, Q. S., Jobe, P. C., Dailey, J. W. (1993). Thalamic deficiency in norepinephrine release detected via intracerebral microdialysis: a synaptic determinant of seizure predisposition in the genetically epilepsy-prone rat. *Epilepsy Res* **14**, 229–236.

Yan, Q. S., Jobe, P. C., Dailey, J. W. (1995). Further evidence of anticonvulsant role for 5-hydroxytryptamine in genetically epilepsy-prone rats. *Br J Pharmacol* **115**, 1314–1318.

CHAPTER 7

Are Anxiety and Depression Two Sides of the Same Coin?

Jana E. Jones

INTRODUCTION

The association between psychiatric disorders and epilepsy can be documented as far back as antiquity (Temkin, 1971). Psychiatric comorbidity in epilepsy received increasingly greater attention and investigation by researchers and clinicians over the past three decades. However, psychiatric disorders in epilepsy are not fully understood, and there is still important work to be completed to increase our knowledge and understanding. Recently, at the second "Curing Epilepsy: Translating Discoveries into Therapies" conference sponsored by the National Institutes of Health (March 29–30, 2007) psychiatric disorders as a comorbidity of epilepsy were highlighted for the first time as part of the research benchmarks or initiatives for the National Institute of Neurological Disorders and Stroke (NINDS). With psychiatric comorbidity identified as a research benchmark, hopefully this significant comorbidity will receive increased study and attention.

Psychiatric symptoms can be directly linked to seizure activity as the symptoms may occur just before, during or after the seizure (peri–ictal, ictal, postictal). Additionally, psychiatric symptoms may not be directly connected with seizure activity (interictal). Psychiatric disorders may appear before and after a diagnosis of epilepsy. Notably, controversy continues to remain regarding the phenomenology of psychiatric disorders in epilepsy and whether or not the symptom profile and semiology is comparable to psychiatric disorders in the general population (i.e., DSM–IV, ICD–10) (Kanner, 2003; Gaitatzis *et al.*, 2004).

Research indicates between 35% and 70% of individuals with epilepsy will meet criteria for a lifetime psychiatric disorder (Silberman *et al.*, 1994; Victoroff *et al.*, 1994; Stefansson *et al.*, 1998; Altshuler *et al.*, 1999; Glosser *et al.*, 2000; Tsopelas *et al.*, 2001). Compared to the general population, other neurological disorders and chronic medical illnesses, adults and children with epilepsy have higher rates of lifetime psychiatric disorders. In a population-based study in Great Britain, Davies *et al.* (2003) reported the rate of any psychiatric disorder was 37% among children with epilepsy compared to 11% among children with diabetes and 9% among controls. Similarly, as part of a 35-year follow-up in a population-based study in Finland, Jalava and Sillanpää (1996) found 23% of adults with epilepsy had a psychiatric disorder compared to 7% in healthy controls. Based on population data, it appears that psychiatric disorders are quite high among individuals with epilepsy.

EPIDEMIOLOGY OF DEPRESSION AND ANXIETY IN EPILEPSY

Depression is believed to be the most common psychiatric disorder among individuals with epilepsy (Hermann *et al.*, 2000; Kanner, 2003). For centuries a link between epilepsy and depression was hypothesized. A poignant example is evidenced by the Hippocratic corpus written around 400 BC. It stated "melancholics ordinarily become epileptics and epileptics melancholics ..." (Lewis, 1934). More recently, Hesdorffer *et al.* (2000) found that individuals with epilepsy were 3.7 times more likely to have a depressive episode prior to the first seizure. In a study in Iceland among children and adults with epilepsy, similar results were reported, revealing individuals with epilepsy were 1.7 times more likely to have a history of major depression prior to the first unprovoked seizure (Hesdorffer *et al.*, 2006). Depression appears to be a risk factor for the onset of seizures, and depression appears to be a significant comorbidity after the seizure disorder is diagnosed.

There continues to be some controversy regarding the exact incidence and prevalence rates of depression in epilepsy due to differing methodologies (self-report vs. psychiatric interview) and samples (population vs. hospital based). In studies utilizing Diagnostic and Statistical Manual of Mental Disorders (DSM) or International Classification of Diseases (ICD) (World Health Organisation, 1992) diagnostic criteria, rates of depressive disorders range from 6% to 64% (see Table 7.1). Swinkels *et al.* (2006) utilized the Composite International Diagnostic Interview (World Health Organization, 1992) to identify psychiatric disorders and found major depression in 21% of individuals with temporal lobe epilepsy (TLE) and 23.4% with extra-TLE. There were no significant differences between the TLE and extra-temporal groups in rates of depression. In a sample of adults with complex partial seizures, Wiegartz *et al.* (1999) reported 31.6% met criteria for a lifetime episode of major depression and 9.2% were identified with a current episode of major depression. This study utilized the Structured Clinical Interview for DSM-IV (SCID-IV) to identify episodes of depressive disorders (First *et al.*, 1996). Using the DSM-III-R criteria, Victoroff *et al.* (1994) identified a lifetime history of a depressive disorder in 64% of the sample and 30% had a lifetime history of major depression. Even among studies utilizing DSM or ICD criteria, there remains variability in prevalence rates of depression in epilepsy.

Like depression, symptoms of anxiety can be exacerbated by seizure activity (peri–ictal, ictal, postictal) and can appear separately or, in other words, interictally. Historically, anxiety and epilepsy have been viewed as highly interconnected. Temkin (1971) documented that even in the 1800s it was believed fright could cause epilepsy. Also similar to depression, methodological issues exist when determining the prevalence and incidence of anxiety disorders. Throughout the literature, hospital- or community-based samples are utilized, making it difficult to make generalizations across studies; uniform methods are frequently not used to assess or measure symptoms of anxiety (symptom checklist vs. interviews), and control groups are frequently omitted (Scicutella, 2001).

In light of these limitations, rates for anxiety disorders are reportedly elevated among individuals with epilepsy and range from 5% to 25% (see Table 7.2). A brief summary of three of these studies follows. Silberman *et al.* (1994) assessed individuals with epilepsy from a tertiary care center using the Schedule of Affective Disorders and Schizophrenia Lifetime Version (SADS-L) (Endicott and Spitzer, 1978), and 15% met criteria for an anxiety disorder and 16% reported symptoms of anxiety on self-report measures. Glosser *et al.* (2000) utilized the SCID to assess individuals with TLE prior to surgery and reported 18% of the total sample met criteria for an anxiety disorder in the previous 12 months. Among Dutch individuals with generalized and complex partial seizures, Swinkels *et al.* (2001) found 30% of the sample met criteria for a lifetime anxiety disorder and 25% meet criteria in the past 12 months. Much like depression, it appears that anxiety disorders are also elevated in epilepsy.

TABLE 7.1   Rates of lifetime depressive disorders

Fiordelli *et al.* (1993)	6%
Jones *et al.* (2005)	21%
Silberman *et al.* (1994)	62%
Swinkels *et al.* (2006)	21%
Victoroff *et al.* (1994)	64%
Wiegartz *et al.* (2000)	32%

TABLE 7.2   Rates of lifetime anxiety disorders

Altshuler *et al.* (1999)	5%
Glosser *et al.* (2000)	18%
Manchanda *et al.* (1996)	11%
Ring *et al.* (1998)	18%
Silberman *et al.* (1994)	16%
Swinkels *et al.* (2001)	30%

Research findings in epilepsy are inconsistent regarding the prevalence rates of depression and anxiety among different seizure types (e.g., TLE vs. generalized epilepsy). Perini *et al.* (1996) found individuals with TLE to have higher rates of affective and personality disorders compared to individuals with juvenile myoclonic epilepsy and individuals with diabetes. In the same year, Manchanda *et al.* (1996) found no significant differences in rates of psychiatric disorders among individuals with localization related vs. generalized seizures. Additionally, in regards to laterality of the seizure focus, there were no significant differences in rates of psychiatric disorders. These two studies reflect the conflicting results reported in the literature and amplify the need for further research.

Depression and anxiety are significant comorbidities in epilepsy. Both psychiatric disorders negatively impact quality of life (Gilliam *et al.*, 2003; Boylan *et al.*, 2004; Johnson *et al.*, 2004). If depression and anxiety are ignored or left untreated there is a greater likelihood for increased rates of suicide and suicidal ideation (Robertson, 1997; Jones *et al.*, 2003). This chapter will review the following topics: (1) co-occurrence of depression and anxiety and common symptomatology; (2) common pathogenic mechanisms; (3) treatment strategies shared by both disorders; and (4) future implications for clinicians and researchers.

ANXIETY AND DEPRESSION IN THE GENERAL POPULATION

The general psychiatric literature provides additional insight and findings related to the discussion of depression, anxiety and the co-occurrence of the two disorders. As clinicians and researchers in epilepsy, we can utilize the knowledge gleaned from the general population about depression and anxiety in order to gain an understanding of the similarities and differences in the presentation of these disorders in epilepsy.

Based on the National Comorbidity Survey-Replication in the United States, lifetime prevalence rates for depressive disorders is 20.8% with major depression being the most prevalent (16.6%), and lifetime anxiety disorders identified in 28.8% with specific phobia (12.5%) and social phobia (12.1%) as the most commonly occurring anxiety disorders (Kessler *et al.*, 2005a, b). Additionally, the European Study of the Epidemiology of Mental Disorders (ESEMeD) project included representative samples from six European countries (Belgium, France, Germany, Italy, The Netherlands, Spain). This project reported lower prevalence rates of depressive disorders (14.0%), major depression (12.8%), anxiety disorders (13.6%), specific phobia (7.7%), and social phobia (2.4%) (Alonso *et al.*, 2004b). These studies confirm what is considered common knowledge: individuals with epilepsy likely have higher prevalence rates of depression and anxiety disorders compared to the general population.

In 1934, Sir Aubrey Lewis described symptoms of anxiety and depression on the same continuum. In contrast, depression and anxiety are currently viewed as separate and distinct disorders in the DSM, fourth edition (DSM-IV) and ICD, tenth edition (ICD-10). However, the possibility that anxiety and depression could be placed on a continuum continues to be studied (Angst, 1997; Boulenger *et al.*, 1997; Levine *et al.*, 2001; Zimmerman and Chelminski, 2003). In tandem with this idea is the significant debate in the literature regarding the extent and impact of subthreshold anxiety and depression (Angst, 1997; Boulenger *et al.*, 1997; Stahl, 1997). The term subsyndromal alludes to a cluster of symptoms that do not reach diagnostic criteria. Subsyndromal symptoms may cause distress and impairment and may require treatment (Ninan and Berger, 2001). In response to the concern that individuals with subthreshold depression and anxiety were presenting to primary care physicians with significant levels of distress and impairment, the ICD-10 and DSM-IV introduced the concept of mixed anxiety and depression. In general clinical practice world wide it appears that 13% have a combined anxiety and depressive disorder (Stein *et al.*, 1995; Sartorius *et al.*, 1996).

There are three major theories or ideas describing the relationship between anxiety and depression (Stahl, 1997; Levine *et al.*, 2001). The first is the traditional theory which essentially states that anxiety and depression are distinct disorders in and of themselves including their treatments (i.e., antidepressants and anxiolytics) (Figure 7.1). This is further amplified by the subcategorization of both anxiety (e.g., generalized anxiety disorder, panic disorder, social phobia, obsessive compulsive disorder) and depression (e.g., major depression, dysthymia, brief intermittent depression). The second is the comorbid theory, which capitalizes on the sheer fact that anxiety and depression often co-exist (Figure 7.1) (Stahl, 1997). The individual presents with two illnesses and will require treatment for both. For example, there is an increased incidence of current or past depressive disorders among individuals with panic disorder. The third is the subsyndromal theory which asserts that some individuals have symptoms of depression and anxiety that are chronic in nature, and these symptoms may or may not always be severe enough to meet criteria for a diagnosis of an anxiety or depressive disorder (Figure 7.1). These subsyndromal symptoms ebb and flow depending on stressors and may escalate over time to meet criteria for a depressive and/or anxiety disorder. Two of the three theories reflect the idea that depression and anxiety are related and may be entangled with one another.

The co-occurrence of anxiety and depression is very commonly recognized in general clinical practice. International epidemiological and clinical studies revealed that comorbid depression and anxiety results in increased distress, impairment, and symptoms resulting in a longer course than either of these disorders alone (Angst, 1997; Kessler *et al.*, 1999; Kessler, 2007). Major depression is often comorbid with other psychiatric conditions, and most commonly co-occurs with symptoms of anxiety. In the Epidemiological Catchment Area program (ECA), 43% of individuals in the community with a depressive disorder also had a comorbid lifetime diagnosis of an anxiety disorder, and 25% with a primary anxiety disorder had a lifetime mood disorder (Regier *et al.*, 1990). Fava *et al.* (2000) reported that

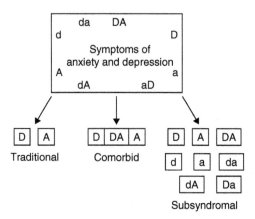

FIGURE 7.1   Three theories on symptoms of anxiety and depression. Syndromal disorders are represented by depressive disorder (D), anxiety disorder (A), and co-occurring depressive and anxiety disorder (DA). Subsyndromal presentation of symptoms include depression (d), anxiety (a), and mixed depression and anxiety (da). Subsyndromal and syndromal mixtures can occur (dA and Da). *Source*: Obtaining permission from author Stahl (1997).

among adults with major depression, anxiety disorders were present in half the sample (50.6%), and the anxiety disorder preceded the onset of depression in 31.4% of the sample. The ESEMeD project analyzed the co-occurrence of psychiatric disorders based on 12-month prevalence rates (Alonso *et al.*, 2004a). They found high pairwise associations between any depressive disorder and any anxiety disorder.

DEPRESSION AND ANXIETY IN EPILEPSY: WHAT DO WE KNOW?

Several recent population-based studies have reported elevated symptoms of depression and anxiety among adults and children with epilepsy (Davies *et al.*, 2003; Strine *et al.*, 2005; Tellez-Zenteno *et al.*, 2005; Kobau *et al.*, 2006). However, in the epilepsy literature the co-occurrence of depression and anxiety has received limited attention. Depression and anxiety disorders are frequently reported in over arching categories (e.g., internalizing disorder, neurosis) with no ability to distinguish the individual prevalence rates of each disorder (Jacoby *et al.*, 1996; Davies *et al.*, 2003; Strine *et al.*, 2003), or depression and anxiety are reported as distinct disorders with no acknowledgement of co-occurrence (Ring *et al.*, 1998; Glosser *et al.*, 2000). Recently, in the 2004 HealthStyles Survey, a large population-based mail survey, Kobau *et al.* (2006) found 16.7% (OR 3.2; 95% CI 1.4–7.4) of individuals with active epilepsy reported both depression and anxiety during the past year. In a multicenter study, among individuals who met criteria for current depressive disorders, 27 of 37 (73%) had a current anxiety disorder (Jones *et al.*, 2005).

Among a sample of 100 children with complex partial seizures and 71 with childhood absence epilepsy, Caplan *et al.* (2005) reported 3.5% of the sample met criteria for both depression and anxiety, and interestingly, there were no co-occurring diagnoses of anxiety and depression among children with absence seizures. Depression and anxiety disorders demand more attention and investigation in order to have a greater understanding of the co-occurrence of these two disorders in epilepsy.

COMMON SYMPTOMATOLOGY IN DEPRESSION AND ANXIETY

Depression and anxiety disorders have overlapping symptomatology. As defined by the DSM-IV-TR (American Psychiatric Association, 2000), major depression and generalized anxiety disorder share four diagnostic symptoms, which include sleep disturbance, difficulty concentrating, restlessness and fatigue. In the DSM-III all anxiety disorders (i.e., social phobia, panic disorder, etc.) were required to occur independently of and symptoms were to be present separate from a depressive episode. However, in the DSM-IIIR (American Psychiatric Association, 1987) this hierarchy was eliminated for all anxiety disorders except generalized anxiety disorder (Zimmerman and Chelminski, 2003). The diagnostic criteria for generalized anxiety disorder states that criteria must be reached independently and outside the presence of a depressive disorder, and this exception is likely related to the fact that both disorders have overlapping symptoms (Zimmermann and Chelminski, 2003; DSM-IV-TR, 2000).

Generalized anxiety disorder was first listed as a separate disorder in the DSM-III (American Psychiatric Association, 1980). According to Kessler *et al.* (1999b), since its introduction generalized anxiety disorder has been noted to be frequently comorbid with other depressive disorders as well as other psychiatric disorders. Using the results from two population-based studies, Kessler *et al.* (1999b) examined the 12-month prevalence rates of individuals with generalized anxiety disorder as the primary disorder who also met criteria for major depression. In the National Comorbidity Survey (NCS) 58.1% had both disorders, and in the Midlife Development in the US Survey, 69.7% met criteria for both depression and anxiety. Interestingly, among individuals with a primary diagnosis of depression, only 17.5% in the NCS and 16.3% in the Midlife Development in the US Survey met criteria for comorbid generalized anxiety disorder and major depression.

In the Early Developmental Stages of Psychopathology Study, a prospective longitudinal study, Bittner *et al.* (2004) found the presence of any anxiety disorder significantly increases the risk for a subsequent major depressive disorder (OR = 2.2, 95%; CI 1.6 – 3.2). Additionally, any of the individual anxiety disorders also increase the risk for a depressive disorder. Generalized anxiety disorder had the largest risk (OR = 4.5, 95%; CI 1.9 – 10.3). Panic disorder (OR = 3.4, 95%; CI 1.2 – 9.0) and agoraphobia (OR = 3.1, 95%; CI 1.4 – 6.7) had the second and third highest risk. Social phobia and specific phobia also had an increased risk for major depression with odds ratio of 2.9 (95%; CI 1.7 – 4.8) and 1.9 (95%; CI 1.2 – 2.8) respectively. Using data from the NCS, Choy *et al.* (2007) found a lifetime prevalence of major depression among individuals with specific phobia to be 40.7%, and in contrast only 14% had major depression without specific phobia. Interestingly, if more than one fear was present the risk of depression was even more elevated and ranged from OR=2.5 (95%; CI 2.0 – 3.1) for two fears to OR = 5.7 (95%; CI 4.3 – 7.6) for more than five fears. After adjusting for lifetime comorbid anxiety disorders, an odds ratio of 1.9 (95%; CI, 1.6 – 2.4) was reported as the risk for major depression if diagnosed with a specific phobia. Additionally, after adjusting for other anxiety disorders, the ECA study reported an odds ratio of 1.7 (95%; CI, 1.6 – 1.8) for depression at 12-month follow-up (Goodwin, 2002).

This frequent co-occurrence adds to speculation and debate over the possibility that generalized anxiety disorder might not be an independent disorder but an actual presentation of major depression or other comorbid psychiatric diagnoses (Kessler *et al.*, 1999b). One central issue in this debate is the extent to which generalized anxiety disorder actually results in functional impairments and to what extent these impairments are related to comorbid depression and other psychiatric disorders (Kessler *et al.*, 1999b).

There is evidence to support the notion that generalized anxiety and depression are two distinct disorders. First, Brown *et al.* (1998) identified separate latent factors of positive and negative affectivity, providing support for the argument that defines the disorders as separate and distinct in spite of the fact that they share four core symptoms. Additionally, Kessler *et al.* (1999b) found impairment ratings of generalized anxiety disorder alone were equal to those of major depression alone even when controlling for other DSM disorders. In conjunction with the above finding, if both disorders are present, the impairment rating is more severe than either of the two disorders alone (Kessler *et al.*, 1999b; Zimmerman and Chelminski, 2003). These results indicate impairments caused by generalized anxiety disorder are likely to be independently significant and not the result of another disorder. Finally, and somewhat contradictory, twin studies have indicated that generalized anxiety disorder and major depression share the same gene; however, epidemiological research indicates there are different sociodemographic predictors for each disorder (Kendler *et al.*, 1992; Skodal *et al.*, 1994). It appears there is evidence to support two distinct disorders (Angst, 1997); however, the exact relationship between anxiety and depression remains to be more clearly delineated.

COMMON PATHOGENIC MECHANISMS IN DEPRESSION, ANXIETY, AND EPILEPSY

There are common pathogenic mechanisms shared by depression, anxiety, and epilepsy. Serotonin (5HT), NE, dopamine, GABA, and glutamate are likely involved in the expression of all three disorders (Jobe *et al.*, 1999; Ressler and Nemeroff,

2000). In depression and anxiety, changes in the noradrenergic and serotonergic systems are implicated as playing a significant role in the expression of these disorders. There appears to be increased noradrenergic activity, contrasting with decreased activity in the serotonergic systems. Pharmacological treatments attempt to modulate this activity. Depending on which medication is introduced, there is a combination of increased or decreased release of the serotonin or NE along with an increase or decrease in receptor activity (Stahl, 1997; Ressler and Nemeroff, 2000). It is suggested that selective serotonin reuptake inhibitors (SSRIs) reduce noradrenergic transmission and increase serotonergic transmission (Ressler and Nemeroff, 2000). If there is a sudden reduction in serotonin there will be an increase in depressive symptoms (Stahl, 1997; Ressler and Nemeroff, 2000). A related hypothesis explaining the effectiveness of SSRIs in treating anxiety indicates that if there is excess serotonin, the receptors downregulate to prevent excessive neuronal activity, reducing symptoms of anxiety; and in depression, an upregulation occurs to enhance neuronal activity (Stahl, 1997). This is a significant simplification of the process, but illustrates the relationship nonetheless.

In epilepsy, if there is a decrease in serotonergic or noradrenergic transmission, kindling of seizures occurs, seizure activity increases, and severity intensifies (Jobe *et al.*, 1999). Increasing serotonin and NE transmission helps to prevent seizures and a reduction will increase the likelihood that a seizure will occur (Jobe *et al.*, 1999). There have been no randomized controlled trials examining the effects of SSRIs on seizure frequency; however, open trials of fluoxetine and citalopram indicated some improvement in seizure frequency (Favale *et al.*, 1995; Specchio *et al.*, 2004). Additionally, there was one controlled trial using imipramine, a tricyclic antidepressant (TCA); it was reported to reduce absence and myoclonic seizures (Fromm *et al.*, 1978).

In a similar fashion, antiepileptic drugs (AEDs) play a dual role with antiepileptic effects and psychotropic properties. All AEDs can have psychiatric side effects (Ketter *et al.*, 1994; Schmitz, 1999; Ettinger, 2006). Valproate, lamotrigine, carbamazepine, and oxcarbazepine have been used as mood stabilizers (Stahl, 2004; Kanner, 2005; Ettinger, 2006). There are two broad categories that can be utilized to classify AEDs – sedating and activating (Ketter *et al.*, 1999). The sedating effects are linked to the GABA inhibitory neurotransmission. These medications result in side effect symptoms of fatigue, cognitive slowing, weight gain, and somnolence. There are possible anxiolytic effects of these drugs as well. The medications included in this category are barbiturates, benzodiazepines, valproate, tiagabine, and vigabatrin. The second category is the activating group, and the resulting symptoms include weight loss, activation, and anxiogenic effects and reduce excitatory glutamate neurotransmission. The medications in this group include felbamate and lamotrigine. Topiramate, zonisamide, and levetiracetam have mixed effects with both activating and sedating profiles (Ketter *et al.*, 1999).

In summary, the common pathogenic mechanisms shared by depression, anxiety, and epilepsy reveal a neurobiological connectedness among the disorders. Serotonin

(5HT), NE, dopamine, GABA, and glutamate are likely involved in the expression of each disorder (Jobe *et al.*, 1999; Ressler and Nemeroff, 2000). It will be important to continue to conduct further investigations of these complex networks to provide a greater understanding of these complicated and interconnected processes.

NEUROANATOMIC IMAGING STUDIES IN DEPRESSION AND ANXIETY

In structural magnetic resonance imaging (MRI) studies of depression, it is reported that there is decreased brain volume in the frontal cortex, hippocampus, striatum, and limbic areas like the subgenual cingulate cortex (Bremner *et al.*, 2000; Bremner, 2002; Bremner *et al.*, 2002; Anand and Shekhar, 2003). Typically, total brain volume is not significantly decreased. Functional studies in depression, including functional MRI (fMRI), single-photon emission-computed tomography (SPECT), and positron emission tomography (PET), have demonstrated decreased activation in the dorsolateral frontal and anterior cingulate cortex with increased activation of the limbic and paralimbic areas (e.g., amygdala, hippocampus, anterior temporal lobes, thalamus, and basal ganglia) (Soares and Mann, 1997; Anand and Shekhar, 2003).

In anxiety, fMRI studies revealed increased activation in the amygdala (Rauch *et al.*, 2000). Studies using emission tomography blood flow showed increased glucose metabolism in the orbitofrontal cortex, anterior cingulate cortex, caudate nucleus, and thalamus in obsessive compulsive disorder (Rauch *et al.*, 2001). Utilizing PET and SPECT in neurochemical studies, it has been demonstrated there is a decrease in benzodiazepine receptors throughout the brain, particularly in the hippocampus and precuneus in those with a panic disorder or generalized anxiety disorder (Malizia *et al.*, 1998). Additionally, magnetic resonance spectroscopy (MRS) studies have revealed decreased GABA levels in panic disorder (Goddard *et al.*, 2001).

There is evidence of brain changes occurring in the limbic-cortical-striatal-pallidal-thalamic circuit in both anxiety and depression. These structures include frontal cortex, hippocampus, thalamus, amygdala, putamen, caudate, and basal ganglia (Sheline, 2003). A complete correspondence between structural impairment in the limbic-cortical-striatal-pallidal-thalamic circuit and resulting depression and anxiety is not confirmed. It is likely that a structural impairment in this circuit results in an increased vulnerability to depression and anxiety (Sheline, 2003).

NEUROANATOMIC IMAGING STUDIES IN EPILEPSY

Epilepsy, particularly focal epilepsy, may involve damage or abnormal activity in brain structures identified as critical for emotional functioning. With quantitative

MRI, Tebartz van Elst *et al.* (1999) studied individuals with TLE and comorbid dysthymia. These individuals with TLE and dysthymia were found to have significantly increased bilateral amygdala volumes, and a positive correlation with left amygdala volumes and depression scores was reported. Quiske *et al.* (2000) examined the relationship between MRI-defined mesial temporal sclerosis (MTS) and Beck Depression Inventory (BDI) scores among 60 individuals with epilepsy. This study found that mean depression scores were significantly elevated in those identified with MTS, and this finding was not related to lateralization of MTS. Similar to the general literature, Bromfield *et al.* (1992), using flourodeoxyglucose (FDG)-PET in individuals with epilepsy, found significant correlations with high depression scores on the Beck Depression Inventory (BDI) and hypometabolism in the inferior frontal regions. In a pilot study, Gilliam *et al.* (2004) found that among 62 individuals with refractory localization related epilepsy, 55 were identified with abnormalities in the temporal lobes (hypometabolism) on FDG-PET, and these same individuals obtained higher scores on the BDI compared to seven individuals with normal scans. Richardson *et al.* (2007) utilized structural MRI volumetrics and FDG-PET to examine the relationship between right and left amygdala and hippocampal volumes and self-reported depression among adults with TLE. Both right and left amygdala volumes increased as depression severity increased; however, there was no increased resting glucose metabolism in this region. As reported in the general psychiatric literature, imaging studies in epilepsy have implicated similar brain regions in depression.

Only one imaging study has been conducted to examine the relationship between interictal anxiety and epilepsy. Satishchandra *et al.* (2003) examined individuals with refractory partial epilepsy and comorbid anxiety. Eight individuals with epilepsy and anxiety, eight individuals with epilepsy without any psychopathology and 15 healthy controls were compared. The anxiety group had an earlier age of onset and longer duration of seizures compared to the no psychopathology group. Comparing the anxiety group to the no psychopathology group, the mean amygdala volumes were larger bilaterally but only statistically significant on the right in the epilepsy group with anxiety ($p<0.05$). Interestingly, when the epilepsy group with no psychopathology was compared to controls the hippocampus and amygdala were significantly smaller on the right ($p<0.05$).

These imaging studies provide further support for the notion that depression, anxiety, and epilepsy have a neurobiological link.

PHARMACOLOGIC AND PSYCHOTHERAPEUTIC TREATMENTS FOR DEPRESSION AND ANXIETY

According to the World Health Organization, by 2020 major depression will be the second leading cause of disability with cardiovascular disease as the leading cause (Michaud *et al.*, 2001). In an editorial, Evans and Charney (2003) stated if an individual has a chronic condition and is depressed there was a poorer prognosis and even increased morbidity and mortality than from the medical diagnosis alone. Additively, co-occurring depression and anxiety is related to poorer prognosis, increased suicide risk, decreased treatment response, and increased functional impairments (Brown *et al.*, 1996; Sherbourne and Wells., 1997; Regier *et al.*, 1998; Kessler *et al.*, 1999a; Kessler *et al.*, 2005a, b). Generalized anxiety disorder and major depression have been demonstrated to negatively impact quality of life more so than depression alone (Mittal *et al.*, 2006). Treatment of depression and anxiety will improve quality of life, reduce suicidality, and decrease functional impairments. Leaving depression and anxiety untreated only leads to poorer outcomes and prognosis.

In the epilepsy literature, a large knowledge gap exists in understanding the efficacy of pharmacologic and psychological treatments for both depression and anxiety. Notably, Kanner (2003) characterized treatment options for depression in epilepsy as remaining "unexplored territory." With this in mind, the pharmacological treatments and psychological treatment options for depression and anxiety will be summarized below.

Pharmacological treatment

There are several factors to consider before initiating any pharmacological treatments. It will be important to rule out any potential factors influencing the presentation of symptoms of depression or anxiety (Kanner, 2003). First, it is important to know if an AED has been discontinued. Ketter *et al.* (1994) demonstrated that AED withdrawal may result in the development of symptoms of depression or anxiety even after a gradual tapering of the antiepileptic medications. Additionally, individuals with ictal anxiety were at higher risk for developing anxiety after AED withdrawal. Second, it is important to assess if an AED with negative psychotropic properties is being used, and if there was a change in the dosing of the AED. The following medications have negative psychotropic properties: phenobarbital, primidone, tiagabine, topiramate, and vigabatrin. If appropriate, an antidepressant can be used to counteract the negative side effects of the AEDs.

When considering the introduction of an antidepressant in an individual with epilepsy on AEDs, there is concern among clinicians regarding the potential to lower the seizure threshold and increase seizure frequency and severity if an antidepressant is utilized. There are potential risks; however, by monitoring plasma serum concentrations, introducing the antidepressant in low doses with small increments, and avoiding using antidepressants with other drugs with proconvulsant properties the overall risks can be significantly minimized (Rosenstein *et al.*, 1993). There are several antidepressant drugs with proconvulsant properties that should be avoided in individuals with epilepsy. These include bupropion, maprotiline, and amoxapine (Kanner, 2003). It is important to treat depression and anxiety in epilepsy while monitoring the potential risks.

Pharmacological treatments for depression

SSRIs are recommended to be the first pharmacological treatment option for depression in epilepsy (Kanner and Balabanov, 2002; Kanner, 2003). SSRIs have fewer side effects and a low risk of increasing seizure frequency or lowering the seizure threshold (Scicutella, 2001). A note of caution should be made, regarding the potential interaction between SSRIs and AEDs since certain SSRIs inhibit certain CYP450 enzymes, resulting in adverse interactions with hepatically metabolized AEDs. These medications include fluoxetine, paroxetine, and fluvoxamine (Scicutella, 2001; Kanner and Balabanov, 2002). Sertraline is frequently recommended due to the minimal pharmacokinetic interactions with AEDs (Barry, 2003; Kanner, 2003). Additionally, Kanner (2003) cautioned that individuals can develop a therapeutic tolerance to sertraline within a few months, resulting in the need to introduce a new SSRI in short order. Citalopram is also a possible treatment option because it does not inhibit the CYP450 enzymes (DeVane, 1999). TCAs are a second line of drugs used to treat depression. However, TCAs have cardiotoxic effects and can easily cause an overdose. In an anecdotal study, Blumer *et al.* (1995) report using low dose imipramine in patients with epilepsy with positive results. There are a number of pharmacological treatment options that appear to be effective in treating depression in epilepsy, but future randomized controlled studies will shed light on the best treatment options for people with epilepsy who have depressive disorders.

Pharmacological treatments for anxiety

A number of SSRIs are approved by the Food and Drug Administration (FDA) in the United States to treat anxiety disorders. Sertraline is approved to treat panic disorder, post traumatic stress disorder and obsessive compulsive disorder, and paroxetine is approved for treatment of panic disorder, obsessive compulsive disorder, generalized anxiety disorder, and social phobia. Obsessive compulsive disorder can also be treated by fluoxetine. In the general population SSRIs are effective, tolerated and safe for the treatment of anxiety disorders. Other drugs approved to treat anxiety disorders include: clonazepam for panic disorder, and alprazolam for generalized anxiety disorder and panic disorder. Benzodiazepines can cause sedation, psychomotor slowing, decreased attention, impaired memory and a risk for addiction. Venlafaxine is approved for use in generalized anxiety disorder (Davidson *et al.*, 1999). It inhibits serotonin and NE reuptake. Venlafaxine has side effects that include nausea, nervousness, and hypertension. Additionally, in a randomized placebo controlled, double blind study, venlafaxine was demonstrated to be efficacious in the treatment of comorbid major depression and generalized anxiety disorder (Silverstone and Salinas, 2001). Buspirone, a partial serotonin agonist and propranolol, a beta blocker, have been used in the treatment of generalized anxiety disorder. No studies have investigated the pharmacological treatment of anxiety disorders in epilepsy, and no studies have examined the treatment options for co-occurring depression and anxiety in epilepsy.

PSYCHOTHERAPEUTIC INTERVENTIONS

There are a number of psychotherapeutic treatments that have been demonstrated to be efficacious for anxiety and depression. Psychotherapy appears to be more effective than placebo, and it is often comparable to pharmacotherapy (Michels, 1997). Empirical studies of cognitive-behavioral therapy (CBT) have increased over the past 15 years with a growing body of evidence indicating their utility and efficacy in the management of depression and anxiety. CBT has been recommended as a first line treatment for depression, generalized anxiety disorder, and panic disorder in adults (Haby et al., 2006). As a result of a meta-analysis Haby et al. (2006) reported the effect size for CBT was 0.68 in all three disorders. This is a moderate to large effect size. In children and adolescents, Kendall (1994) and his colleagues (Kendall and Southam-Gerow, 1996; Kendall et al., 1997) have demonstrated the efficacy of CBT in reducing symptoms of anxiety in randomized controlled trials, with treatment gains maintained to 1-year and 3-year follow-ups (Kendall, 1994; Kendall and Southam-Gerow, 1996; Kendall et al., 1997).

Despite awareness of the efficacy of psychotherapeutic interventions for depression and anxiety, they have rarely been studied in epilepsy. In a recent Cochrane review of psychological treatments in epilepsy, Ramaratnam et al. (2005) concluded that due to the limited number of studies and methodological concerns there is not enough evidence to endorse psychological treatments in epilepsy. The authors conducted a focused review of randomized controlled trials of psychological treatments in epilepsy. There were three studies that identified anxiety as measured outcome (Sultana, 1987; Helgeson et al., 1990; Olley et al., 2001). The results of the studies were mixed; one study found no change (Helgeson et al., 1990), and the other two studies found significant reductions in symptoms of anxiety (Sultana, 1987; Olley et al., 2001). The interventions used were psycho-educational programs and relaxation plus behavioral therapy. Depression was the outcome measure in six studies with three studies using an educational approach (Helgeson et al., 1990; Olley et al., 2001; May and Pfafflin, 2002), two utilized CBT (Davis, 1984; Tan and Bruni, 1986) and one study used relaxation plus behavior therapy (Sultana, 1987). Similarly, the results of the interventions were mixed with half reporting no change following the intervention (Tan and Bruni, 1986; Sultana, 1987; May and Pfafflin, 2002) and the other half reporting improvement in depressive symptoms (Davis, 1984; Helgeson et al., 1990; Olley et al., 2001).

There is significant discussion in the epilepsy literature regarding the under-recognition and under-treatment of depression, and more recently, anxiety disorders; however, very few psychological intervention studies have been conducted even though it has been demonstrated in the general psychiatric and psychological literature that there are efficacious treatment modalities for both depression and anxiety. Armed with this knowledge, we need to begin to study these treatment modalities using randomized controlled trials in individuals with epilepsy to identify and develop efficacious treatment options in epilepsy.

DISCUSSION AND IMPLICATIONS

Epidemiological data consistently indicate that depression and anxiety occur more frequently among individuals with epilepsy compared to the general population and other medical conditions. The relationship between depression and anxiety requires further study to examine the complexities and interconnectedness of the two disorders. Depression and anxiety frequently co-occur both in the general population, and it appears, in epilepsy. Depression, anxiety and epilepsy appear to share related neurobiological mechanisms, and the interplay between the three disorders is not clear. It has been noted time and time again in the epilepsy literature that psychiatric disorders, more specifically depression and anxiety, are frequently under-recognized and under-treated. Yet, there are very few randomized controlled studies for either pharmacological or psychotherapeutic interventions for depression, anxiety or co-occurring depression and anxiety in epilepsy. As researchers and clinicians in epilepsy there is much for us to learn. Psychiatric comorbidity in epilepsy is wrought with complex neurobiological systems that are only just beginning to be understood.

There are a number of implications for future research. First and importantly, we must begin to aggressively study treatment options, both pharmacological and psychotherapeutic, in order to reduce the negative impact of the disorders on the lives of individuals with epilepsy. Second, we must continue to conduct neuroanatomical studies to obtain more knowledge about the neurobiological link between epilepsy, anxiety, and depression. Third, we must continue to explore the neuropathogenic mechanisms of all three disorders, and the potential implications of the dynamic relationship on the expression of these disorders. This knowledge could inform the development of medications to treat depression, anxiety, and epilepsy. Finally, it will be important to identify environmental and psychosocial risk factors for anxiety and depression. It is clear that neurobiological and genetic mechanisms are influenced by environmental factors and their expression is based on exposure to particular stimuli, and thus these factors cannot be ignored. Although clinicians and researchers have been discussing psychiatric comorbidity in epilepsy for several decades, there is much more work to be done.

REFERENCES

Alonso, J., Angermeyer, M. C., Bernert, S., Bruffaerts, R., Brugha, T. S., Bryson, H., *et al.* (2004a). 12-month comorbidity patterns and associated factors in Europe: results from the European study of the epidemiology of mental disorders (ESEMeD) project. *Acta Psychiatr Scand Suppl* **109** (Suppl 420), 28–37.

Alonso, J., Angermeyer, M. C., Bernert, S., Bruffaerts, R., Brugha, T. S., Bryson, H., *et al.* (2004b). Prevalence of mental disorders in Europe: results from the European study of the epidemiology of mental disorders (ESEMeD) project. *Acta Psychiatr Scand Suppl* **109** (Suppl 420), 21–27.

Altshuler, L., Rausch, R., Delrahim, S., Kay, J., Crandall, P. (1999). Temporal lobe epilepsy, temporal lobectomy, and major depression. *J Neuropsychiatry Clin Neurosci* **11**(4), 436–443.

American Psychiatric Association (1980). *Diagnostic and statistical manual of mental disorders*, 3rd edition. (1980). Washington, DC; American Psychiatric Association.

American Psychiatric Association (1987). *Diagnostic and statistical manual of mental disorders*, 3rd edition, revised. Washington, DC: American Psychiatric Association.

American Psychiatric Association (2000). *Diagnostic and statistical manual of mental disorders*, 4th edition, text revision. Washington, DC: American Psychiatric Association.

American Psychiatric Association (2000). *Diagnostic and Statistical Manual of Mental Disorders*, 4th edition, *text revision*. Washington, DC: American Psychiatric Association.

Anand, A., Shekhar, A. (2003). Brain imaging studies in mood and anxiety disorders: special emphasis on the amygdala. *Ann NY Acad Sci* **985**, 370–388.

Angst, J. (1997). Depression and anxiety: implications for nosology, course, and treatment. *J Clin Psychiatry* **58**(Suppl 8), 3–5.

Barry, J. J. (2003). The recognition and management of mood disorders as a comorbidity of epilepsy. *Epilepsia* **44**(Suppl 4), 30–40.

Bittner, A., Goodwin, R. D., Wittchen, H. U., Beesdo, K., Hofler, M., Lieb, R. (2004). What characteristics of primary anxiety disorders predict subsequent major depressive disorder? *J Clin Psychiatry* **65**(5), 618–626. quiz 730.

Blumer, D., Montouris, G., Hermann, B. (1995). Psychiatric morbidity in seizure patients on a neurodiagnostic monitoring unit. *J Neuropsychiatry Clin Neurosci* **7**(4), 445–456.

Boulenger, J. P., Fournier, M., Rosales, D., Lavallee, Y. J. (1997). Mixed anxiety and depression: from theory to practice. *J Clin Psychiatry* **58**(Suppl 8), 27–34.

Boylan, L. S., Flint, L. A., Labovitz, D. L., Jackson, S. C., Starner, K., Devinsky, O. (2004). Depression but not seizure frequency predicts quality of life in treatment-resistant epilepsy. *Neurology* **62**(2), 258–261.

Bremner, J. D., Narayan, M., Anderson, E. R., Staib, L. H., Miller, H. L., Charney, D. S. (2000). Hippocampal volume reduction in major depression. *Am J Psychiatry* **157**(1), 115–118.

Bremner, J. D. (2002a). Structural changes in the brain in depression and relationship to symptom recurrence. *CNS Spectr* **7**(2), 129–130. 135–129.

Bremner, J. D., Vythilingam, M., Vermetten, E., Nazeer, A., Adil, J., Khan, S., *et al.* (2002b). Reduced volume of orbitofrontal cortex in major depression. *Biol Psychiatry* **51**(4), 273–279.

Bromfield, E. B., Altshuler, L., Leiderman, D. B., Balish, M., Ketter, T. A., Devinsky, O., *et al.* (1992). Cerebral metabolism and depression in patients with complex partial seizures. *Arch Neurol* **49**(6), 617–623.

Brown, C., Schulberg, H. C., Madonia, M. J., Shear, M. K., Houck, P. R. (1996). Treatment outcomes for primary care patients with major depression and lifetime anxiety disorders. *Am J Psychiatry* **153**(10), 1293–1300.

Brown, T. A., Chorpita, B. F., Barlow, D. H. (1998). Structural relationships among dimensions of the DSM-IV anxiety and mood disorders and dimensions of negative affect, positive affect, and autonomic arousal. *J Abnorm Psychol* **107**(2), 179–192.

Caplan, R., Siddarth, P., Gurbani, S., Hanson, R., Sankar, R., Shields, W. D. (2005). Depression and anxiety disorders in pediatric epilepsy. *Epilepsia* **46**(5), 720–730.

Choy, Y., Fver A. J., Goodwin, R. D. (2007). Specific phobia and comorbid depression: a closer look at the National Comorbidity Survey data. *Compr Psychiatry* **48**(2), 132–136.

Composite International Diagnostic Interview (CIDI). (1993). World Health Organization, Geneva.

Davidson, J. R., DuPont, R. L., Hedges, D., Haskins, J. T. (1999). Efficacy, safety, and tolerability of venlafaxine extended release and buspirone in outpatients with generalized anxiety disorder. *J Clin Psychiatry* **60**(8), 528–535.

Davies, S., Heyman, I., Goodman, R. (2003). A population survey of mental health problems in children with epilepsy. *Dev Med Child Neurol* **45**(5), 292–295.

Davis, G. R., Armstrong, H. E., Jr., Donovan, D. M., Temkin, N. R. (1984). Cognitive-behavioral treatment of depressed affect among epileptics: preliminary findings. *J Clin Psychol* **40**(4), 930–935.

DeVane, C. L. (1999). Metabolism and pharmacokinetics of selective serotonin reuptake inhibitors. *Cell Mol Neurobiol* **19**(4), 443–466.

Endicott, J., Spitzer, R. L. (1978). A diagnostic interview: the schedule for affective disorders and schizophrenia. *Arch Gen Psychiatry* **35**(7), 837–844.

Ettinger, A. B. (2006). Psychotropic effects of antiepileptic drugs. *Neurology* **67**(11), 1916–1925.

Evans, D. L., Charney, D. S. (2003). Mood disorders and medical illness: a major public health problem. *Biol Psychiatry* **54**(3), 177–180.

Fava, M., Rankin, M. A., Wright, E. C., Alpert, J. E., Nierenberg, A. A., Pava, J., et al. (2000). Anxiety disorders in major depression. *Compr Psychiatry* **41**(2), 97–102.

Favale, E., Rubino, V., Mainardi, P., Lunardi, G., Albano, C. (1995). Anticonvulsant effect of fluoxetine in humans. *Neurology* **45**(10), 1926–1927.

Fiordelli, E., Beghi, E., Bogliun, G., Crespi, V. (1993). Epilepsy and psychiatric disturbance. A cross-sectional study. *Br J Psychiatry* **163**, 446–450.

First, M. B., Gibbon, M., Spitzer, R. L., Williams, J. (1996). *Structured Clinical Interview for DSM-IV Axis I Disorders – Research Version*. New York: Biometrics Research Department, New York State Psychiatric Institute.

Fromm, G. H., Wessel, H. B., Glass, J. D., Alvin, J. D., Van Horn, G. (1978). Imipramine in absence and myoclonic-astatic seizures. *Neurology* **28**(9 Pt 1), 953–957.

Gaitatzis, A., Trimble, M. R., Sander, J. W. (2004). The psychiatric comorbidity of epilepsy. *Acta Neurol Scand* **110**(4), 207–220.

Gilliam, F., Hecimovic, H., Sheline, Y. (2003). Psychiatric comorbidity, health, and function in epilepsy. *Epilepsy Behav* **4**(Suppl 4), S26–S30.

Gilliam, F. G., Santos, J., Vahle, V., Carter, J., Brown, K., Hecimovic, H. (2004). Depression in epilepsy: ignoring clinical expression of neuronal network dysfunction? *Epilepsia* **45**(Suppl 2), 28–33.

Glosser, G., Zwil, A. S., Glosser, D. S., O'Connor, M. J., Sperling, M. R. (2000). Psychiatric aspects of temporal lobe epilepsy before and after anterior temporal lobectomy. *J Neurol Neurosurg Psychiatry* **68**(1), 53–58.

Goddard, A. W., Mason, G. F., Almai, A., Rothman, D. L., Behar, K. L., Petroff, O. A., et al. (2001). Reductions in occipital cortex GABA levels in panic disorder detected with 1H-magnetic resonance spectroscopy. *Arch Gen Psychiatry* **58**(6), 556–561.

Goodwin, R. D. (2002). Anxiety disorders and the onset of depression among adults in the community. *Psychol Med* **32**(6), 1121–1124.

Haby, M. M., Donnelly, M., Corry, J., Vos, T. (2006). Cognitive behavioural therapy for depression, panic disorder and generalized anxiety disorder: a meta-regression of factors that may predict outcome. *Australian and New Zealand Journal of Psychiatry* **40**, 9–19.

Helgeson, D. C., Mittan, R., Tan, S. Y., Chayasirisobhon, S. (1990). Sepulveda epilepsy education: the efficacy of a psychoeducational treatment program in treating medical and psychosocial aspects of epilepsy. *Epilepsia* **31**(1), 75–82.

Hermann, B. P., Seidenberg, M., Bell, B. (2000a). Psychiatric comorbidity in chronic epilepsy: identification, consequences, and treatment of major depression. *Epilepsia* **41**(Suppl 2), S31–S41.

Hesdorffer, D. C., Hauser, W. A., Annegers, J. F., Cascino, G. (2000). Major depression is a risk factor for seizures in older adults. *Ann Neurol* **47**(2), 246–249.

Hesdorffer, D. C., Hauser, W. A., Olafsson, E., Ludvigsson, P., Kjartansson, O. (2006). Depression and suicide attempt as risk factors for incident unprovoked seizures. *Ann Neurol* **59**(1), 35–41.

ICD-10 Classification of Mental and Behavioral Disorders. 1992. Geneva: WHO.

Jacoby, A., Baker, G. A., Steen N., Potts, P., Chadwick, D.W. (1996). The clinical course of epilepsy and its psychosocial correlates: findings from a UK Community study. *Epilepsia* **37**(2), 148–161.

Jalava, M., Sillanpää, M. (1996). Concurrent illnesses in adults with childhood-onset epilepsy: a population-based 35-year follow-up study. *Epilepsia* **37**(12), 1155–1163.

Jobe, P. C., Dailey, J. W., Wernicke, J. F. (1999). A noradrenergic and serotonergic hypothesis of the linkage between epilepsy and affective disorders. *Crit Rev Neurobiol* **13**(4), 317–356.

Johnson, E. K., Jones, J. E., Seidenberg, M., Hermann, B. P. (2004a). The relative impact of anxiety, depression, and clinical seizure features on health-related quality of life in epilepsy. *Epilepsia* **45**(5), 544–550.

Jones, J. E., Hermann, B. P., Barry, J. J., Gilliam, F. G., Kanner, A. M., Meador, K. J. (2003). Rates and risk factors for suicide, suicidal ideation, and suicide attempts in chronic epilepsy. *Epilepsy Behav* **4**(Suppl 3), S31–S38.

Jones, J. E., Hermann, B. P., Barry, J. J., Gilliam, F., Kanner, A. M., Meador, K. J. (2005a). Clinical assessment of Axis I psychiatric morbidity in chronic epilepsy: a multicenter investigation. *J Neuropsychiatry Clin Neurosci* **17**(2), 172–179.

Kanner, A. M. (2003). Depression in epilepsy: prevalence, clinical semiology, pathogenic mechanisms, and treatment. *Biol Psychiatry* **54**(3), 388–398.

Kanner, A. M. (2005). Depression in epilepsy: a neurobiologic perspective. *Epilepsy Curr* **5**(1), 21–27.

Kanner, A. M., Balabanov, A. (2002). Depression and epilepsy: how closely related are they? *Neurology* **58**(8 Suppl 5), S27–S39.

Kendall, P. C. (1994). Treating anxiety disorders in children: results of a randomized clinical trial. *J Consult Clin Psychol* **62**(1), 100–110.

Kendall, P. C., Southam-Gerow, M. A. (1996). Long term follow-up of a cognitive-behavioral therapy for anxiety-disordered youth. *J Consult Clin Psychol* **64**(4), 724–730.

Kendler, K. S., Neale, M. C., Kessler, R. C., Heath, A. C., Eaves, L. J. (1992). Major depression and generalized anxiety disorder. Same genes, (partly) different environments? *Arch Gen Psychiatry* **49**(9), 716–722.

Kendall, P. C., Flannery-Schroeder, E., Panichelli-Mindel, S. M., Southam-Gerow, M., Henin, A., Warman, M. (1997). Therapy for youths with anxiety disorders: a second randomized clinical trial. *J Consult Clin Psychol* **65**(3), 366–380.

Kessler, R. C. (2007). The global burden of anxiety and mood disorders: putting the European study of the epidemiology of mental disorders (ESEMeD) findings into perspective. *J Clin Psychiatry* **68**(Suppl 2), 10–19.

Kessler, R. C., DuPont, R. L., Berglund, P., Wittchen, H. U. (1999). Impairment in pure and comorbid generalized anxiety disorder and major depression at 12 months in two national surveys. *Am J Psychiatry* **156**(12), 1915–1923.

Kessler, R. C., Berglund, P., Borges, G., Nock, M., Wang, P. S. (2005a). Trends in suicide ideation, plans, gestures, and attempts in the United States, 1990–1992 to 2001–2003. *JAMA* **293**(20), 2487–2495.

Kessler, R. C., Chiu, W. T., Demler, O., Merikangas, K. R., Walters, E. E. (2005b). Prevalence, severity, and comorbidity of 12-month DSM-IV disorders in the National Comorbidity Survey Replication. *Arch Gen Psychiatry* **62**(6), 617–627.

Ketter, T. A., Malow, B. A., Flamini, R., White, S. R., Post, R. M., Theodore, W. H. (1994). Anticonvulsant withdrawal-emergent psychopathology. *Neurology* **44**(1), 55–61.

Ketter, T. A., Post, R. M., Theodore, W. H. (1999). Positive and negative psychiatric effects of antiepileptic drugs in patients with seizure disorders. *Neurology* **53**(5 Suppl 2), S53–S67.

Kobau, R., Gilliam, F., Thurman, D. J. (2006). Prevalence of self-reported epilepsy or seizure disorder and its associations with self-reported depression and anxiety: results from the 2004 HealthStyles Survey. *Epilepsia* **47**(11), 1915–1921.

Levine, J., Cole, D. P., Chengappa, K. N. R., Gershon, S. (2001). Anxiety disorders and major depression, together or apart. *Depress Anxiety* **14**, 94–104.

Lewis, A. (1934). Melancholia: a historical review. *J Ment Sci* **80**, 1–42.

Malizia, A. L., Cunningham, V. J., Bell, C. J., Liddle, P. F., Jones, T., Nutt, D. J. (1998). Decreased brain GABA(a)-benzodiazepine receptor binding in panic disorder: preliminary results from a quantitative PET study. *Arch Gen Psychiatry* **55**(8), 715–720.

Manchanda, R., Schaefer, B., McLachlan, R. S., Blume, W. T., Wiebe, S., Girvin, J. P., et al. (1996). Psychiatric disorders in candidates for surgery for epilepsy. *J Neurol Neurosurg Psychiatry* **61**(1), 82–89.

May, T. W., Pfäfflin, M. (2002). The efficacy of an educational treatment program for patients with epilepsy (MOSES): results of a controlled, randomized study. Modular service package epilepsy. *Epilepsia* **43**(5), 539–549.

Michaud, C. M., Murray, C. J., Bloom, B. R. (2001). Burden of disease – implications for future research. *JAMA* **285**(5), 535–539.

Michels, R. (1997). Psychotherapeutic approaches to the treatment of anxiety and depressive disorders. *J Clin Psychiatry* **58**(Suppl 13), 30–32.

Mittal, D., Fortney, J. C., Pyne, J. M., Edlund, M. J., Wetherell, J. L. (2006). Impact of comorbid anxiety disorders on health-related quality of life among patients with major depressive disorder. *Psychiatr Serv* **57**(12), 1731–1737.

Ninan, P. T., Berger, J. (2001). Symptomatic and syndromal anxiety and depression. *Depress Anxiety* **14**, 79–85.

Olley, B. O., Osinowo, H. O., Brieger, W. R. (2001). Psycho-educational therapy among Nigerian adult patients with epilepsy: a controlled outcome study. *Patient Educ Couns* **42**(1), 25–33.

Perini, G. I., Tosin, C., Carraro, C., Bernasconi, G., Canevini, M. P., Canger, R., et al. (1996). Interictal mood and personality disorders in temporal lobe epilepsy and juvenile myoclonic epilepsy. *J Neurol Neurosurg Psychiatry* **61**(6), 601–605.

Quiske, A., Helmstaedter, C., Lux, S., Elger, C. E. (2000). Depression in patients with temporal lobe epilepsy is related to mesial temporal sclerosis. *Epilepsy Res* **39**(2), 121–125.

Ramaratnam, S., Baker, G. A., Goldstein, L. H. (2005). Psychological treatments for epilepsy. *Cochrane Database Syst Rev* (4). CD002029.

Rauch, S. L., Whalen, P. J., Shin, L. M., McInerney, S. C., Macklin, M. L., Lasko, N. B., et al. (2000). Exaggerated amygdala response to masked facial stimuli in posttraumatic stress disorder: a functional MRI study. *Biol Psychiatry* **47**(9), 769–776.

Rauch, S. L., Dougherty, D. D., Cosgrove, G. R., Cassem, E. H., Alpert, N. M., Price, B. H., et al. (2001). Cerebral metabolic correlates as potential predictors of response to anterior cingulotomy for obsessive compulsive disorder. *Biol Psychiatry* **50**(9), 659–667.

Regier, D. A., Narrow, W. E., Rae, D. S. (1990). The epidemiology of anxiety disorders: the Epidemiologic Catchment Area (ECA) experience. *J Psychiatr Res* **24**(Suppl 2), 3–14.

Regier, D. A., Rae, D. S., Narrow, W. E., Kaelber, C. T., Schatzberg, A. F. (1998). Prevalence of anxiety disorders and their comorbidity with mood and addictive disorders. *Br J Psychiatry* (Suppl 34), 24–28.

Ressler, K. J., Nemeroff, C. B. (2000). Role of serotonergic and noradrenergic systems in the pathophysiology of depression and anxiety disorders. *Depress Anxiety* **12**(Suppl 1), 2–19.

Richardson, E. J., Griffith, H. R., Martin, R. C., Paige, A. L., Stewart, C. C., Jones, J., et al. (2007). Structural and functional neuroimaging correlates of depression in temporal lobe epilepsy. *Epilepsy Behav* **10**(2), 242–249.

Ring, H. A., Moriarty, J., Trimble, M. R. (1998). A prospective study of the early postsurgical psychiatric associations of epilepsy surgery. *J Neurol Neurosurg Psychiatry* **64**(5), 601–604.

Robertson, M. M. (1997). Suicide, parasuicide, and epilepsy. In *Epilepsy: A Comprehensive Textbook* (J. Engel, T. A. Pedley, eds), pp. 2141–2151. Philadelphia, PA: Lippincott-Raven.

Rosenstein, D. L., Nelson, J. C., Jacobs, S. C. (1993). Seizures associated with antidepressants: a review. *J Clin Psychiatry* **54**(8), 289–299.

Sartorius, N., Ustun, T. B., Lecrubier, Y., Wittchen, H. U. (1996). Depression comorbid with anxiety: results from the WHO study on psychological disorders in primary health care. *Br J Psychiatry* (Suppl 30), 38–43.

Satishchandra, P., Krishnamoorthy, E. S., van Elst, L. T., Lemieux, L., Koepp, M., Brown, R. J., et al. (2003). Mesial temporal structures and comorbid anxiety in refractory partial epilepsy. *J Neuropsychiatry Clin Neurosci* **15**(4), 450–452.

Schmitz, B. (1999). Psychiatric syndromes related to antiepileptic drugs. *Epilepsia* **40**(Suppl 10), S65–S70.

Scicutella, A. (2001). Anxiety disorders in epilepsy. In *Psychiatric Issues in Epilepsy: A Practical Guide to Diagnosis and Treatment* (A. B. Ettinger, A. M. Kanner, eds), pp. 95–109. Philadelphia, PA: Lippincott, Williams & Wilkins.

Sheline, Y. I. (2003). Neuroimaging studies of mood disorder effects on the brain. *Biol Psychiatry* **54**(3), 338–352.

Sherbourne, C. D., Wells, K. B. (1997). Course of depression in patients with comorbid anxiety disorders. *J Affect Disord* **43**(3), 245–250.

Silberman, E. K., Sussman, N., Skillings, G., Callanan, M. (1994). Aura phenomena and psychopathology: a pilot investigation. *Epilepsia* **35**(4), 778–784.

Silverstone, P. H., Salinas, E. (2001). Efficacy of venlafaxine extended release in patients with major depressive disorder and comorbid generalized anxiety disorder. *J Clin Psychiatry* **62**(7), 523–529.

Skodal, A. E., Schwartz, S., Dohrenwend, B. P., Levav, I., Shrout, P. E. (1994). Minor depression in a cohort of young adults in Israel. *Arch Gen Psychiatry* **51**(7), 542–551.

Soares, J. C., Mann, J. J. (1997). The functional neuroanatomy of mood disorders. *J Psychiatr Res* **31**(4), 393–432.

Specchio, L. M., Iudice, A., Specchio, N., La Neve, A., Spinelli, A., Galli, R., et al. (2004). Citalopram as treatment of depression in patients with epilepsy. *Clin Neuropharmacol* **27**(3), 133–136.

Stahl, S. M. (1997). Mixed depression and anxiety: serotonin1A receptors as a common pharmacologic link. *J Clin Psychiatry* **58**(Suppl 8), 20–26.

Stahl, S. M. (2004). Anticonvulsants as mood stabilizers and adjuncts to antipsychotics: Valproate, lamotrigine, carbamazepine, and oxcarbazepine and actions at voltage-gated sodium channels. *J Clin Psychiatry* **65**(6), 738–739.

Stefansson, S. B., Olafsson, E., Hauser, W. A. (1998). Psychiatric morbidity in epilepsy: a case controlled study of adults receiving disability benefits. *J Neurol Neurosurg Psychiatry* **64**(2), 238–241.

Stein, M. B., Kirk, P., Prabhu, V., Grott, M., Terepa, M. (1995). Mixed anxiety-depression in a primary-care clinic. *J Affect Disord* **34**(2), 79–84.

Strine, T. W., Kobau, R., Chapman, D. P., Therman, D. J., Price, P., Balluz, L. S. (2005). Psychological distress, comorbidities, and health behaviors among US adults with seizures: results from the 2002 National Health Interview Survey. *Epilepsia* **46**(7), 1133–1139.

Sultana, S. M. (1987). A study on the psychological factors and the effect of psychological treatment in intractable epilepsy. Unpublished doctoral dissertation. University of Madras, India.

Swinkels, W. A., Kuyk, J., de Graaf, E. H., van Dyck, R., Spinhoven, P. (2001). Prevalence of psychopathology in Dutch epilepsy inpatients: a comparative study. *Epilepsy Behav* **2**(5), 441–447.

Swinkels, W. A., van Emde Boas, W., Kuyk, J., van Dyck, R., Spinhoven, P. (2006). Interictal depression, anxiety, personality traits, and psychological dissociation in patients with temporal lobe epilepsy (TLE) and extra-TLE. *Epilepsia* **47**(12), 2092–2103.

Tan, S. Y., Bruni, J. (1986). Cognitive-behavior therapy with adult patients with epilepsy: a controlled outcome study. *Epilepsia* **27**(3), 225–233.

Tebartz van Elst, L., Woermann, F. G., Lemieux, L., Trimble, M. R. (1999). Amygdala enlargement in dysthymia – a volumetric study of patients with temporal lobe epilepsy. *Biol Psychiatry* **46**(12), 1614–1623.

Tellez-Zenteno, J. F., Matijevic, S., Wiebe, S. (2005). Somatic comorbidity of epilepsy in the general population in Canada. *Epilepsia* **46**(12), 1955–1962.

Temkin, O. (1971). *The falling sickness* 2nd edition. Baltimore, MD: The Johns Hopkins Press.

Tsopelas, N. D., Saintfort, R., Fricchione, G. L. (2001). The relationship of psychiatric illnesses and seizures. *Curr Psychiatry Rep* **3**(3), 235–242.

Victoroff, J. I., Benson, F., Grafton, S. T., Engel, J., Jr., Mazziotta, J. C. (1994). Depression in complex partial seizures. Electroencephalography and cerebral metabolic correlates. *Arch Neurol* **51**(2), 155–163.

Wiegartz, P., Seidenberg, M., Woodard, A., Gidal, B., Hermann, B. (1999). Co-morbid psychiatric disorder in chronic epilepsy: recognition and etiology of depression. *Neurology* **53**(5 Suppl 2), S3–S8.

World Health Organization (1992). *ICD-10 Classification of Mental and Behavioral Disorders.* Geneva: World Health Organization.

World Health Organization (1993). *Composite International Diagnostic Interview (CIDI).* Geneva: World Health Organization.

Zimmerman, M., Chelminski, I. (2003). Generalized anxiety disorder in patients with major depression: is DSM-IV's hierarchy correct? *Am J Psychiatry* **160**(3), 504–512.

CHAPTER 8

Does Psychosis of Epilepsy Differ from Primary Psychotic Disorders?

Kousuke Kanemoto, Yukari Tadokoro and Tomohiro Oshima

INTRODUCTION

With good reason, most contemporary physicians specializing in epilepsy may well wonder why this disease used to be regarded as one of the major psychoses along with mood disorders and schizophrenia. According to population-based studies (Krohn, 1961; Helgason, 1964; Gudmundsson, 1966; Zielinski, 1974; Jalava and Sillanpaa, 1996; Bredkjaer *et al.* 1998; Qin *et al.*, 2005), the incidence of psychosis among patients with epilepsy ranges from 2% to 6% at most. Further, in our prospective study (Tadokoro *et al.*, 2007), only 7 (2.3%) out of 302 patients with epilepsy (average follow-up period 4 years) newly developed interictal or postictal psychoses. Overall, except for early studies in mental hospitals or specialized centers (Gibbs, 1951; Stantage, 1973; Shukla *et al.*, 1979), recent reports, nearly unanimously, agree that the overwhelming majority of individuals with epilepsy have never experienced psychotic episodes, which inevitably leads to the long-disputed question if there exists an intrinsic relationship between epilepsy and psychosis. A paper published in 1963 by Slater and colleagues of the venerable Maudsley Hospital (Slater and Beard, 1963) represents a milestone in this discussion. They concluded that the incidence of comorbidity between epilepsy and psychosis is far above the level of chance. Together with the report of Gibbs (1951), most fundamental clinical features of interictal psychosis among populations with epilepsy, such as a close link with temporal lobe epilepsy (TLE) and a long interval (greater than 10 years) until the onset of psychosis, have been suggested in this epoch-making paper, which many authors later confirmed (Flor-Henry, 1969; Bruens, 1971; Stantage and Fenton, 1975; Jensen and Larsen, 1979; Shukla *et al.*, 1979; Perez and Trimble, 1980; Parnas *et al.*, 1982; Sengoku *et al.*, 1983; Edeh and Toone, 1987; Kristensen and Sindrup, 1987; Mendez *et al.*,

1993; Onuma *et al.*, 1995; Adachi *et al.*, 2000). Indeed, those phenomena are rather rare among patients with epilepsy, though most recent large scale population-based studies (Jalava and Sillanpaa 1996; Bredkjaer *et al.*, 1998; Qin *et al.*, 2005) have found a higher prevalence of psychotic disorders in patients with epilepsy than in the general population. Less widely noticed, Slater and colleagues suggested that this link between epilepsy and psychosis is non-specific, and that psychosis in patients with epilepsy occurs not as a result of epileptic activity itself, but rather indirectly as a sequel to non-specific damage to a vulnerable portion of brain tissue. In other words, psychosis in epilepsy is nothing but a sign of a damaged strategic portion of the cortex (equal to the temporal lobe in Slater's context), impairment of which prepares brain milieu predisposed to psychosis.

This idea is in sharp contrast with that of Landolt (Landolt, 1953; Landolt 1958; Landolt 1963), another pioneer of the modern concept of epileptic psychosis. As an experienced psychiatrist, who dedicated his life to people with mental illness, Landolt spent nearly his entire life within a psychiatric ward. In the midst of such devoted activity, he found a paradoxical normalization of the electro-encephalogram (EEG) in patients with epilepsy while psychotic episodes became manifest. Strikingly, simultaneous with the disappearance of psychotic symptoms, epileptic seizures and accompanying EEG abnormalities recurred. In an ensuing study, Tellenbach extended Landolt's concept by focusing on the clinical side of the seesaw phenomenon between seizures and psychosis, and proposed the concept of alternative psychosis (Tellenbach, 1965). While Slater's psychosis should be understood properly as symptomatic psychosis resulting from temporal lobe insufficiency rather than epileptic psychosis, alternative psychosis is a genuine epileptic psychosis, because of the essential link with epileptic activity. To consider this critical but unsolved discrepancy of opinions concerning the nosological positioning of epileptic psychosis in the present report, attention is initially focused mainly on the clinical picture. After reviewing phenomenological and demographic differences between the two conditions, we will return to this potential schism between two founders of the modern concepts of epileptic psychosis.

WHICH TYPE OF EPILEPTIC PSYCHOSIS SHOULD BE COMPARED WITH WHICH PRIMARY PSYCHOTIC DISORDER?

With regard to the wide-ranging clinical manifestations of epileptic psychoses and primary psychotic disorders, it is important, first of all, to determine which should be compared. Table 8.1 demonstrates the incidence of subcategories of psychosis occurring in patients with epilepsy who visited the Kansai Regional Epilepsy Center from 1983 to 1999 (Kanemoto *et al.*, 2001), definitions of which are listed in Table 8.2. Chronic epileptic psychosis (CP), acute interictal psychosis (AIP), and postictal psychosis (PIP) together constituted 95% of all patients who had

TABLE 8.1   Subcategories of epileptic psychoses ($n = 200$)

Postictal psychosis	26.5% (53/200)
AIP as well	2.0% (4/200)
evolving into CP	1.0% (2/200)
Acute interictal psychosis	52.0% (104/200)
evolving into CP	11.5% (23/200)
Chronic psychosis	29.5% (59/200)
Others	5.0% (10/200)

Source: Data recalculated from Kanemoto *et al.* (2001).
AIP: Acute interictal psychosis; CP: Chronic psychosis.

TABLE 8.2   Definitions of subcategories of epileptic psychoses

- Postictal psychosis: Psychosis that follows immediately after 1 or generally multiple seizures (mostly complex partial or secondarily generalized), occurring within 1 week of the last seizure.
- Acute interictal psychosis: Psychosis that develops when seizures have ceased or reduced significantly in frequency (alternative psychosis) or when seizures are unrelated to any recent increase in seizure activity.
- Chronic epileptic psychosis: Any psychotic state lasting for more than 6 months in patients with epilepsy.

experienced psychotic episodes at any time in their history. The others include psychotic episodes newly occurring postoperatively, psychotic episodes corresponding to diffuse spike and wave discharges in EEG findings, and other miscellaneous psychotic episodes of an obscure nature. In agreement with the current data, previous studies have shown that an overwhelming majority of epileptic psychosis can be explained by the first three major subcategories. For example, CP occupied 33.3% ($n = 102$) in the series of Adachi *et al.* (personal communication) and CP plus AIP 75.4% ($n = 58$) in the series of Matsuura *et al.* (2004) which agree well with the figures shown in Table 8.2.

As for primary psychotic disorders, one of the most widely used criteria of psychiatric diagnosis, DSM-IV (American Psychiatric Association, 1994) subdivides primary psychotic disorders mainly into five subcategories: nuclear schizophrenia, schizophreniform disorder (SFD), brief psychotic disorder (BPD), delusional disorder, and schizoaffective disorder. In DSM-IV, the pivotal reference point on which the first three psychotic disorders are divided is time criteria. On the other hand, the determinant features of the last two categories are the contents of the psychotic experience. Figure 8.1 is a rough sketch of the interrelationships between subcategories of primary psychotic disorders in DSM-IV and corresponding epileptic psychoses. Since the duration of psychotic episodes might be a potent determinant of the contents of psychotic episodes at the same time,

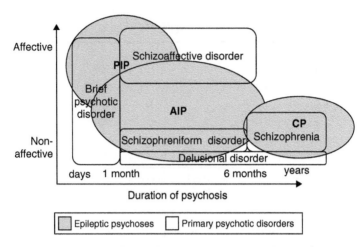

FIGURE 8.1   Duration of illness of primary psychotic disorders and corresponding epileptic psychoses.

comparisons would be more unbiased if the durations matched, at least roughly. Along that line, good matches to AIP and CP ideally would be SFD and schizophrenia, respectively. If the same rule is applied, PIP should be preferentially compared with BPDs, because most episodes of PIP remit within 1 month (Adachi *et al.*, Levin, 1952; Logsdail and Toone 1988; So *et al.*, 1990; Savard *et al.*, 1991; Devinsky *et al.*, 1995; Umbricht *et al.*, 1995; Kanemoto 1996, 2002; Kanner *et al.*, 1996; Lancman 1999). While transient psychotic episodes are more often encountered than chronic ones in consecutive series of patients with epilepsy, as shown above, this trend seems to be reversed in populations with primary psychotic disorders (Perala *et al.*, 2007). Although this could be more apparent than real as a result of a failure to recruit some patients with transient psychotic episodes into the group with primary psychotic disorders, nevertheless, a simple comparison between functional and epileptic psychosis without further specification would reveal only a difference between subacute and chronic courses of psychotic illnesses, instead of an aimed comparison. Actually, such matching based on time criteria between subcategories of primary psychotic disorders and epileptic psychoses can only be done with difficulty at present, because psychopathological observations devoted to SFD and BPD are mostly skipped in the recent literature. Under the rubric of acute and transient psychotic disorders (ATPD) of ICD-10 (Pull *et al.*, 1984; World Health Organization, 1992), a comparatively greater number of psychopathological descriptions are found (Marneros *et al.*, 2005). However, as already noted, a dichotomy of transient psychosis with illness duration of 1 and 6 months in the DSM criteria conforms better to the well-recognized division of transient epileptic psychoses into PIP and AIP. Thus, reference to a mixture of literature based on DSM and ICD is preferable especially for this review of transient psychotic episodes.

TABLE 8.3   Clinical features of CP in comparison with other organic psychoses

- Predominance of Schneiderian first-rank symptoms
- Less frequent visual hallucinations
- Less frequent rapidly changing higher cognitive function except for epileptic seizures

Note: Diffuse Lewy body disease (DLBD) and Systemic lupus erythematosus (SLE) psychoses as model organic psychoses. CP: Chronic psychosis.

TABLE 8.4   CP with and without preceding AIP episodes

	CP without preceding AIP episodes ($n = 36$)	AIP evolving into CP ($n = 24$)
TLE[a]	41.7% ($n = 15$)	82.6% ($n = 19$)
Others	58.3% ($n = 21$)	17.4% ($n = 4$)

Notes: $\chi^2 = 9.634, p = 0.002$
Source: Data recalculated from Kanemoto *et al.* (2001).
[a]TLE was diagnosed here only if complex focal seizures except for those of apparent frontal origin were registered, or if typical auras with limbic characteristics such as *déjà vu* and ictal fear preceded seizures with impaired consciousness, including secondarily generalized seizures. CP: Chronic psychosis; AIP: Acute interictal psychosis; TLE: Temporal lobe epilepsy.

Chronic epileptic psychosis vs. schizophrenia

Superficially, the comments of previous authors on the psychopathological similarities that appear to exist between CP and functional schizophrenia are conflicting (Gibbs, 1951; Hill, 1953; Pond, 1957; Flor-Henry, 1969; Taylor, 1972; Stantage and Fenton, 1975; Jensen and Larsen, 1979; Perez and Trimble, 1980; Toone *et al.*, 1980; Parnas *et al.*, 1982; Sherwin *et al.*, 1982; Edeh and Toone, 1987; Stevens, 1988; Roberts *et al.*, 1990; Trimble, 1991; Mace, 1993; Bruton *et al.*, 1994; Kanemoto *et al.*, 2001; Matsuura *et al.*, 2004). However, this diversity of opinions seems to be more apparent than real in some aspects. First, the problem of control subjects should be mentioned. When looking at other representative organic psychoses such as those occurring in diffuse Lewy body disease (Del Ser *et al.*, 2000; Ballard *et al.*, 2001) and SLE psychosis (World Health Organization, 1992; Iverson, 1993; Velakoulis *et al.*, 2006; Wekking, 1993; Yu *et al.*, 2006), the similarities between CP and schizophrenia supervenes (Table 8.3). On the other hand, dissimilarities are obvious when CP is directly compared with schizophrenia (Table 8.4) (Slater and Beard 1963; Perez and Trimble 1980; Toone *et al.*, 1980; Tadokoro *et al.*, 2007). For example, visual hallucinations are often assumed to be more frequently encountered in patients with CP when schizophrenia is adopted as a control, while the reverse is true when compared with other organic psychoses. Further, the important question that should be asked is whether the psychopathological difference, if any, between CP and schizophrenia is really qualitative. In other words, some authors suspect that the apparent psychopathological characteristics of CP are nothing more than artifacts resulting from a less severe course of illness. Matsuura et al. (2004) were the first to ask this question seriously. Thereafter, after recruiting all consenting patients with interictal psychosis and schizophrenia during the same study period, we also confirmed significantly lower scores in the group with epileptic psychosis using assessments with PANSS (Positive and Negative Syndrome Scale). Matsuura et al. concluded that the differences in symptom profiles in both groups were quantitative rather than qualitative, and that a psychopathological distinction between both groups could not be maintained. However, their conclusion requires further amplification, because epileptic psychosis was treated as a unified entity in their analysis (so was CP and AIP in ours), which inevitably biases epileptic psychoses toward lower scores.

The heterogeneity of CP, especially in association with TLE, should also be addressed, because the reported proportion of TLE varies greatly among published studies (Slater and Beard, 1963; Bruens, 1971; Shukla et al., 1979; Perez and Trimble 1980; Parnas et al., 1982; Adachi et al., 2000; Kanemoto et al., 2001; Adachi et al., 2002). In Table 8.5, patients who had CP without preceding AIP episodes (18%) and those who had AIP episodes evolving into CP (11%) in our series are compared. Strikingly, the correlation between each group with TLE showed a statistically significant difference. While the second group was mainly associated with TLE, accompanied epilepsy types varied considerably in the first group. In our series, CP without preceding AIP episodes showed a more insidious onset and took a clinical course scarcely distinguishable from functional schizophrenia in a substantial number of patients. Such an observation regarding the first CP group leads to another fundamental question of whether a personal history of epilepsy increases the susceptibility to nuclear schizophrenia. This question dates back to the former "psychodynamic hypothesis" proposed by Pond (1957), which is echoed later in the holistic concepts of Schmidt and Wolf (1989) and Wolf (1991). A recent study by Qin et al. (2005), one of the largest population-based studies to date to investigate schizophrenia-like psychosis in epilepsy, seems to answer this question in the affirmative, though a number of limitations such as preclusion of ambulatory patients from the study, lack of direct contact with the patients, and the non-critical acceptance of epilepsy types diagnosed by physicians with presumably extremely diverse backgrounds, diminish the validity of their results in regard to other subsidiary questions. In addition, the obscure nature of the allegedly generalized type of epilepsy, which remained unanswered whether it included secondarily

TABLE 8.5   Clinical features of CP in comparison with schizophrenia

- Better pre-morbid personality
- Well-retained affect (less negative symptoms)
- Deteriorates finally to organic type of impairment rather than a typically schizophrenic defect state
- Paranoid features predominate

generalized seizures or not and whether this concept should be understood in the frame of epileptic seizures or epileptic syndromes, makes their results, especially those concerning epilepsy types, unreliable. However, such shortcomings are shared by all previous population-based studies on this topic (Krohn 1961; Helgason 1964; Gudmundsson 1966; Zielinski 1974; Jalava and Sillanpaa 1996; Bredkjaer *et al.*, 1998; Qin *et al.*, 2005). Nevertheless, even with these limitations, their answer to the main question remains valid and is supported by several other studies as well (Jalava and Sillanpaa, 1996; Bredkjaer *et al.*, 1998). Separated from epilepsy, a variety of types of insults to brain tissue, such as head trauma (Zhang and Sachdev, 2003; David and Prince, 2005) and intrauterine infection (Kendell and Kemp, 1989), irrespective of localization, are known to increase the risk of later development of schizophrenia. It is plausible that epilepsy could be such a non-specific contributor to developing schizophrenia. If that is the case, some proportion of CP, presumably the first group in our series, should be regarded as genuine schizophrenia that only becomes manifested with epilepsy as a precipitating factor.

In contrast, when discussing the second CP group, CP following repeated AIP episodes, true epileptic psychosis may be in question. In this situation, seizures may somehow modify the brain over the long term, which results in some structural change that leads to the genesis of psychosis. Indeed, it remains unsettled whether seizures themselves cause permanent changes to brain tissue, however, there is some evidence of plastic regenerative changes in response to repeated occurrences of seizures (Sutula, 2004) and brief psychotic episodes (Umbricht *et al.*, 1995), with a proportion of patients developing CP on follow-up. It is important to repeat here that, in our series, this development from AIP into CP was rather specific to patients with TLE. In a review of data from depth EEG studies of patients who underwent lobectomy procedures, Mace (1993) suggested a unique electrophysiological interrelationship between limbic epilepsy and schizophrenia. As illustrated in Figure 8.2, both schizophrenia and CP are reported to exhibit spike discharges alongside the whole limbic circuit, including the septal and amygdalo-hippocampal regions. However, spiking is most marked in the septal area in schizophrenia and in the amygdalo-hippocampal region in TLE without psychosis. Mace postulated that CP mimics the distribution of spiking seen in schizophrenia, though it remains to be distinguished from that of schizophrenia. It follows that both schizophrenia and TLE are disorders of the limbic circuit, with only different components that are preferentially disturbed. Thus, CP represents a shifted pattern of disturbance within the limbic circuit toward schizophrenia.

Acute interictal psychosis vs. schizophreniform disorder

A literature search revealed no previous studies that directly compared AIP either with SFD (DSM-IV) or ATPD (ICD-10), however, most of the features listed in Table 8.5 as characteristics of epileptic psychoses, in contrast to those of schizophrenia, are shared in common with SFD as well as ATPD. Based on the present

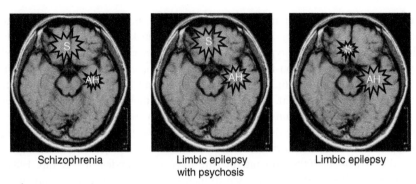

| Schizophrenia | Limbic epilepsy with psychosis | Limbic epilepsy |

 Spiking;   S Septal area; AH Amygdalo-Hippocampal area

FIGURE 8.2   Hypothesis of unique electrophysiological interrelationships between limbic epilepsy and schizophrenia (Mace, 1993). *Source:* Data from Kendrich and Gibbs (1957) and Heath (1986) are schematized.

TABLE 8.6   Delusion/hallucination ratios in patients with epileptic psychosis and schizophrenia

Schizophrenia	Interictal psychosis
($n = 13$)	($n = 46$)
0.37 (SD = 0.17)	0.49 (SD = 0.11)

Note: $F = 7.37, p = 0.009; t = 2.32, p = 0.035$. The delusion/hallucination ratios presented here were calculated using PANSS scores. Scores of hallucinatory behavior are the numerators, while those of delusion plus suspiciousness are the denominators.

Source: Recalculated from personal data in a prospective study in the process of submission by Tadokoro *et al.* (2007)

knowledge, no phenomenological differences between AIP and SFD for both symptomatology and course of illness appear to exist, except for the relative predominance of paranoid complaints. In Table 8.6, personal data in this respect are presented. However, as mentioned above, this could also be an artifact merely as a result of the difference in duration of psychosis.

In Table 8.7, AIP with and without complete remission at the time of examination were demonstrated as a function of linking with TLE. Paradoxically, AIP with complete remission exhibited a less related link with TLE, just like CP without preceding AIP episodes. This is in sharp contrast to the high proportion of TLE in patients with AIP that eventually evolves into CP. Even if a massive inclusion of extra-temporal and generalized epilepsies in CP without preceding AIP episodes can be understood under the rubric of the "epilepsy as a non-specific risk factor

TABLE 8.7   AIP with and without complete remission

	AIP with complete remission ($n = 81$)	AIP evolving into CP ($n = 23$)
TLE[a]	49.4% ($n = 40$)	82.6% ($n = 19$)
Others	50.6% ($n = 41$)	17.4% ($n = 4$)

Note: $\chi^2 = 8.056, p = 0.005$.
Source: Recalculated data from Kanemoto et al. (2001).
[a]TLE was diagnosed here only if complex focal seizures except for those of apparent frontal origin were registered, or if typical auras with limbic characteristics such as déjà vu and ictal fear preceded seizures with impaired consciousness, including secondarily generalized seizures. CP: Chronic psychosis; AIP: Acute interictal psychosis; TLE: Temporal lobe epilepsy.

TABLE 8.8   Incidence of drug-induced and alternative psychosis

Schizophrenia ($n = 54$)	Psychotic disorders ($n = 78$)	Non-schizophrenic
Drug-induced	7 (13.0%)	31 (39.7%)
Alternative	8 (14.8%)	33 (42.3%)

Note: A statistically significant difference was seen between the groups ($p < 0.01$).
Source: Cited from Kanemoto et al. (2001).

for schizophrenia" hypothesis, this cannot be applied easily to AIP with complete remission, because the resulting clinical picture including the transient nature differs decisively from schizophrenia. Most episodes in so-called drug-induced epileptic psychosis as well as the controversial alternative psychosis fall into this category. While alternative and drug-induced psychosis tend to be regarded as very rare entities in epilepsy centers mainly dedicated to epilepsy surgery (Mace, 1993), they stand out in tertiary epilepsy centers, where exhaustive pharmacotherapy is routinely attempted for intractable epilepsy for an extended period of time by physicians (Wolf, 1991; Adachi et al., 2000; Kanemoto et al., 2001). As shown in Table 8.8, more than one third of the group of patients with non-schizophrenic epileptic psychosis in our previous series, mostly those with AIP, were composed of the alternative or drug-induced type (Kanemoto et al., 2001). It should also be noted that drug-induced psychosis and alternative psychosis are closely interrelated. Indeed, two thirds of drug-induced psychotic episodes (30 of 45 episodes) became manifest in the form of alternative psychosis.

Usually, drug-induced or alternative psychosis appears not within days, but rather within weeks and lasts for more than 1 month and less than half a year, which agrees well with the concept of SFD. Since alternative psychosis has an intrinsic relationship to seizure activity, though the specific involvement of relevant antiepileptics such as ethosuximide (Wolf, 1991; Schmitz and Trimble, 1998), vigabatrin (Sander et al., 1991), and zonisamide (Kanemoto et al., 2001) may modify the clinical picture, it seems to be a good candidate as a genuine epileptic psychosis.

Postictal psychosis vs. brief psychotic disorder (or bouffée délirante)

Bouffée délirante is a traditional term used in French psychiatry for primary psychotic disorder of short duration (Ferrey and Zebdi, 1999; Pichot, 1986; Pillmann et al., 2003; Pull et al., 1984). The clinical features listed by Pichot (Pichot, 1986), such as sudden onset, "bolt from the blue", delusions characterized by numerous, diverse, and protean delusional themes without recognizable structure and cohesiveness, sometimes accompanied by hallucinations, clouding of consciousness associated with emotional instability, and a rapid return to the premorbid level of functioning, all show strong phenotypic similarities to PIP. However, bouffée délirante, which fulfills these strict criteria, is reported to represent only 2.4% of all patients with non-organic psychotic disorders (Pull et al., 1984), in contrast to the continuing popularity of that diagnosis among French clinicians (Ferrey and Zebdi, 1999).

As mentioned in the introduction section, based on the duration of psychosis, BPD is in good accordance with PIP just like bouffée délirante. However, little has been reported about the precise psychopathological nature of this DSM-IV category. Thus, more definite conclusions await precise case control studies between these groups in the future, though some speculation can be made based on previous reports of PIP, BPD, ATPD, and bouffée délirante as well as our own data. Episodes in the nuclear group of PIP show a very stereotyped course of illness (Logsdail and Toone 1988; Savard et al., 1991; Devinsky et al., 1995; Umbricht et al., 1995; Kanemoto et al., 1996; Kanner et al., 1996; Oshima et al., 2006). Following clustered complex partial seizures or secondarily generalized seizures, mental function returns to the premorbid level of functioning, except for a subtle feeling of derealization in some patients. Following this intervening lucid interval, which lasts for 1–3 days on average (Adachi et al., 2007), a rapidly increasing emotional lability with a mostly hypomanic nature comes to the fore, culminating into polymorphous psychosis without a recognizable structure or cohesiveness within a few days. The most notable difference between nuclear PIP and BPD/bouffée délirante lies in how the symptoms develop, rather than what kind of symptoms are presented. A delicate inquiry during the lucid interval often reveals a unique feeling of detachment from the surrounding environment with perfectly preserved cognitive function. In contrast to bouffée délirante or BPD, the initial hypomanic state is strikingly short (usually from 12 to 36 hours) and tends to remain purely affective, uncontaminated by various psychotic complaints such as protean delusional themes and visual hallucinations. It should be noted, however, that the "bouffée délirante-type" clinical picture soon predominates during the ensuing stage shortly after the initial hypomanic stage. In the nuclear group of PIP, initial detached and ensuing affective stages, mainly with elevated moods, are so characteristic and regularly repeating that pertinent intervention during the affective stage could often prevent PIP from developing fully into frank psychosis (Lancman, 1999).

In our recent study (Oshima *et al.*, 2006), we focused on a variant of PIP that showed a different clinical picture from nuclear PIP. Typically, episodes of this variant of PIP widely outnumber those of the nuclear type and often occur without an intervening lucid interval. A noted similarity between the contents of illusory experiences during PIP and those during aura was also suggested. In this variant, complex partial seizures can occur even in the midst of PIP. This variant of PIP often overlaps with peri-ictal psychosis (Hermann *et al.*, 1982; Wieser *et al.*, 1985; Kanemoto, 1997; Takeda *et al.*, 2001), in which psychotic symptoms develop gradually and parallel to increases in seizure frequency (Trimble and Schmitz, 1997). Wieser's hypothesis, which assumes circumscribed limbic status epilepticus limited within the limbic circuit as an underlying pathogenesis, may explain this variant of PIP very well. A strikingly close association of PIP with TLE also supports this, which has been confirmed in previous series (Logsdail and Toone, 1988; Devinsky *et al.*, 1995; Umbricht *et al.*, 1995), as well as in our series: 41 (77.4%) out of 53 patients with PIP also had TLE (Kanemoto, 2002).

CONCLUDING REMARKS: PLURAL EPILEPTIC PSYCHOSES SHOULD BE UNDERSTOOD UNDER PLURAL HYPOTHESES

Does the psychosis of epilepsy differ from a primary psychotic disorder? Historically, several answers have been presented. As we discussed in the preceding sections, variations of the answer in the negative may be summarized as follows: the coexistence of the two conditions in a single individual is only accidental; epilepsy increases the vulnerability to a primary psychotic disorder; and the two conditions develop from a common etiological factor, but independently. Since Slater's report 40 years ago, the first answer seems to be denied by several studies including large scale epidemiological investigations (Jalava and Sillanpaa, 1996; Bredkjaer *et al.*, 1998; Qin *et al.*, 2005). When focused on CP, the increased incidence of psychosis among patients with epilepsy has tended to be explained by one of the latter two answers, both of which presuppose an enhanced susceptibility to schizophrenia in patients with epilepsy. As mentioned in the introduction section, the corollary to these is that epileptic psychosis is nothing but a genuine schizophrenia, which at most may become modified by the presence of epilepsy. Simply stated, Slater seemed to prefer the third alternative and advocated the view that damage to the medial temporal structures, irrespective of genetic or exogenous origin (such as epilepsy), is associated with schizophrenia. Mace pointed out that this simple causal link appears to be increasingly untenable in light of subsequent developments in psychiatry and neurology. However, Slater's fundamental idea outlived decades of dispute, and became modified and innovated by Taylor's as well as Stevens' more sophisticated versions of the same line of explanation. While Taylor stressed the ontogenetic rather than anatomical impact of a common pathology leading to both epilepsy and psychosis (Taylor, 1971, 1975), Stevens proposed that the critical factor causing predisposion to schizophrenic psychoses among some individuals with organically induced psychosis (including epilepsy) is the reactive occurrence of abnormal neuronal regeneration, especially when it coincides with a vulnerable period (Stevens, 1989, 1991). These more sophisticated versions of the common etiology hypothesis offered some prospect of reconciliation between the second and the third line of explanations. It is important to note here that the leading authors who have discussed CP nearly unanimously presumed that CP is an organically induced schizophrenia in essence.

If we conclude that the essence of epilepsy lies in seizure activity, then the question can be paraphrased as follows: Does psychosis occurring in patients with epilepsy have an intrinsic relationship with seizure activity? While answers in the negative predominate when discussing CP, some kind of intrinsic relationship between seizure and psychosis is difficult to deny with regard to transient epileptic psychosis, especially PIP and alternative psychosis. Very recently, Sachdev (2007) provided a unifying hypothesis related to the pathogenesis of alternative and postictal psychoses, and suggested that the brain's inhibitory processes in response to seizures plays a key role in the development of the psychosis underlying the two conditions. Twenty years before Sachdev, this emphasis on the excessive inhibitory process as a cause of psychosis was suggested by Stevens (1986). Wolf (1991) attempted to explain transient epileptic psychosis in a different way, but also as a phenomenon intrinsically related to seizures, and also contended that some epileptic psychoses, especially alternative psychosis, result directly from a deviant seizure that has spread from an original epileptogenic area. Originally, some areas of the reticular activating system were proposed as the destination for this deviant seizure spread. The shifting peak of seizure activity within the limbic circuit depicted in Figure 8.3 may be regarded as a variant of this deviant seizure spread hypothesis. In this case, however, the septal area was considered as the destined area. Finally, Wieser's hypothesis of limbic status epilepticus (Wieser *et al.*, 1985) should also be categorized as an affirmative answer to the initial question.

In Table 8.9, the major theories for epileptic psychoses and suitable models for respective theories are concisely summarized. To repeat, previous theories can be divided into two groups. In one group (Table 8.9a, b), psychotic episodes occurring concurrently with epilepsy are regarded as essentially nothing but primary psychotic disorders, while in the other (Table 8.9c, d, e), a direct casual relationship between seizure activity and resultant psychosis is presumed. Figure 8.3 is a bird's eye view, in which subcategories of epileptic psychoses are allocated as a function of association with seizures and TLE. A uniting theory covering all the epileptic psychoses does not seem to exist, and lack of awareness of the limited applicability of a single explanation or theory may lead to serious confusion. Answers to the initial question, "Does epileptic psychosis differ from a primary psychotic disorder?", cannot be given wholly, but only separately one by one. As depicted in Figure 8.3, the answers are shaded from the most notable "Yes" of PIP

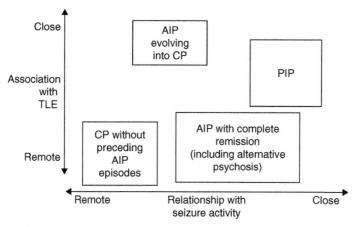

Figure 8.3   Subcategories of epileptic psychosis as a function of their associations with TLE and seizure activity. CP: Chronic psychosis; AIP: Acute interictal pyschosis; PIP: Post-ictal pyschosis; TLE: Temporal lobe epilepsy.

TABLE 8.9   Major hypotheses for epileptic psychoses and corresponding models

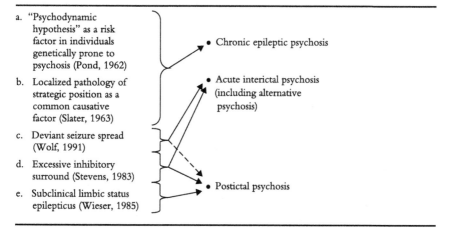

to the approximate "No" of CP without preceding AIP, based on how close the association with seizures is. Evidently, plural epileptic psychoses should be understood under plural theories. It is possible that complex interrelationships between various sites within the limbic circuit and excessive inhibitory spread will serve as promising key concepts in future studies, at least for transient epileptic psychoses.

It is interesting that as far back as the mid-19th century, Farlet named PIP as a true epileptic psychosis (Farlet, 1860/1861). The true value of that long-forgotten statement is now becoming more apparent.

The first author is indebted to Dr. Naoto Adachi for his valuable advice and Mrs. Lumina Ogawa for her help in the literature searches.

REFERENCES

Adachi, N., Matsuura, M., Okubo, Y., Oana, Y., Takei, N., Kato, M., Hara, T., Onuma, T. (2000). Predictive variables for interictal psychosis in epilepsy. *Neurology* **55**, 1310–1314.

Adachi, N., Matsuura, M., Hara, T., *et al.* (2002). Psychoses and epilepsy: are interictal and postictal psychoses distinct clinical entities? *Epilepsia* **43**, 1574–1582.

Adachi, N., Ito, M., Kanemoto, K., Akanuma, N., Okazaki, M., Ishida, S., Sekimoto, M., Kato, M., Kawasaki, J., Tadokoro, Y., Oshima, T., Onuma, T. (2007). Duration of postictal psychotic episodes. *Epilepsia* **48**, 1531–1537.

American Psychiatric Association (1994). *Diagnostic and statistical manual of mental disorders* 4th edition. Washington, DC: American Psychiatric Association.

Ballard, C. G., O'Brien, J. T., Swann, A. G., Thompson, P., Neill, D., McKeith, I. G. (2001). The natural history of psychosis and depression in dementia with Lewy bodies and Alzheimer's disease: persistence and new cases over 1 year of follow-up. *J Clin Psychiatry* **62**, 46–49.

Bredkjaer, S. R., Mortensen, P. B., Parnas, J. (1998). Epilepsy and non-organic no-affective psychosis. National epidemiologic study. *Br J Psychiatry* **172**, 235–238.

Bruens, J. H. (1971). Psychosis in epilepsy. *Psychiatrica, Neurologica, Neurochirurgica* **74**, 174–192.

Bruton, C. J., Stevens, J. R., Frith, C. D. (1994). Epilepsy, psychosis, and schizophrenia: clinical and neuropathologic correlations. *Neurology* **44**, 34–42.

David, A. S., Prince, M. (2005). Psychosis following head injury: a critical review. *J Neurol Neurosurg Psychiatry* **76**(Suppl 1), 53–60.

Del Ser, T., McKeith, I., Anand, R., Cicin-Sain, A., Ferrara, R., Spiegel, R. (2000). Dementia with Lewy bodies: findings from an international multicentre study. *Int J Geriatr Psychiatry* **15**, 1034–1045.

Devinsky, O., Abramson, H., Alper, K., FitzGerald, L. S., Perrine, K., Calderon, J., Luciano, D. (1995). Postictal psychosis: a case control series of 20 patients and 150 controls. *Epilepsy Res* **20**, 247–253.

Edeh, J., Toone, B. (1987). Relationship between interictal psychopathology and the type of epilepsy. *Br J Psychiatry* **151**, 95–101.

Farlet, J., De l'état des épileptiques. Archives Générales de Médecine 1860, 1861; 16:661–699;17:461–491;18:423–443.

Ferrey, G., Zebdi, S. (1999). Evolution et prognostique des troubles psychotiqueu aigus ('bouffée délirante polymorphe'). *Encephale* **25**(Special 3), 26–32.

Flor-Henry, P. (1969). Psychosis and temporal lobe epilepsy. *Epilepsia* **10**, 363–395.

Gibbs, F. A. (1951). Ictal and non-ictal psychiatric disorders in temporal lobe epilepsy. *J Nerv Ment Dis* **113**, 522–528.

Gudmundsson, G. (1966). Epilepsy in Iceland. *Acta Neurol Scand* **25**(Suppl), 1.

Heath, R. G. (1986). Studies with deep electrodes in patients intractably ill with epilepsy and other disorders In: What is epilepsy? (eds Trimble, M. R., Reynolds, E. H.). pp. 126–138. Edinburgh Churchill, Linvingstone.)

Helgason, T. (1964). Epidemiology of mental disorders in Iceland: Psychoses. *Acta Psychiatry Scand* **173**(Suppl), 67–95.

Hermann, B. P., Dikmen, S., Schwartz, M. S., Karnes, W. E. (1982). Interictal psychopathology in patients with ictal fear: a quantitative investigation. *Neurology* **32**, 7–11.

Hill, D. (1953). Psychiatric disorders of epilepsy. *The Mental Press* **229**, 473–475.

Iverson, G. L. (1993). Psychopathology associated with systemic lupus erythematosus: a methodological review. *Semin Arthritis Rheum* **22**, 242–251.

Jalava, M., Sillanpaa, M. (1996). Concurrent illness in adults with childhood-onset epilepsy: a population-based 35-year follow-up study. *Epilepsia* **37**, 1155–1163.

Jensen, I., Larsen, J. K. (1979). Psychoses in drug-resistant temporal lobe epilepsy. *J Neurol Neurosurg Psychiatry* **42**, 256–265.

Kanemoto, K. (1997). Periictal Capgras syndrome following clustered ictal fear. *Epilepsia* **38**, 847–850.

Kanemoto, K. (2002). Postictal psychoses, revisited. In *The Neuropsychiatry of epilepsy* (M. Trimble, B. Schmitz, eds), pp. 117–131. Cambridge: Cambridge University Press.

Kanemoto, K., Kawasaki, J., Kawai, I. (1996). Postictal psychosis: a comparison with acute interictal and chronic psychoses. *Epilepsia* **37**, 551–556.

Kanemoto, K., Tsuji, T., Kawasaki, J. (2001). Re-examination of interictal psychoses; based on DSM IV psychosis classification and international epilepsy classification. *Epilepsia* **42**, 98–103.

Kanner, A. M., Stagno, S., Kotagel, P., Morris, H. H. (1996). Postictal psychiatric events during prolonged video-electroencephalographic monitoring studies. *Arch Neurol* **53**, 258–263.

Kendell, R. E., Kemp, I. W. (1989). Maternal influenza in the etiology of schizophrenia. *Arch Gen Psychiatry* **46**, 878–882.

Kendrich, J. R., Gibbs, F. A. (1957). Origin, spread and neurological treatment of the psychomotor type of seizure discharge. *J Neurosurg* **14**, 270–284.

Kristensen, O., Sindrup, E. H. (1987). Psychomotor epilepsy and psychosis. I. Physical Aspects. *Acta Neurol Scand* **57**, 361–369.

Krohn, W. (1961). A study of epilepsy in northern Norway, its frequency and character. *Acta Psychiat Scand* **150**(Suppl), 215–225.

Lancman, M. E. (1999). Psychosis and peri-ictal confusional states. *Neurology* **53**(Suppl 2), S33–38.

Landolt, H. (1958). Serial EEG investigations during psychotic episodes in epileptic patients and during schizophrenic attacks. In *Lectures on Epilepsy* (A. M. Lorenz De Haas, ed.), pp. 91–133. Elsevier: Amsterdam.

Landolt, H. (1953). Einige klinisch-elektroencephalographische Korrelationen bei epileptischen Dämmerzuständen. *Nervenarzt* **24**, 479.

Landolt, H. (1963). Die Dammer- und Verstimmungszustande bei Epilepsie und ihre Elektroencephalographie. *Dtsch Z Nervenheilkunde* **185**, 411–430.

Levin, S. (1952). Epileptic clouded states. A review of 52 cases. *J Nerv Ment Dis* **116**(4), 215–225.

Logsdail, S. J., Toone, B. K. (1988). Postictal psychosis: a clinical and phenomenological description. *Br J Psychiatry* **152**, 246–252.

Mace, C. J. (1993). Epilepsy and schizophrenia. *Br J Psychiatry* **163**, 439–445.

Marneros, A., Pillmann, F., Haring, A., Balzuweit, S., Blöink, R. (2005). Is the psychopathology of acute and transient psychotic disorder different from schizophrenic and schizoaffective disorders? *Eur Psychiatry* **20**, 315–320.

Matsuura, M., Adachi, N., Oana, Y., Okubo, Y., Kato, M., Nakano, T., Takei, N. (2004). A polydiagnostic and dimensional comparison of epileptic psychoses and schizophrenia spectrum disorders. *Schzophrenia Research* **69**, 189–201.

Mendez, M. F., Grau, R., Doss, R. C., Taylor, J. L. (1993). Schizophrenia in epilepsy: seizure and psychosis variables. *Neurology* **43**, 1073–1077.

Oshima, T., Tadokoro, Y., Kanemoto, K. (2006). A prospective study of postictal psychoses with emphasis on the periictal type. *Epilepsia* **47**, 2131–2134.

Onuma, T., Adachi, N., Ishida, S., Katou, M., Uesugi, S. (1995). Prevalence and annual incidence of psychoses in patients with epilepsy. *Epilepsia* **36**(suppl 3), S218.

Parnas, J., Korsgaard, S., Krautwald, O., Jensen, P. S. (1982). Chronic psychosis in epilepsy. *Acta Psychiat Scand* **66**, 282–293.

Perala, J., Suvisaari, J., Saarni, S. I., Kuoppasalmi, K., Isometsa, E., Pirkola, S., Partonen, T., Tuulio-Henriksson, A., Hintikka, J., Kieseppa, T., Harkanen, T., Koskinen, S., Lonnqvist, J. (2007). Lifetime prevalence of psychotic and bipolar I disorders in a general population. *Arch Gen Psychiatry* **64**, 19–28.

Perez, M. M., Trimble, M. R. (1980). Epileptic psychosis-diagnostic comparison with process schizophrenia. *Br J Psychiatry* **141**, 256–261.

Pichot, P. (1986). The concept of 'bouffée délirante' with special reference to the Scandinavian concept of reactive psychosis. *Psychopathology* **19**, 35–43.

Pillmann, F., Haring, A., Balzuweit, S., Blöink, R., Marneros, A. (2003). Bouffée délirante and ICD-10 acute and transient psychoses: a comparative study. *Aust New Z J Psychiatry* **37**, 327–333.

Pond, D. A. (1957). Psychiatric aspects of epilepsy. *Journal of the Indian Medical Profession* **3**, 1441–1451.

Pond, D. A. (1962) Discussion remark. Proceedings of the Royal Society of Medicine **55**, 316.

Pull, C. B., Pull, M. C., Pichot, P. (1984). Des critères empiriques français pour les psychoses. I. Position du problème et méthodologie. *Encéphale* **10**, 119–123.

Qin, P., Xu, H., Laursen, T. M., Vestergaard, M., Mortensen, P. B. (2005). Risk for schizophrenia and schizo-phrenia-like psychosis among patients with epilepsy: population based cohort study. *BMJ* **331**, 23.

Roberts, G. W., Done, D. J., Bruton, C., Crow, T. J. (1990). A "mock-up" of schizophrenia: temporal lobe epilepsy and schizophrenia-like psychosis. *Biol Psychiatry* **28**, 127–143.

Sachdev, P. S. (2007). Alternating and postictal psychoses: review and a unifying hypothesis. *Schizophrenia Bull* **33**, 1029–1037.

Sander, J. W., Hart, Y. M., Trimble, M. R., Shorvon, S. D. (1991). Vigabatrin and psychosis. *J Neurology Neurosurg Psychiatry* **54**, 435–439.

Savard, G., Andermann, F., Olivier, A., Remillard, G. M. (1991). Postictal psychosis after partial complex seizures: a multiple case study. *Epilepsia* **32**, 225–231.

Schmidt, B., Wolf, P. (1989). *Psychosis in epilepsy: Frequency and relation to different types of epilepsy.* New Delhi, India: Presented at the World Congress of Epilepsy.

Schmitz, B., Trimble, M. R. (1998). Epilogue. In *Forced normalization and alternative psychoses of epilepsy* (M. R. Trimble, B. Schmitz, eds), pp. 221–227. Petersfield: Wrightson Biomedical Publishing.

Sengoku, A., Yagi, K., Seino, M., Wada, T. (1983). Risks of occurrence of psychoses in relation to the types of epilepsies and epileptic seizures. *Folia Psychiat Neurol Jpn* **37**, 221–225.

Sherwin, I., Peron-Magnan, P., Bancaud, J., Boris, A., Talairach, J. (1982). Prevalence of psychosis as a function of the laterality of the epileptogenic lesion. *Arch Neurol* **39**, 621–625.

Shukla, G. D., Srivastava, O. N., Katiyar, B. C., Joshi, V., Mohan, P. K. (1979). Psychiatric manifestations in temporal lobe epilepsy: a controlled study. *Br J Psychiatry* **135**, 411–417.

Slater, E., Beard, A. W. (1963). The schizophrenia-like psychoses of epilepsy. *Br J Psychiatry* **109**, 95–150.

So, N. K., Savard, G., Andermann, A., Olivier, A., Quensney, L. F. (1990). Acute postictal psychosis: a stereo EEG study. *Epilepsia* **31**, 188–193.

Stantage, K. (1973). Schizophreniform psychosis among epileptics in a mental hospital. *Br J Psychiatry* **123**, 231–232.

Stantage, K., Fenton, G. W. (1975). Psychiatric symptom profile of patients with epilepsy: a controlled investigation. *Psychological Medicine* **5**, 152–160.

Stevens, J. R. (1983) Psychosis and epilepsy. *Ann Neurol* **14**, 348–8.

Stevens, J. R. (1988). Epilepsy, psychosis, and schizophrenia. *Schizophrenia Research* **1**, 79–89.

Stevens, J. R. (1991). Psychosis and the temporal lobe. Neurobehavioral Problems in Epilepsy: Advances in Neurology, **55** (D. Smith, D. Treiman, M. Trimble, eds), pp. 76–96. New York: Raven Press.

Stevens, J. R. (1986). Abnormal reinnervation as a basis for schizophrenia: a hypothesis. *Arch Gen Psychiatry* **49**, 238–243.

Sutula, T. P. (2004). Mechanisms of epilepsy progression: current theories and perspectives from neuro-plasticity in adulthood and development. *Epilepsy Res* **60**, 161–167.

Tadokoro Y, Oshima T, Shimizu H, Kanemoto K. (2007). Interictal psychoses in comparison with schizophrenia; a prospective study. *Epilepsia* **48**, 2345–2351.

Takeda, Y., Inoue, Y., Tottori, T., Mihara, T. (2001). Acute psychosis during intracranial EEG monitoring: close relationship between psychotic symptoms and discharges in amygdala. *Epilepsia* **42**, 719–724.

Taylor, D. C. (1972). Mental state and temporal lobe epilepsy. A correlative account of 100 patients treated surgically. *Epilepsia* **13**, 727–765.

Taylor, D. C. (1971). Ontogenesis of chronic epileptic psychoses: A re-analysis. *Psychol Med* **1**, 247–253.

Taylor, D. C. (1975). Factors influencing the occurrence of schizophrenia-like psychosis in patients with temporal lobe epilepsy. *Psychol Med* **5**, 249–254.

Tellenbach, H. (1965). Epilepsie als Anfallsleiden und als Psychose. *Nervenarzt* **36**, 190–202.

Toone, B. K., Garralda, M. E., Ron, M. A. (1980). The psychoses of epilepsy and the functional psychoses. *Br J Psychiatry* **137**, 245–249.

Trimble, M. R. (1991). *The Psychoses of Epilepsy*. New York: Raven Press.

Trimble, M. R., Schmitz, B. (1997). The psychoses of epilepsy/Schizophrenia. In *Epilepsy: A Comprehensive Textbook* (J. Engel, Jr., T. A. Pedley, eds), pp. 2071–2081. Philadelphia, PA: Lippincott-Raven Publishers.

Umbricht, D., Degreef, G., Barr, W. B., Lieberman, J. A., Pollack, S., Schaul, N. (1995). Postictal and chronic psychoses in patients with temporal lobe epilepsy. *Am J Psychiatry* **152**, 224–231.

Velakoulis, D., Wood, S. J., Wong, M. T., *et al.* (2006). Hippocampal and amygdale volumes according to psychosis stage and diagnosis: a magnetic resonance imaging study of chronic schizophrenia, first-episode psychosis, and ultra-high-risk individuals. *Arch Gen Psychiatry* **63**, 139–149.

Wekking, E. M. (1993). Psychiatric symptoms in systemic lupus erythematosus: an update. *Psychosom Med* **55**, 219–228.

Wieser, H. G., Hailemariam, S., Regard, M. (1985). Unilateral limbic epileptic status activity: stereo-EEG, behavioral, and cognitive data. *Epilepsia* **26**, 19–29.

Wolf, P. (1991). Acute behavioral symptomatology at disappearance of epileptiform EEG abnormality: paradoxical or forced normalization. In *Neurobehavioral Problems in Epilepsy* (D. Smith, D. Treiman, M. R. Trimble, eds), pp. 127–142. New York: Raven Press.

World Health Organization (1992). The ICD-10 Classification of Mental and Behavioural Disorders: Clinical and diagnostic guidelines. Geneva.

Yu, H. H., Lee, J. H., Wang, L. C., Yang, Y. H., Chiang, B. L., Wekking, E. M. (2006). Neuropsychiatric manifestations in pediatric systemic lupus erythematosus: a 20-year study. *Lupus* **15**, 651–657.

Zhang, Q., Sachdev, P. S. (2003). Psychotic disorder and traumatic brain injury. *Curr Psychiatry Rep* **5**, 197–201.

Zielinski, J. J. (1974). *Epidemiology and Medical–Social Problems of Epilepsy in Warsaw*. Washington, DC: US Government Printing Office.

CHAPTER 9

Are the Psychoses of Epilepsy a Neurological Disease?

E. S. Krishnamoorthy and R. Seethalakshmi

INTRODUCTION

A close association between psychoses and epilepsy has been known since 1854 confirmed by the fact that the early mental hospitals in Europe had special wards dedicated to epilepsy. The early proponents of psychosis of epilepsy (POE) included Kraepelin (1918) who believed that the "dementia" resulting from epilepsy was different from "dementia praecox" or schizophrenia. This was further affirmed by Vorkastner (1918) and Krapf (1928) who clarified that schizophrenia-like symptoms might follow epilepsy and that these need to be differentiated from true schizophrenia. Contrary to this, Glaus (1931) and Gruhle (1963) opined that the schizophrenic symptoms of epilepsy could be true independent schizophrenia. In 1860, with the description of alternating psychosis, this relationship underwent some radical thinking, and sadly petered out until the 1950s when the term schizophrenia-like psychosis of epilepsy (SLPE) was coined by Slater *et al.* (1963).

Throughout its history, however, POE has been largely considered an epileptic equivalent; both Falret (1854) and Morel (1873) employed terms such as "larval epilepsy", "epileptic mania", and "grands maux intellectuals". This chapter will attempt to justify a neurological diagnosis of the psychoses of epilepsy. This will be argued in two parts – in the first, we will endeavor to establish POE as a distinctive entity different from functional psychoses with schizophrenia as the gold standard. The second part of the discussion will entail neurological explanations of POE.

THE PSYCHOSES OF EPILEPSY

Among the most systematic descriptions of POE is the categorization based on the temporal relationship of the psychosis to the ictus.

1. Ictal psychosis: This can be defined as a nonconvulsive status epilepticus with symptoms resembling psychosis; ictal psychosis is associated with concurrent electrographic ictal patterns. The psychosis is usually brief, lasting from hours to days. The manifestation may not always be truly psychotic – consciousness is altered during the episode, insight is usually maintained and delusions and hallucinations are often absent during lucid periods, although there have been several reports of hallucinations (visual and auditory, and somesthetic), paranoia and bizarre psychotic-like catatonic behavior (Kaplan, 2002) during active episodes of ictus.

2. Postictal psychosis (PIP): Postictal psychotic episodes are brief psychotic episodes that occur in close proximity to seizure clusters. PIP can also be associated with a recent exacerbation in seizure frequency that may be related to withdrawal of anticonvulsants. Between the last seizure and the psychosis there is usually a nonpsychotic lucid period, which can range from 12 to 72 h to a week. Similar to ictal psychoses, consciousness may be clouded. Unlike ictal psychosis, however, in PIP true psychotic symptoms are dominant throughout the episode. These are pleomorphic in nature and include persecutory, grandiose, referential, somatic, and religious delusions, catatonic features, and hallucinations. The presentation may additionally be colored with affective symptoms including both manic and depressive symptoms. Schneiderian first rank symptoms that are considered pathognomonic of schizophrenia have also been described rarely. These symptoms usually resolve within a few days.

3. Interictal psychosis.

Chronic (also called SLPE): The principle characteristics of the "schizophrenia-like psychoses of epilepsy' were first described as follows (Toone, 1991):

1. Among patients with epilepsy, a schizophrenia-like syndrome occurs more commonly than chance would predict.

2. While the condition shares many of the characteristics of schizophrenia, many experts feel that it is distinguishable from that condition based on certain clinical features: the relative preservation of affect, the relative absence of negative symptoms and cognitive deterioration; the preservation of the personality in between episodes and the near total absence of hebephrenic features.

3. Temporal lobe epilepsy (TLE) is overrepresented.

4. A family history of schizophrenia occurs with no greater frequency than in the general population.

5. The onset of seizures almost invariably precedes the development of psychotic symptoms, usually by an interval of several years.

6. Within the various combinations of epilepsy subtypes and psychosis subtypes, a schizophrenia-like syndrome is usually associated with a dominant temporal lobe focus.

The phenomenon of forced normalization epitomizes the antagonistic relationship between psychosis and seizures. It can be defined as "the phenomenon characterized by the fact that, with the occurrence of psychotic states, the electroencephalogram (EEG) becomes more normal or entirely normal, as compared with previous and subsequent EEG findings." A recent diagnostic guideline suggests behavioral change including psychosis accompanied by over 50% reduction of EEG ictal activity or caregiver report of complete cessation of seizure activity (Krishnamoorthy and Trimble, 1999; Krishnamoorthy et al., 2002).

Relationship between schizophrenia (psychosis) and POE

Schizophrenia is considered the archetypal psychotic illness. The other primary psychotic illnesses include brief psychotic disorder, and schizophreniform disorder and schizoaffective disorder which differs from schizophrenia in being predominantly disorders of mood. The question "does POE warrant the distinction of being a separate psychotic entity?" begs answering, given that schizophrenia despite being a discrete nosological entity in psychiatry is acknowledged as having a number of subtypes with widespread phenomenological correlates. To that end, we will examine the epidemiological evidence for an association between epilepsy and psychosis and phenomenological evidence for distinctions between POE and schizophrenia.

Epidemiological issues

Schizophrenia is estimated to have a prevalence of 0.5–1% in the general population; the prevalence of psychotic disorders is likely to be much higher. Epilepsy has a point prevalence of 0.4–1% in the general population, and the lifetime risk of having at least one unprovoked seizure is 5–10%. A number of studies have suggested prevalence rates of schizophrenia within an epileptic population varying between 3% and 7% (Toone, 2000). The overall evidence suggests that schizophrenia-like psychosis is 6–12 times more likely to occur in people with epilepsy than in the general population. Further, when compared with other neurological disorders (e.g., migraine), the prevalence of schizophrenia in people with epilepsy has been found to be nine times higher (Mendez et al., 1993). A standardized incidence ratio of 1.48 for all epilepsy and 2.35 for psychomotor epilepsy argues strongly for epilepsy as a risk factor for schizophrenia, establishing beyond any reasonable doubt that epilepsy is a risk factor for the development of psychosis.

Schizophrenic-like psychosis of epilepsy (SLPE) is a unique disorder and not an artifact of random association. About 12.5–25% of POE is usually PIP (Adachi et al., 2002); 6.4–10% individuals with epilepsy can develop PIP (Sachdev, 1998). Forced normalization or the alternative psychosis of epilepsy is considerably less frequent with only 3 among 697 individuals demonstrating POE in Schmitz's (1998) study.

Phenomenological variations between POE and schizophrenia

SLPE primarily differs from schizophrenia in the severity of symptoms – quantitatively rather than qualitatively (Toone *et al.*, 1982; Tadokoro *et al.*, 2007). Although SLPE and schizophrenia seem alike in their phenomenology, subtle differences have been identified. SLPE is characterized by an increased frequency of persecutory and referential delusions while the incidences of catatonia and negative symptoms have been observed to be higher in schizophrenia. The higher preponderance of affective symptoms and the pleomorphic nature of SLPE also make it different from schizophrenia. In PIP, individuals report grandiose and religious delusions; on the other hand perceptual delusions and second-person hallucinations are less common. Additionally, individuals with SLPE do not demonstrate "affective flattening" and asociality that are common features in schizophrenics; establishment of interpersonal rapport and empathy has been found to be easier as compared to schizophrenics, especially in the earlier stages.

The age at onset in schizophrenia is usually 15–25 years; while in SLPE it has been noted that the psychosis emerges approximately 10–14 years after the onset of the seizure disorder. Individuals with schizophrenia often have greater deterioration in premorbid functioning than individuals with SLPE. These individuals have also been noted to have paranoid or schizoid premorbid personalities. On the contrary, individuals with SLPE have personalities characterized by obstinacy, egocentricity, and aggressiveness. An increased family history of schizophrenia as compared to the general population has not been noted in individuals with SLPE, again distinguishing this from schizophrenia.

The course of psychosis in SLPE is often varied with episodes presenting with different symptoms. Additionally, affected patients have a better prognosis as compared to schizophrenics and usually respond to lower doses of antipsychotic medications. Indeed, in both PIP and SLPE, adjustments in the dosage of the antiepileptic drug (AED) may help in alleviation of symptoms. This is unlike "schizophrenia" where antipsychotics form the first line of management. Besides, SLPE is characterized by a lesser degree of cognitive impairment in terms of memory and executive function as compared to schizophrenia; some impairment is similar.

Neuropathological findings relating to medial temporal defects in SLPE have been contradictory (Sachdev, 1998). Neuroimaging data on the other hand have suggested some interesting evidence (van Elst *et al.*, 2002; Rüsch *et al.*, 2004). SLPE has been associated with increased cerebral volume loss and amygdala enlargement as compared to matched patients without psychosis and healthy volunteers. In contrast to patients with schizophrenia, hippocampal volumes are preserved in patients with POE. Cerebral volume loss is more pronounced in patients with SLPE compared with patients with PIP. This finding has been corroborated in histopathological studies that have demonstrated lesser loss of CA1 neurons in the hippocampus in individuals with POE. EEG studies have also demonstrated similar spike discharges in medial temporal and frontal structures in both schizophrenia and SLPE. Both SLPE and schizophrenia have been reported to have higher than normal levels of dopa decarboxylase activity, possibly the result of suppressed tonic release of dopamine in striatum because of low corticostriatal glutamatergic input (Reith *et al.*, 1994). A recent magnetization transfer study (Flügel *et al.*, 2006) suggested that SLPE is associated with cortical magnetization transfer ratio (MTR) reductions in the left middle and superior temporal gyri. While superior temporal reductions have been reported in schizophrenia, this entity has more widespread abnormalities including involvement of the frontotemporal regions and medial temporal regions (Foong *et al.*, 2001). These cortical reductions on MTR are indicative of focal disruption of cortical neurons or dendrites suggesting neurodevelopmental defects. It has been suggested that the functional rather than structural abnormality in SLPE may be left-sided temporal.

TABLE 9.1   Schizophrenia and POE: A comparative analysis

Schizophrenia		POE	
		SLPE	PIP
Etiology	Biological and psychological factors	The epileptic process?	The epileptic process?
Family history of schizophrenia	Present	Absent	Absent
Phenomenology			
Age at onset	15–25 years	Related to age at onset of seizure disorder	Related to the age of onset of seizure disorder?
Premorbid/risk factors	Obstetric complications	Left-sided focus, temporal lobe epilepsy, left-handedness, female sex, presence of a gangliogliomas and focal lesions	Similar to SLPE; the predilection for clustering of seizures is a unique risk factor for POE
Premorbid personality	Paranoid or schizoid traits	Obstinacy, aggressive tendencies, egocentricty, religiosity, affective changes, features of the Geschwind syndrome	Pleomorphic paranoid-affective features, explosive behavior (DSH/self mutilation, etc.); bewilderment
Symptoms	Catatonia and negative symptoms including asociality	Persecutory and grandiose delusions	Grandiose and religious delusions
Treatment	Antipsychotic	AED optimization; antipsychotics	Benzodiazepines; AED optimization; short course of antipsychotics if required
Prognosis	Usually chronic deteriorating	Responds well to medication; remains fairly well integrated overall	Usually one-time; may be recurrent in a proportion; well between episodes; occasionally leads to SLPE

IS POE A NEUROLOGICAL DISEASE?

There is clear epidemiological evidence of an increased prevalence of POE in individuals with complex partial seizures (described in the previous section). Flor-Henry (1969a, b) suggested a preponderance of schizophrenia-like symptoms in dominant TLE. Epidemiological studies have identified other risk factors including a severe form of epilepsy involving multiple seizure types, a history of status epilepticus, and resistance to drug treatment. POE has also been associated with comparatively more stigmata of neurological damage including "morbid antecedents such as birth trauma and neurological signs" (Flor-Henry, 1969a, b).

Development of POE

It has been postulated that behavioral changes secondary to seizures could be either a positive effect of seizures – the epilepsy activating the areas mediating the development of psychotic symptomatology (e.g. ictal psychosis), or a negative effect – temporary dysfunction leading to inability to perform behaviors or disinhibition of behaviors such as PIP (Sachdev, 1998).

From an etiopathogenic perspective, it has been hypothesized that POE may be the result of:

1. Neurodevelopmental defect leading to cortical dysgenesis that results in both seizures and psychosis. This has been supported by the increased frequency of alien tissue tumors such as gangliogliomas as compared to mesial temporal sclerosis in individuals with POE. Cryptic insults to the vulnerable brain may lead to further synaptic reorganization and possibly psychosis.
2. Diffuse brain damage leading to seizures and psychosis. This theory borrows from neuropathological evidence in schizophrenia suggesting that psychosis is the result of diffuse brain damage rather than a single area.

PIP has been associated with bitemporal foci lending credence to long-term potentiation (LTP) from recurrent limbic kindling leading to the development of secondary epileptogenic foci. Individuals with PIP have also been shown to have an increased incidence of alien tissue such as gangliogliomas; and have been found to lack in Ammon's Horn sclerosis. As stated above, SLPE has been associated with left-sided temporal deficits by some investigators. These specific regions were first identified by Flor-Henry (1969a, b) and later corroborated by Perez and Trimble (1980); after all, the temporal lobe has far-reaching connecting circuits extending into almost all affected parts of the brain and this could thus explain the presentation of psychosis in individuals with chronic seizure disorder. Yet, the pathogenic role of the left temporal lobe in SLPE is being increasingly questioned by some authors who have suggested involvement of the basal ganglia and periventricular regions moreso than temporal lobe lesions (Sachdev, 1998).

Various mechanisms have been suggested to explain the development of POE at the neurophysiological level.

Limbic kindling

Of the many neurobiological hypotheses that have been put forward in trying to explain the evolution of POE, kindling is perhaps the most controversial, yet fascinating potential explanation (Krishnamoorthy and Trimble, 1999). Kindling is relatively well accepted as an animal model of epilepsy. Kindling is somewhat less well accepted as an animal model of psychoses. The exploration of its relevance to the POE interface is therefore of interest. Kindling defines a mechanism by which repeated (daily) brief, high-frequency trains of electrical pulses to limbic and cortical areas produce a change in response to the stimulus, such that the latter elicits a motor convulsion that outlasts the stimulus train. Thus, when a current of low amplitude but high pulse frequency is applied to the amygdala, a region of great susceptibility, there is a gradual stepwise progression of behavioral and EEG responses that finally culminate in a full motor seizure. Pharmacologic kindling and behavioral sensitization have been used to explain the development of psychosis in animal models (Stevens and Livermore, 1978; Sato et al., 1990). Based on the animal data (extensively reviewed elsewhere), it has been postulated that the end point for pharmacologic kindling may be a particular form of affective expressions or behaviors, unlike electrical kindling that terminates in a motor seizure (Pollock, 1987; Smith and Darlington, 1996).

Role of neurotransmitters

A number of neurotransmitters have been thought to have a role in POE, of which the principal often cited neurotransmitter is dopamine (Trimble, 1991). Dopamine is an inhibitory neurotransmitter that reduces the hyperexcitability associated with seizures and is also psychogenic in nature. Another hypothesis that has gained credence over the years is the Glutamate-GABA hypothesis. When the affective spectrum of psychoses (to which POE have phenomenological affinity) are considered as a whole, GABAergic preponderance (Benes and Berretta, 2001) and glutamatergic deficit (Meldrum, 2000) are hypothesized as being the common denominator. The significance of course is the underlying antagonism between seizures and psychosis that this implies; an epileptogenic state is the classic neurobiological oxymoron – glutamatergic preponderance and GABAergic deficit. Thus, enhanced glutamatergic excitation is a well known epileptogenic mechanism, particularly at the level of the N-methyl-d-aspartate glutamate (NMDA) receptor. In psychoses, on the other hand, an endogenous antagonist at the NMDA receptor, N-acetyl-aspartyl-glutamate, appears to have enhanced activity. This hypothesized dysfunction of glutamatergic transmission also interdigitates with the traditional dopamine hypothesis of schizophrenia. Presynaptic dopamine receptors on corticostriatal and limbic glutamatergic terminals provide a negative regulation of glutamate release.

Loss of γ-aminobutyric acid (GABA) inhibition is another potential epileptogenic factor; AEDs that increase GABA levels are associated with the development of a psychopathologic state in up to 10% of patients including mood changes, agitation, and even psychotic symptoms of a paranoid nature (Trimble, 1998).

Channelopathy

A recent hypothesis (Krishnamoorthy *et al.*, 2002) that attempts to explain alternative psychosis or forced normalization gains support from evidence of abnormal ion channels in episodic neurological disorders. Ion channels provide the basis for regulation of excitability in the CNS. Mutations in ion channels have been implicated in various epilepsy syndromes including autosomal dominant nocturnal frontal lobe epilepsy (neuronal nicotinic acetylcholine receptor), benign familial neonatal convulsions (potassium channels), generalized epilepsy with febrile seizures plus (sodium channels or the GABA(A) receptor), and episodic ataxia type-1, which is associated with epilepsy in a few patients (another type of potassium channel). There is a viewpoint gaining ground that paroxysmal psychiatric disorders (bipolar disorder for example) are also a product of malfunctioning ion channels (Amann and Grunze, 2005). Given that ion channels are expressed in the limbic regions of the brain and that AEDs have an influence on both ion channels and behavior, one may speculate on the role that ion channels have in the development of POE, in particular paroxysmal forms such as the forced normalization/alternative psychosis state, where both seizures and psychotic symptoms appear to be affected by common underlying factors. Unfortunately, at this point, these hypotheses remain unproven.

Excitation–inhibition imbalance (Sachdev, 2007)

In a recent paper, Sachdev (2007) puts forward an interesting hypothesis. Epilepsy can be described as a state in which there is an imbalance in neuronal excitation and inhibition. Seizure termination is an active inhibition resulting from fast inhibitory postsynaptic potential (IPSP) through $GABA_A$ receptors, a hyperpolarizing potential through $GABA_B$ receptors and after hyperpolarization produced by calcium-activated potassium currents. The hyperpolarizing pumps can produce prolonged inhibition of neuronal activity in an attempt to maintain homeostasis in the brain, leading to the various postictal states. He concludes that both postictal psychosis and alternative psychosis (forced normalization) are states akin to Todd's paresis; states of postictal cortical inactivation with ongoing subcortical or mesial temporal epileptic activity and cortical inhibition referred to as an "inhibitory surround" in response to ongoing seizures.

Other factors that may have a role in the evolution of psychosis are AEDs (Krishnamoorthy and Trimble, 1999), folic acid deficiency (Levi and Waxman, 1975), laterality of lesion (Flor-Henry, 1969a, b) and the presence of alien tissue in the brain (Taylor, 1975). The predominance of temporal lobe epilepsy occurring with POE also indicates impairment or defect in these or linked structures supported by the evolving literature implicating mesial temporal structures in the development of psychopathology (Krishnamoorthy, 2007).

CONCLUSION

POE thus can be considered an important neuropsychiatric sequel of a common neurological condition – epilepsy. Although psychodynamic factors such as the presence of a chronic illness and the sudden cessation of seizures may play some role in the development of interictal psychopathology, the majority of the explanations for POE seem to be neurological in nature and related to the ictal process. While our current understanding about the precise nature of these neurobiological processes is limited, this should not deter us from exploring them further given the plethora of molecular, cellular and imaging techniques that are now available to us. That the psychosis of epilepsy is a neurological disease is almost a certainty. Whether this will lead to a better understanding of the neurobiology of psychosis overall and of psychopathology in general is the moot point!

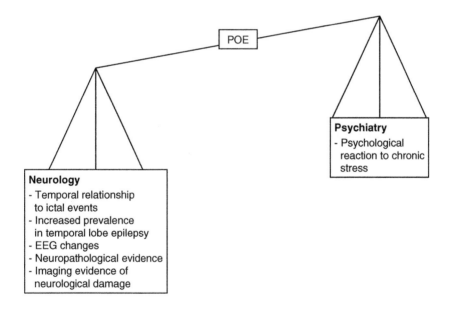

REFERENCES

Adachi, N., Matsuura, M., Hara, T., Oana, Y., Okubo, Y., Kato, M., Onuma, T. (2002). Psychoses and epilepsy: Are interictal and postictal psychoses distinct clinical entities? *Epilepsia* **43**(12), 1574–1582.

Amann, B., Grunze, H. (2005). Neurochemical underpinnings in bipolar disorder and epilepsy. *Epilepsia* **46**, 26–30.

Benes, F. M., Berretta, S. (2001). GABAergic interneurons: implications for understanding schizophrenia and bipolar disorder. *Neuropsychopharmacology* **25**(1), 1–27.

Falret, J. P. (1854). Memoire sur la folie ciculaire. *Bull Acad Med (Paris)* **19**, 382–400.

Flor-Henry, P. (1969a). Psychosis and temporal lobe epilepsy. *Epilepsia* **10**, 363–365.

Flor-Henry, P. (1969b). Schizophrenic-like reactions and affective psychoses associated with temporal lobe epilepsy: etiological factors. *Am J Psychiatry* **126**, 3.

Flügel, D., Cercignani, M., Symms, M. R., Koepp, M. J., Foong, J. (2006). A magnetization transfer imaging study in patients with temporal lobe epilepsy and interictal psychosis. *Biol Psychiatry* **59**(6), 560–567.

Foong, J., Symms, M. R., Barker, G. J., Maier, M., Woermann, F. G., Miller, D. H., *et al.* (2001). Neuropathological abnormalities in schizophrenia: evidence from magnetization transfer imaging. *Brain* **124**, 882–892.

Glaus, A. (1931). Uber combinationron schizophrenie and epilepsie. *Zentralbe Neurol. Psychiatri* **135**, 450–550.

Gruhle, H. W. (1936). Uber den wahn bei Epilepsie. *Z Gesampte Neurol Psychiatry* **154**, 395–399.

Kaplan, P. W. (2002). Behavioral manifestations of nonconvulsive status epilepticus. *Epilepsy Behav* **3**, 122–139.

Kraepelin, B. (1918). *Dementia Praecox*. (R. M. Barclay, ed.). Edinburgh: Livingstone.

Krapf, E. (1928). Epilepsie und schizophrenie. *Arch Psychiatr Nervenkr* **83**, 547–586.

Krishnamoorthy, E. S. (2007). A differential role for the hippocampus and amygdala in neuropsychiatric disorders. *J Neurol Neurosurg Psychiatry* **78**(11), 1165–1166.

Krishnamoorthy, E. S., Trimble, M. R. (1999). Forced normalization: clinical and therapeutic relevance. *Epilepsia* **40**(Suppl 10), S57–sS64.

Krishnamoorthy, E. S., Trimble, M. R., Sander, J. W. A. S., Kanner, A. M. (2002). Forced normalization at the interface between epilepsy and psychiatry. *Epilepsy Behav* **3**, 3–8.

Levi, R. N., Waxman, S. (1975). Schizophrenia, epilepsy, cancer, methionine and folate metabolism: Pathogenesis of schizophrenia. *Lancet* **11**, 11–13.

Meldrum, B. S. (2000). Glutamate as a neurotransmitter in the brain: review of physiology and pathology. *J Nutr* **130**, 1007S–s1015S.

Mendez, M. F., Grau, R., Doss, R. C., *et al.* (1993). Schizophrenia in epilepsy: seizure and psychosis variables. *Neurology* **43**, 1073–1077.

Morel, B. A. (1873). Discussion sur l'Epilepsie larvee. *Ann Med Psychol* **9**, 155–163. (series 5)

Perez, M. M., Trimble, M. R. (1980). Epileptic psychosis – diagnostic comparison with process schizophrenia. *Br J Psychiatry* **137**, 245–249.

Pollock, D. C. (1987). Models for understanding the antagonism between seizures and psychosis. *Prog Neuro-Psychopharmacol Biol Psychiatr* **11**, 483–504.

Reith, J., Benkelfat, C., Sherwin, A., Yasuhara, Y., Kuwabara, H., Andermann, F., Bachneff, S., Cumming, P., Diksic, M., Dyve, S. E., Etienne, P., Evans, A. C., Lal, S., Shevell, M., Savard, G., Wong, D. F., Chouinard, G., Gjedde, A. (1994). Elevated dopa decarboxylase activity in living brain of patients with psychosis. *Proc Natl Acad Sci USA* **91**, 11651–11654.

Rüsch, N., van Elst, L. T., Baeumer, D., Ebert, D., Trimble, M. R. (2004). Absence of cortical gray matter abnormalities in psychosis of epilepsy: a voxel-based MRI study in patients with temporal lobe epilepsy. *J Neuropsychiatry Clin Neurosci* **16**, 148–155.

Sachdev, P. (1998). Schizophrenia-like psychosis and epilepsy: the status of the association. *Am J Psychiatry* **155**, 325–336.

Sachdev, P. S. (2007). Alternating and postictal psychoses: review and a unifying hypothesis. *Schizophr Bull* **33**(4), 1029–1037.

Sato, M., Racine, R. J., McIntyre, D. C. (1990). Kindling: basic mechanisms and clinical validity. *Electroencephal Clin Neurophysiol* **76**, 459–472.

Schmitz, B. (1998). Forced normalization: history of a concept. In *Forced normalization and alternative psychosis of epilepsy* (M. R. Trimble, B. Schmitz, eds), pp. 7–24. Petersfield: Wrightson Biomedical Publishing.

Slater, E., Beard, A. W., Glithero, E. (1963). The schizophrenia-like psychoses of epilepsy, i–v. *Br J Psychiatry* **109**, 95–150.

Smith, P. F., Darlington, C. L. (1996). The development of psychosis in epilepsy: a re-examination of the kindling hypothesis. *Behav Brain Res* **75**(1-2), 59–66.

Stevens, J. R., Livermore, A. (1978). Kindling of the mesolimbic dopamine system: animal model of psychosis. *Neurology* **28**, 36–46.

Tadokoro, Y., Oshima, T., Kanemoto, K. (2007). Interictal psychoses in comparison with schizophrenia – a prospective study. *Epilepsia* **48**(12): 2345–2351.

Taylor, D. C. (1975). Factors influencing the occurrence of schizophrenia-like psychosis in patients with temporal lobe epilepsy. *Psychol Med* **5**, 249–254.

Toone, B. K. (1991). The psychoses of epilepsy. *J Roy Soc Med* **84**, 457–459.

Toone, B. K. (2000). The psychoses of epilepsy. *J Neurol Neurosurg Psychiatry* **69**(1), 1–3.

Toone, B. K., Garralda, M. E., Ron, M. A. (1982). The psychoses of epilepsy and the functional psychoses: a clinical and phenomenological comparison. *Br J Psychiatry* **141**, 256–261.

Trimble, M. R. (1991). *The Psychoses of Epilepsy*. New York: Raven Press.

Trirnble, M. R. (1998). Forced normalization and the role of anticonvulsants. In *Forced normalization and alternative psychoses of epilepsy* (M. R. Trimble, B. Schmitz, eds), pp. 169–178. Petersfield: Wrightson Biomedical Publishing.

van Elst, L. T., Baeumer, D., Lemieux, L., Woermann, F. G., Koepp, M., Krishnamoorthy, S., Thompson, P. J., Ebert, D., Trimble, M. R. (2002). Amygdala pathology in psychosis of epilepsy: a magnetic resonance imaging study in patients with temporal lobe epilepsy. *Brain* **125**(1), 140–149.

Vorkastner, W. (1918). *Neurologie Psychiatric Psychologie and Ihren Grenzgebieten.*

CHAPTER 10

Is ADHD in Epilepsy the Expression of a Neurological Disorder?

David W. Dunn and William G. Kronenberger

INTRODUCTION

Children with epilepsy have substantially more problems with behavior and difficulties with school performance than siblings or normal population controls. Problems with attention and attention deficit hyperactivity disorder (ADHD) are found consistently in the evaluation of both emotional troubles and academic delays. In this chapter there are two primary questions. Is ADHD in epilepsy the expression of a neurological disorder? Is ADHD in epilepsy different from ADHD in children without epilepsy? In this chapter we will review studies of the prevalence of problems of attention in children with epilepsy and will explore possible reasons for difficulties with attention in this population. We also will address the controversy about the most appropriate treatment for the child with ADHD and epilepsy.

ATTENTION DEFICIT HYPERACTIVITY DISORDER

ADHD is one of the most extensively studied disorders in child and adolescent psychiatry. ADHD affects 4–12% of children and 3–5% of adults. The prevalence is higher in males with a range of ratios of 3:1 to 10:1, although the male-to-female ratio is smaller for ADHD, predominantly inattentive type. ADHD is classified as predominantly inattentive type if there are at least six symptoms of inattention but less than six symptoms of hyperactivity–impulsivity, predominantly hyperactive–impulsive type if there are six or more symptoms of hyperactivity and impulsivity and less than six symptoms of inattention, and combined type if there are six or more symptoms in each category. In clinical samples, ADHD, combined type is most common. ADHD is often comorbid with other conditions. In at least half of the children with ADHD, additional diagnoses of oppositional defiant disorder, conduct disorder, anxiety disorders, mood disorders, Tourette's disorder, and/or learning disorder may be found.

ADHD is a familial disorder with an estimated heritability of 0.76 (Biederman and Faraone, 2005). The genetic studies of ADHD suggest multiple genes conferring vulnerability for ADHD. Studies have particularly implicated dopamine receptor and dopamine transporter genes. Biological factors associated with ADHD are maternal smoking and alcohol use, prematurity, and fetal hypoxia. Neurological causes of ADHD include lead poisoning and traumatic brain injury. Psychosocial adversity may contribute to the severity of ADHD but does not seem to be a sole etiological factor.

Both the dopaminergic and the noradrenergic systems are involved in attention. Dopamine and norepinephrine influence working memory, and norepinephrine is important in orienting and vigilance. Vigilance and orienting are centered in the posterior cerebral cortex where the parietal lobe functions in disengagement and the superior colliculi in directing attention to a new stimulus. The anterior cingulate cortex, orbitofrontal cortex, and prefrontal cortex play a critical role in executive function. The cerebellar vermis may also influence attention.

Imaging studies in ADHD have shown abnormalities in multiple areas of the central nervous system (CNS). The most consistent findings are reduction in the volume of the dorsolateral prefrontal cortex, caudate, and cerebellum (Castellanos, 2001). One longitudinal study found enlargement of the lateral ventricles, decreased volume of the cerebral and cerebellar volumes, and smaller caudate volumes early but not later in adolescence (Castellanos *et al.*, 2002). Functional magnetic imaging and proton magnetic imaging have demonstrated reduced metabolism in the prefrontal cortex, more in the nondominant side.

Recent treatment algorithms of ADHD suggest starting with stimulant medication (Biederman and Faraone, 2005; Pliszka *et al.*, 2006). This is based on the impressive effect size (0.91–0.095) for stimulant medication. Atomoxetine is also effective with an effect size of 0.6–0.7. Other agents that may be effective are bupropion, modafinil, tricyclic antidepressants, risperidone, the alpha-2 agonists and monoamine oxidase (MAO) inhibitors.

ADHD IN CHILDREN WITH EPILEPSY

If ADHD in children with epilepsy is the same as ADHD in children without epilepsy, there should be a higher rate of ADHD in boys with epilepsy compared to girls with epilepsy. Here the data are not consistent. Two older studies found more inattention and hyperkinetic behavior in boys with epilepsy than in girls (Ounsted, 1955; Stores *et al.*, 1978). In one, the sample seemed to fit ADHD plus oppositional disorder or conduct disorder or possibly early onset juvenile bipolar disorder. In contrast, three recent studies found no gender difference (Dunn *et al.*, 2003; Hesdorffer *et al.*, 2004; Gonzalez-Heydrich *et al.*, 2007). These three recent studies used standardized testing and DSM criteria for the diagnosis of ADHD. The evidence from these studies suggests that ADHD in children with epilepsy differs from that seen in the typical psychiatric clinic sample.

PREVALENCE OF ATTENTION PROBLEMS AND ADHD IN EPILEPSY SAMPLES

The prevalence figures for ADHD or problems with attention in children with epilepsy can be used to compare children with epilepsy to those with ADHD without epilepsy. These prevalence rates also may help determine etiological factors for ADHD in children with epilepsy. If ADHD is found more often in subgroups of children with epilepsy, this may suggest possible causes for problems with attention.

Prevalence rates for ADHD in children with epilepsy vary widely from 0% to 77%. Variations in prevalence may be partially the results of differences in measures of ADHD or may be the result of assessing different samples of children with epilepsy. Several studies reported symptoms of inattention, hyperactivity, or impulsivity without using Diagnostic and Statistical Manual (DSM) criteria. In a population-based survey, 28.1% of children with epilepsy were described as hyperactive and 39% as impulsive compared to 4.9% and 11% of normal population controls (Carlton-Ford *et al.*, 1995; McDermott *et al.*, 1995). Teachers reported inattention in 42% of children with epilepsy in the report by Holdsworth and Whitmore (1974) and inattention or hyperactivity in 58% of the children described by Sturniolo and Galletti (1994).

In general, those studies that utilized standard measures found slightly lower rates. Reviewing 15 studies of ADHD in children or adolescents with epilepsy, we found prevalence rates from 0% to 50%. In three studies, prevalence rates were less than 20% (Ounsted, 1955; Davies *et al.*, 2003; Freilinger *et al.*, 2006), two were 20–30% (Thome-Souza *et al.*, 2004; Keene *et al.*, 2005), five were 30–40% (Hoare and Mann, 1994; Hempel *et al.*, 1995; Semrud-Clikeman and Wical, 1999; McCusker *et al.*, 2002; Dunn *et al.*, 2003), and two were higher than 40% (Salpekar *et al.*, 2005; Hanssen and Bauer *et al.*, 2007). Two studies described change over time. Williams *et al.* (2002a) noted rates dropping from 31% to 21–24% and Borgatti *et al.* (2004) found rates increasing 21–42%. Though not all studies used impairment criteria, the more common prevalence figures of 30–40% are 3–4 times higher than rates found in the general population, again suggesting a difference in the ADHD found in children with epilepsy compared to that found in the general population.

Children with ADHD without epilepsy are more likely to have ADHD, combined type. ADHD, predominantly hyperactive–impulsive type is least common. ADHD, predominantly inattentive type, is seen more in population studies than clinical samples and may be more prevalent in girls with ADHD (Biederman and

Faraone, 2005). Separation of ADHD in subtypes occurs infrequently in studies of children with epilepsy. Dunn *et al.* (2003) found a prevalence of 37.7% for ADHD in a clinical sample of children with epilepsy. They reported ADHD, combined type in 11.4%, ADHD, predominantly inattentive type in 24%, and ADHD, predominantly hyperactive–impulsive type in 2.4%. Williams *et al.* (2002a) noted inattention in 31% and hyperactivity–impulsivity in 31% of children with epilepsy at baseline. At follow up, a mean of 7 months later, 27% had inattention and 24% hyperactivity–impulsivity. Gonzalez-Heydrich *et al.* (2007) found more ADHD, combined type (58%) than ADHD, predominantly inattentive type (42%) in their clinical sample of children with epilepsy. They noted that the sample was drawn from a medication trial for ADHD in children with epilepsy and may have been biased toward patients with the more disruptive symptoms of hyperactivity and impulsivity. The reports by Dunn *et al.* (2003) and Williams *et al.* (2002a) suggest that the inattentive form of ADHD is more common in children with epilepsy than seen in children without seizures, but the report by Gonzalez-Heydrich *et al.* (2007) finds similar distributions of subtypes of ADHD. More research is needed before definite conclusions can be made.

Comorbidity in children with ADHD without epilepsy is common. However, only two groups have specifically commented on additional disorders in children with epilepsy and ADHD. Caplan *et al.* (2004) reported ADHD plus other disruptive behavior disorders in 17% of children with complex partial seizures compared to 6% of controls. In addition, they found ADHD with comorbid anxiety or depressive disorders in another 23% of children with epilepsy compared to 3% of controls. Gonzalez-Heydrich *et al.* (2007) found a comorbid condition in 67% of children with epilepsy and ADHD. They noted an anxiety disorder in 36% and oppositional defiant disorder in 31%. Though the data are limited, the rate of comorbid psychiatric conditions in children with epilepsy and ADHD is similar to the rate seen in children with ADHD without epilepsy.

Neuropsychological deficits are common in both groups of children. Sánchez-Carpintero and Neville (2003), in a review of attention in children with epilepsy, found consistent abnormalities in sustained attention and some evidence of deficits in divided attention. Selective attention was less often impaired. Widespread deficits in executive function have been found by Høie *et al.* (2006) in children with epilepsy. Impairment was found in children with all epilepsy syndromes except benign childhood epilepsy with centrotemporal spikes. No differences have been reported in the neuropsychological assessments of children with epilepsy and ADHD compared to children with ADHD alone.

Risk factors for ADHD in children with epilepsy

The risk factors for psychopathology in children with epilepsy can be divided into three major categories-neurological problems, seizure-related problems, and psychosocial problems. CNS problems in addition to epilepsy significantly increase the risk of behavioral problems. This has been shown by epidemiological studies that find children with complicated epilepsy to have twice the rate of behavioral problems seen in children with uncomplicated seizures. Seizure-related variables are age of onset, duration of epilepsy, recurrent seizures, and antiepileptic drugs (AEDs). Psychosocial variables are family functioning and child attitude toward illness. These same risk factors may be important in ADHD in children with epilepsy.

The association of neurological factors with ADHD in children with epilepsy has been shown by the increased risk of ADHD in children with complicated epilepsy compared to those with uncomplicated seizures. Davies *et al.* (2003) found a prevalence of ADHD of 12% in children with complicated epilepsy compared to none in children with uncomplicated seizures. The study with the highest reported rate of hyperactivity (77%) was completed in children with intractable epileptic encephalopathies (Ferrie *et al.*, 1997). In contrast, the prevalence of ADHD in children with epilepsy and intellectual disability (mental retardation) is not higher than found in children with normal intelligence. Steffenburg *et al.* (1996) noted ADHD in 7% of children with epilepsy and intellectual disability and Thome-Souza *et al.* (2004) found ADHD in 44% of children with epilepsy and IQ <70 compared to 56% in children with IQ >70. The lower rate of ADHD in the population of children with epilepsy and intellectual disability may be due to the presence of more autistic disorders that normally preclude the diagnosis of ADHD. There are few data on the association of an abnormal neurological examination or abnormalities on neuroimaging with ADHD in children with epilepsy. Williams *et al.* (2002a) did find that normal magnetic resonance imaging (MRI) was associated with fewer symptoms of ADHD on follow up examination.

Another way of assessing the role of neurological factors is to evaluate children with new-onset seizures. Problems present at the time of a first seizure could not be attributed to recurrent seizures, AEDs, or the psychosocial stresses of epilepsy. Instead, they are more likely secondary to a common underlying neurological dysfunction that causes both seizures and attention deficits. Three studies have in part shown increased rates of ADHD in children with new-onset seizures. In a population-based study, Hesdorffer *et al.* (2004) found an association of the inattentive form of ADHD with new-onset seizures. The history of ADHD, predominantly inattentive type was 2.5-fold more common in children with seizures than in controls. In contrast, the increased risk of seizures was not seen in those with ADHD, combined type. Similarly, Austin *et al.* (2001), in a study of children with first recognized seizures, used the CBCL to assess behavior in the 6 months prior to the first recognized seizures. They divided the seizure group into those with prior unrecognized seizures and those with no prior seizures. They found more elevated attention problems scores in children with new-onset seizures than seen in sibling controls. The prevalence of children in the clinical range for attention problems was 15.8% in the children with prior unrecognized seizures, 8.1% in those with apparently true new-onset seizures, and 2.2–3.4% in siblings. In contrast, Oostrom *et al.* (2002) found only partial differences between children with new-onset seizures and controls. When compared to controls, the children with new-onset epilepsy had more errors on a reaction time measure and more errors of omission on a sustained attention task, but execution time and motor speeds were not different.

Seizure-related factors also may be important. Imaging studies in children with ADHD without epilepsy would suggest that children with an epileptic focus in the frontal cortex are at greater risk. However, seizure type and seizure focus have not been consistently associated with ADHD in children with epilepsy. Small case series of children with frontal lobe epilepsy have noted frequent symptoms of ADHD but in another case series, children with generalized seizures were more likely to have symptoms of ADHD than those with focal seizures (Hempel *et al.*, 1995; Sherman *et al.*, 2000; Hernandez *et al.*, 2003). There were no significant differences by seizure type or focus in the children reported by Bennett-Levy and Stores (1984), Carlton-Ford *et al.* (1995), Williams *et al.* (2002a), or Dunn *et al.* (2003). Studies of children with benign focal epilepsy with centrotemporal spikes have found more attention problems in the children with bilateral or right-sided temporal spikes (Piccirilli *et al.*, 1994).

Some authors have suggested that transient cognitive impairment may occur in children with epileptiform discharges not associated with seizures. However, Pressler *et al.* (2005) were only able to demonstrate general improvement in behavior after reduction of spike number, not specific improvement in attention. In a very well-controlled study, Aldenkamp and Arends (2004) used video-EEG monitoring during neuropsychological testing to evaluate the effect of seizure discharges and seizures on attention and processing speed. They found that non-convulsive seizures were associated with impaired attention and slow processing speed and frequent epileptiform discharges with slow processing speed. Additional support for the importance of seizure-related factors comes from the new-onset study of Austin *et al.* (2001). They found that, at the time of initial diagnosis, the children with prior unrecognized seizures had almost twice the rate of problems with attention as the children with true first seizures.

Children with recurrent seizures may be at increased risk for ADHD compared to those with controlled seizures. Hermann *et al.* (1989) found that in girls, intractable epilepsy was associated with hyperactivity. McCusker *et al.* (2002) found an association between ADHD and seizure frequency.

Treatment of seizures can contribute to symptoms of ADHD (Kwan and Brodie, 2001; Bourgeois, 2004; Loring and Meador, 2004). Problems with attention and hyperactivity have occurred most commonly with the barbiturates and, to a lesser extent, with phenytoin, carbamazepine, and valproic acid. Of the newer AEDs, topiramate has been associated with impaired attention as well as language difficulties. Hyperactivity was reported as a side effect of gabapentin in children with developmental delay and epilepsy. Hyperactivity has been seen with vigabatrin.

Psychosocial factors have not been implicated in the etiology of ADHD, but may contribute to the severity of symptoms. In children with epilepsy, psychosocial factors have been associated with symptoms of ADHD. McDermott *et al.* (1995) found that hyperactivity was associated with poverty and urban residence. Carlton-Ford *et al.* (1995) noted that the association between impulsivity and epilepsy was eliminated after controlling for learning disability and family processes. McCusker *et al.* (2002) reported an association between both family cohesion and family conflict and ADHD in children with intractable epilepsy. Oostrom *et al.* (2002) described a stronger association of attention problems with family factors than seizure variables.

Problems with attention in children with epilepsy do not appear to be a nonspecific reaction to a chronic illness. Rodenburg *et al.* (2005) in a recent meta-analysis of 46 studies of behavioral problems in children with epilepsy found that attention problems, thought problems, and social problems were more specific to epilepsy, whereas depression, withdrawn behavior, somatic complaints, and aggressive behavior were a more generic response to chronic illness. They found smaller effect sizes when comparing children with epilepsy to siblings. They suggested that family factors may be more related to behavior problems in general than specifically to attention, thought, or social problems in children with epilepsy.

Treatment of problems of attention in children with epilepsy

Is treatment of ADHD in children with epilepsy different from treatment in children without epilepsy? Pharmacological treatment of ADHD in children without epilepsy starts with stimulant medication. If stimulants are not effective or cause side effects, next choices are atomoxetine, antidepressants such as bupropion or tricyclic antidepressants, and adrenergic agents such as clonidine and guanfacine.

In children with epilepsy, stimulants may be more problematic. In the product information from the pharmaceutical companies there are warnings against the use of stimulants in patients with seizures. Presumably there may be lowering of the seizure threshold with the use of stimulant medication. The data to support this concern are limited. Feldman *et al.* (1989) treated children with ADHD and well-controlled epilepsy with methylphenidate 0.3 mg/kg/dose and found improvement in attention without adverse effect on seizure control. Gross-Tsur *et al.* (1997) found that children with ADHD and controlled seizures had no breakthrough seizures when treated with methylphenidate 0.3 mg/kg/day for 2 months. In the small group with active seizures, 2 improved and 3 experienced worsening seizure control. Gucuyener *et al.* (2003) noted no new seizures in 62 children with ADHD and an abnormal EEG that were treated with methylphenidate for 12 months and no increase in mean seizure frequency in the 57 children with ADHD and epilepsy that received methylphenidate 0.3–1 mg/kg/day. In this second group, 5 of 57 had an increase in seizure frequency. The data seem to suggest that stimulants are safe and effective in children with controlled seizures and probably though less definitely safe in those with active seizures.

The response of children with epilepsy and ADHD may be different from that seen in children with ADHD alone. Though children with epilepsy and ADHD respond to stimulant medication, stimulants seem to be less efficacious in these children. Semrud-Clikeman and Wical (1999) treated children with complex partial seizures and ADHD and children with ADHD alone with methylphenidate. On a computer performance task, the children with ADHD alone normalized after methylphenidate whereas the children with epilepsy and ADHD improved from 3.5 standard deviations below normal to 1.5 standard deviations below normal. Gonzalez-Heydrich *et al.* (2004) found that 63% of children with epilepsy responded to methylphenidate and 24% to amphetamine preparations.

Atomoxetine is often employed if stimulants are not effective or cause unacceptable side effects. Atomoxetine is a norepinephrine reuptake inhibitor that is effective in the treatment of ADHD. It has not been extensively studied in children with epilepsy. Hernández and Barragán (2005) used atomoxetine in 17 children with epilepsy and ADHD and found an improvement in symptoms of ADHD without an increase in seizure frequency. In a review of the clinical trials of atomoxetine, Wernicke *et al.* (2006) found that new-onset seizures occurred at a rate of 2/1,000/year. Assessing the reports for new seizures after the introduction of atomoxetine on the market, they found a rate of new seizures of 8.2/100,000, which is below expected incidence rates.

In children with ADHD without epilepsy, bupropion and tricyclic antidepressants are used as alternative agents. However, both bupropion and the tricyclic antidepressants may be problematic in children with epilepsy. Bupropion carries a high relative risk for seizures at high does and an intermediate risk at moderate or low doses. The tricyclic antidepressants have an intermediate risk at moderate to high dose (Alldredge, 1999). Drug interactions between valproic acid and tricyclics may lead to elevated serum levels of tricyclic antidepressants and interaction between barbiturates, phenytoin, and carbamazepine and tricyclics may cause decreased serum levels of tricyclics.

Comorbidity

Is there sufficient evidence to conclude that the co-occurrence of ADHD and epilepsy exceeds that expected by chance and that the co-occurrence of these disorders is not an artifact of measurement? Many studies of ADHD and epilepsy are based on clinical samples from comprehensive epilepsy centers. There could be a referral bias with those children having both epilepsy and psychiatric problems seen in these clinics. Though the lack of ADHD in the epidemiological study of Davies *et al.* (2003) might suggest this is the case, the population-based surveys reported by Carlton-Ford *et al.* (1995) and McDermott *et al.* (1995) and the community series by Austin *et al.* (2001) show the increased prevalence of ADHD in children with epilepsy is not the result of a referral bias.

Could the comorbidity be an artifact of measurement? If problems with attention are an ictal manifestation of seizures or occur only as a side effect of medication there would be no reason to consider epilepsy and ADHD comorbid conditions. Here the evidence for true comorbidity seems strong. Problems with attention have been documented at the onset of epilepsy prior to adverse effects of recurrent seizures and before the introduction of AEDs. In addition, Williams *et al.* (2002b) have demonstrated differences in symptoms when comparing ADHD ratings in children with absence seizures to those in children with ADHD.

A final question is more difficult to answer. Is the co-occurrence of ADHD and epilepsy an example of a common syndrome complex with both disorders sharing a genetic or biologic cause or does one disorder predispose or provoke the other? Hesdorffer *et al.* (2004) hypothesized a common underlying etiology suggesting a possible common dysfunction in the central norepinephrine system. The increased risk of ADHD in children with additional neurological disturbance would favor a common underlying cause for both disorders. The increased rate of epileptiform abnormalities on EEGs in children with ADHD also is evidence for a common underlying condition. In contrast, the prominence of inattentive symptoms might suggest that seizures and AEDs adversely affect neuropsychological functioning resulting in symptoms of ADHD (Noeker and Haverkamp, 2003). It may be that just as epilepsy is a heterogeneous condition, ADHD and epilepsy may be the results of multiple causes that vary by the individual patient.

CONCLUSIONS

ADHD in children with epilepsy is more common than ADHD in the general population. There do seem to be significant differences between ADHD and epilepsy and ADHD alone. Comparing ADHD with epilepsy to ADHD alone, there does not seem to be gender differences, symptoms of inattention appear to be more common, and the response to stimulant medication is less robust. The presence of symptoms of ADHD prior to or at the onset of epilepsy and the increased prevalence in children with epilepsy and additional CNS dysfunction suggests a common underlying etiology. However, the association of ADHD with prior unrecognized seizures, more frequent seizures, and AED adverse effects suggest that seizure-related factors contribute to the presence or severity of symptoms of ADHD. Appropriate treatment should begin with assessment of seizure control and AEDs followed by pharmacotherapy with stimulants or atomoxetine.

REFERENCES

Aldenkamp, A., Arends, J. (2004). The relative influence of epileptic EEG discharges, short nonconvulsive seizures, and type of epilepsy on cognitive function. *Epilepsia* **45**, 54–63.

Alldredge, B. K. (1999). Seizure risk associated with psychotropic drugs: clinical and pharmacokinetic considerations. *Neurology* **53**(Suppl. 2), S68–S75.

Austin, J. K., Harezlak, J., Dunn, D. W., Huster, G. A., Rose, D. F., Ambrosius, W. T. (2001). Behavior problems in children prior to first recognized seizures. *Pediatrics* **107**, 115–122.

Bennett-Levy, J., Stores, G. (1984). The nature of cognitive dysfunction in school children with epilepsy. *Acta Neurol Scand* **69**(suppl 99), 79–86.

Biederman, J., Faraone, S. V. (2005). Attention-deficit hyperactivity disorder. *Lancet* **366**, 237–248.

Borgatti, R., Piccinelli, P., Montirosso, R., *et al.* (2004). Study of attentional processes in children with idiopathic epilepsy by Conners' continuous performance test. *J Child Neurol* **19**, 509–515.

Bourgeois, B. F. D. (2004). Determining the effects of antiepileptic drugs on cognitive function in pediatric patients with epilepsy. *J Child Neurol* **19**(Suppl 1), S15–S24.

Caplan, R., Siddarth, P., Gurbani, S., *et al.* (2004). Psychopathology and pediatric complex partial seizures: seizure-related, cognitive, and linguistic variables. *Epilepsia* **45**, 1273–1281.

Carlton-Ford, S., Miller, R., Brown, M., Nealeigh, N., Jennings, P. (1995). Epilepsy and children's social and psychological adjustment. *J Health Soc Behav* **36**, 285–301.

Castellanos, F. X. (2001). Neuroimaging studies of ADHD. In *Stimulant Drugs and ADHD* (M. V. Solanto, A. F. T. Arnsten, F. X. Castellanos, eds), pp. 243–258. New York: Oxford University Press.

Castellanos, F. X., Lee, P. P., Sharp, W., *et al.* (2002). Developmental trajectories of brain volume abnormalities in children and adolescents with attention-deficit/hyperactivity disorder. *JAMA* **288**, 1740–1748.

Davies, S., Heyman, I., Goodman, R. (2003). A population survey of mental health problems in children with epilepsy. *Dev Med Child Neurol* **45**, 292–295.

Dunn, D. W., Austin, J. K., Harezlak, J., Ambrosius, W. T. (2003). ADHD and epilepsy in childhood. *Dev Med Child Neurol* **45**, 50–54.

Feldman, H., Crumrine, P., Handen, B. L., *et al.* (1989). Methylphenidate in children with seizures and attention-deficit disorder. *Am J Dis Child* **143**, 1081–1086.

Ferrie, C. D., Madigan, C., Tilling, K., Maisey, M. N., Marsden, P. K., Robinson, R. O. (1997). Adaptive and maladaptive behaviour in children with epileptic encephalopathies: correlation with cerebral glucose metabolism. *Dev Med Child Neurol* **39**, 588–595.

Freilinger, M., Reisel, B., Reiter, E., *et al.* (2006). Behavioral and emotional problems in children with epilepsy. *J Child Neurol* **21**, 939–945.

Gonzalez-Heydrich, J., Hsin, O., Hickory, M., *et al.* (2004). Comparisons of response to stimulant preparations in pediatric epilepsy (abstract). *AACAP 2004 Sci Proc* **31**, 107–108.

Gonzalez-Heydrich, J., Dodds, A., Whitney, J., *et al.* (2007). Psychiatric disorders and behavioral characteristics of pediatric patients with both epilepsy and attention-deficit hyperactivity disorder. *Epilepsy Behav* **10**, 384–388.

Gross-Tsur, V., Manor, O., van der Meere, J., *et al.* (1997). Epilepsy and attention deficit hyperactivity disorder: Is methylphenidate safe and effective? *J Pediatr* **130**, 670–674.

Gucuyener, K., Erdemogly, A. K., Senol, S., *et al.* (2003). Use of methylphenidate for attention-deficit hyperactivity disorder in patients with epilepsy or electroencephalographic abnormalities. *J Child Neurol* **18**, 109–112.

Hanssen-Bauer, K., Heyerdahl, S., Eriksson, A. (2007). Mental health problems in children and adolescents referred to a national epilepsy center. *Epilepsy Behav* **10**, 255–262.

Hempel, A. M., Frost, M. D., Ritter, F. J., Farnham, S. (1995). Factors influencing the incidence of ADHD in pediatric epilepsy patients. *Epilepsia* **36**(Suppl 4), 122. (Abstract)

Hermann, B. P., Whitman, S., Dell, J. (1989). Correlates of behavior problems and social competence in children with epilepsy, aged 6–11. In *Childhood Epilepsies: Neuropsychological, Psychosocial and Intervention Aspects* (B. Hermann and M. Seidenberg, eds), pp. 143–157. New York: John Wiley and Sons.

Hernández, A. J. C., Barragán, P. E. J. (2005). Efficacy of atomoxetine treatment in children with ADHD and epilepsy. *Epilepsia* **46**(Suppl 6), 241.

Hernandez, M.-T., Sauerwein, H. C., Jambaqué, I., *et al.* (2003). Attention, memory, and behavioral adjustment in children with frontal lobe epilepsy. *Epilepsy Behav* **4**, 522–536.

Hesdorffer, D. C., Ludvigsson, P., Olafsson, E., *et al.* (2004). ADHD as a risk factor for incident unprovoked seizures and epilepsy in children. *Arch Gen Psychiatry* **61**, 731–736.

Hoare, P., Mann, H. (1994). Self-esteem and behavioural adjustment in children with epilepsy and children with diabetes mellitus. *J Psychosom Res* **38**, 859–869.

Høie, B., Mykletun, A., Waaler, P. E., Skeidsvoll, H., Sommerfelt, K. (2006). Executive functions and seizure-related factors in children with epilepsy in western Norway. *Dev Med Child Neurol* **48**, 519–525.

Holdsworth, L., Whitmore, K. (1974). A study of children with epilepsy attending ordinary schools. 1: their seizure patterns, progress and behaviour in school. *Dev Med Child Neurol* **16**, 746–758.

Keene, D. L., Manion, I., Whiting, S. *et al.* (2005). A survey of behavior problems in children with epilepsy. *Epilepsy Behav* **6**, 581–586.

Kwan, P., Brodie, M. J. (2001). Neuropsychological effects of epilepsy and antiepileptic drugs. *Lancet* **357**, 216–222.

Loring, D. W., Meador, K. J. (2004). Cognitive side effects of antiepileptic drugs in children. *Neurology* **62**, 872–877.

McCusker, C. G., Kennedy, P. J., Anderson, J., *et al.* (2002). Adjustment in children with intractable epilepsy: importance of seizure duration and family factors. *Dev Med Child Neurol* **44**, 681–687.

McDermott, S., Mani, A., Krishnaswami, S. (1995). A population-based analysis of specific behavior problems associated with childhood seizures. *J Epilepsy* **8**, 110–118.

Noeker, M., Haverkamp, F. (2003). Neuropsychological deficiencies as a mediator between CNS dysfunction and inattentive behavior in childhood epilepsy. *Dev Med Child Neurol* **45**, 717–718.

Oostrom, K. J., Schouten, A., Kruitwagen, C. L. J. J., *et al.* (2002). Attention deficits are not characteristic of school children with newly diagnosed epilepsy. *Epilepsia* **43**, 301–310.

Ounsted, C. (1955). The hyperkinetic syndrome in epileptic children. *Lancet* **2**, 303–311.

Piccirilli, M., D'Alessandro, P., Sciarma, T. *et al.* (1994). Attention problems in epilepsy: possible significance of the epileptic focus. *Epilepsia* **35**, 1091–1096.

Pliszka, S. R., Crismon, M. L., Hughes, C. W., *et al.* (2006). The Texas children's medication algorithm project: revision of the algorithm for pharmacotherapy of attention-deficit/hyperactivity disorder. *J Am Acad Child Adolesc Psychiatry* **45**, 642–657.

Pressler, R. M., Robinson, R. O., Wilson, G. A., Binnie, C. D. (2005). Treatment of interictal epileptiform discharges can improve behavior in children with behavioral problems and epilepsy. *J Pediatr* **146**, 112–117.

Rodenburg, R., Stams, G. J., Meijer, A. M., *et al.* (2005). Psychopathology in children with epilepsy: a meta-analysis. *J Pediatr Psychol* **30**, 453–468.

Salpekar, J. A., Cushner-Weinstein, S., Conry, J. A., *et al.* (2005). Attention deficit hyperactivity disorder symptoms in pediatric epilepsy patients (abstract). *Epilepsia* **46**(Suppl. 8), 80.

Sánchez-Carpintero, R., Neville, B. G. R. (2003). Attentional ability in children with epilepsy. *Epilepsia* **44**, 1340–1349.

Semrud-Clikeman, M., Wical, B. (1999). Components of attention in children with complex partial seizures with and without ADHD. *Epilepsia* **40**, 211–215.

Sherman, E. M. S., Armitage, L. L., Connolly, M. B., Wambera, K. M., Esther, S. (2000). Behaviors symptomatic of ADHD in pediatric epilepsy: relationship to frontal lobe epileptiform abnormalities and other neurological predictors (abstract). *Epilepsia* **41**(Suppl.7), 191.

Steffenburg, S., Gillberg, C., Steffenburg, U. (1996). Psychiatric disorder in children and adolescents with mental retardation and active epilepsy. *Arch Neurol* **23**, 904–912.

Stores, G., Hart, J., Piran, N. (1978). Inattentiveness in school children with epilepsy. *Epilepsia* **19**, 169–175.

Sturniolo, M. G., Galletti, F. (1994). Idiopathic epilepsy and school achievement. *Arch Dis Child* **70**, 424–428.

Thome-Souza, S., Kuczynski, E., Assumpção, F., *et al.* (2004). Which factors may play a pivotal role on determining the type of psychiatric disorder in children and adolescents with epilepsy? *Epilepsy Behav* **5**, 988–994.

Wernicke, J. F., Chilcott, K. E., Jin, L., *et al.* (2006). Seizure risk in patients with ADHD treated with atomoxetine (abstract). *Neuropediatrics* **37**(Suppl 1), S165.

Williams, J., Lange, B., Phillips, T., *et al.* (2002a). The course of inattentive and hyperactive–impulsive symptoms in children with new onset seizures. *Epilepsy Behav* **3**, 517–521.

Williams, J., Sharp, G. B., DelosReyes, E., *et al.* (2002b). Symptom differences in children with absence seizures versus inattention. *Epilepsy Behav* **3**, 245–248.

CHAPTER 11

Are Psychogenic Non-epileptic Seizures an Expression of "Neurologic" Pathology?

Markus Reuber

INTRODUCTION

Seventy years ago C. G. Jung felt there was "little doubt nowadays about the psychogenesis of hysteria and other neuroses, although 30 years ago some brain enthusiasts still vaguely suspected that at bottom there was something organically wrong in the neuroses" (Jung, 1939). Apparently, Jung's statement was premature. Far from settled, the issue of "organicity" continues to agitate clinicians and researchers dealing with "hysteria" or its modern day equivalents. Only a decade ago, Devinsky lamented the "quagmire," which continues to surround the diagnosis of psychogenic non-epileptic seizures (PNES) (Devinsky, 1998). Any exploration of the nosology and etiology of "psychogenic," "medically unexplained," "conversion," "dissociative," "somatoform," "somatized" or "functional" symptoms including PNES continues to face the risk of getting bogged down in an area in which strongly held expert opinion is more easy to come by than evidence from well-conducted studies. This state of affairs is no reflection of the lack of ambition or scientific rigor of the clinical scientists active in this field. It is better explained by the genuine complexity of the field. In this chapter, I will attempt to contain the risks associated with a discussion of this topic by setting out clear definitions of the terms I intend to use and by stating the aims and limitations of my approach. The section "DEFINITIONS" will be devoted to this. The section "'NEUROLOGIC' FACTORS" will examine to what extent individual "neurologic," "physical" or "organic" factors explain PNES. This will be followed by sections on "PSYCHOGENIC FACTORS" and "CONSTITUTIONAL AND DEVELOPMENTAL FACTORS". In the section "A MULTIFACTORIAL ETIOLOGIC MODEL FOR PNES" I will draw conclusions from the preceding three sections, propose an integrative scheme accommodating "neurologic," "psychogenic" and constitutional etiologic factors, and state my view on this particular controversy.

DEFINITIONS

Psychogenic non-epileptic seizures

In a conventional definition PNES have been characterized as "episodes of altered movement, emotion, sensation, or experience, similar to those seen in epilepsy, but which have purely emotional causes" (Lesser, 1996a). For the purpose of this discussion, I will need to clarify a few points. Firstly, I will need to change the last subclause to "…, but which are not caused by epileptic activity"; otherwise this chapter would make little sense. Secondly, I should point out that I will take a very narrow view of the association of attack manifestations and epileptic discharges. This means that I will not consider subjective symptoms or behavioral changes as fully "explained" by ictal epileptic activity in the brain if they were not synchronous and the ictal changes did not affect an appropriate part of the brain. I will therefore discuss postictal symptoms which are not associated with ictal electroencephalographic changes as (at least potentially) "psychogenic". Thirdly, I would like to address the issue of the "similarity" of epileptic seizures and PNES. There is a wealth of evidence from video-electroencephalogram (video-EEG) studies that PNES actually look quite different from epileptic seizures (Reuber and Elger, 2003). What is more, panic attacks or visual flashbacks of disturbing memories are rarely classified as PNES although focal epileptic seizures could produce very similar symptoms. In reality, the attacks I am writing about here are not so much defined by the fact that they look or feel "similar" to manifestations of epilepsy, but that patients, patients' peers and doctors mistake the experiences as epileptic or struggle to distinguish them from epileptic seizures. This does not mean that it is impossible to define PNES in the prototypic, polythetic way we are familiar with from the definition of mental disorders (criteria A, B, C…) or medical conditions (such as the diagnostic criteria for migraine). It does, however, mean that the definition of PNES has to incorporate the domain of patients', peers' and doctors' illness perceptions. Although it is recognized that illness perceptions or representations, including the name of the condition (Peters et al., 1998), its cause (Scambler, 1994), and to what extent the patient is responsible for it (Finerman and Bennet, 1995) are relevant to the acceptance of treatment and clinical outcomes (e.g., (Horne, 1999)), their inclusion in the definition of a condition is unusual – and the inclusion of peers' or doctors' (mis-) perceptions is unique. For this reason, it is not surprising that a definition incorporating illness perceptions does not place PNES in a single diagnostic category in the current diagnostic manuals (International Classification of Diseases-tenth edition (ICD-10), Diagnosis and Statistical Manual-fourth edition (DSM-IV)).

I should point out that the consideration of illness perceptions in the definition of PNES means that, in explicit contrast to Charcot's statement that hysteria is the same condition, everywhere and always (Charcot, 1889), we need to restrict the use of the term PNES to the "here" (in developed countries) and "now" (in the last century or so).

Having been so particular about what I mean by PNES, I am rather less concerned with the label for the phenomenon. An incomplete list of labels used over the last 150 years includes pseudoseizures, non-epileptic attack disorder, non-epileptic events; and dissociative, conversion, psychogenic, functional and hysterical seizures. For the purpose of this discussion, I think that the meaning of these terms is sufficiently similar for them to be regarded as synonyms.

"Neurologic"

My difficulties with the distinction of "neurologic" vs. "psychogenic" or "constitutional and developmental" factors will become apparent in the following sections. However, for the purpose of this discussion I will operationally define states as "neurologic" if they are primarily related to medically explained structural changes, functional states or conditions. This definition is intended to embrace certain types of neuropathology (e.g., hippocampal sclerosis), medically defined disorders (e.g., types of epilepsy or parasomnia), abnormal physiological states (e.g., biochemical or blood gas changes, endocrine disturbances) and medically explicable states of episodic normal brain functioning (e.g., startle or sleep). It is implicit in this use of the term "neurologic" that there should not be a significant volitional component in any seizure manifestations, but that the subjective or observable seizure phenomenology should be explainable as a direct chemical, physiological or mechanical consequence of the "neurologic" state or condition.

"Psychogenic"

In contrast, I will use the term "psychogenic" to signify that the seizure manifestations are primarily unrelated to medically explained structural changes, functional states or conditions. Instead "psychogenic" factors are considered as affecting the brain through a reaction to mental distress or as a learned response. I will also discuss illness representations under this heading. A prototypic (but incomplete) list of examples of "psychogenic" factors would include remote or recent traumatic experiences, a challenging environment in adulthood, and inextricable conflicts and beliefs about illness and emotions. Reactive distress could also be triggered by medically explained "neurologic" states or conditions (such as simple partial seizures or presyncopal states). "Psychogenic" factors could include social or financial rewards for illness behavior. However, in line with the current diagnostic manuals

(ICD-10, DSM-IV), I would make a distinction between situations in which such factors act "subconsciously" and deliberate acts (such as those seen in factitious disorder or malingering).

Constitutional or developmental

I will include factors under this heading which are inherited, acquired in childhood or result from complex gene–environment interaction in childhood and early adolescence. By the time individuals reach adulthood, these factors have become relatively stable and remain so throughout adult life. I appreciate that it is difficult to make a clear distinction between certain personality traits (which I would place in this category) and traumatic aspects of childhood experience which may be intimately linked with their development.

This discussion

Although I will have to refer to much of the literature on different aspects of the etiology of PNES, this discussion focuses on the question of "organicity". I will not address the questions whether PNES are best conceptualized as a dissociative or somatoform disorder, or indeed, whether PNES warrant the status of a separate axis-I disorder (as the ICD-10 and the DSM-IV suggest), or are merely a manifestation of other disorders. I will also not refer to the literature on PNES in children (in whom etiological formulations are likely to be different), because I have no clinical experience with this patient group.

"NEUROLOGIC" FACTORS

Epilepsy

Discussing epileptic discharges as a possible etiologic factor in the generation of PNES touches directly on the question whether PNES actually exist. Given that I have defined PNES as attacks in which subjective and behavioral manifestations are not related to epileptic discharges, a link with undetected epileptic activity would degrade PNES to misdiagnosed epileptic seizures. However, epilepsy plays a prominent role in the PNES literature and cannot be ignored here.

Whilst the percentage of patients with concurrent epilepsy has varied from 3.6% to 58% in different PNES series (Reuber *et al.*, 2003a), it is generally accepted that the prevalence of epilepsy is greater amongst patients with PNES than in the general population (Lesser *et al.*, 1983). In our own series, the clinical history suggested additional epilepsy in 119 of 329 PNES patients (36.2%). Of these patients,

90 (27.4%) also had interictal EEG changes. If we required the documentation of ictal EEG changes for the diagnosis of additional epilepsy, 58 (17.6%) would be classed as having PNES with concurrent epilepsy (Reuber et al., 2003a). Several authors have noted that PNES are (almost) always preceded by epilepsy (Rabe, 1970; Devinsky et al., 1996; Henry and Drury, 1997; Glosser et al., 1999; Devinsky et al., 2001). Occasionally, PNES stop after epileptic seizures have been fully controlled (Reuber et al., 2002c). It is therefore perhaps not surprising that epilepsy is the most commonly suspected neurologic explanation of PNES in clinical practice. In many cases this suspicion is simply a matter of misdiagnosis. Most studies suggest that physicians with less experience tend to overdiagnose epilepsy (Smith et al., 1999; Zaidi et al., 2000). However, the difficulties with the distinction of epileptic seizures and PNES are not fully explained by lack of diagnostic experience. There are some genuinely gray areas. Invasive EEG-recordings have demonstrated that ictal behavior which would be considered typical of a non-epileptic seizure can follow on from ictal EEG changes likely to be causing a simple partial seizure (Wilner and Bream, 1993; Devinsky and Gordon, 1998; Papacostas et al., 2006). Given that only 10–20% of simple partial seizures produce identifiable changes in scalp EEG-recordings (Bare et al., 1994; Devinsky et al., 1988), this observation raises the possibility that a non-epileptic "elaboration" of epileptic symptoms is commoner than the small number of well-documented cases suggests. Even complex partial seizures may not be associated with scalp EEG changes. Given that the semiology of these seizures may include emotionally charged screams, bilateral motor activity with retained consciousness, ictal speech arrest with unimpaired postictal recollection or ictal responsiveness with postictal amnesia (Ebner et al., 1995; Salanova et al., 1995), the distinction of PNES from these seizure types poses a particular challenge. To make matters worse, it has been reported that such epileptic seizures may start after traumatic life events ("pathoplastic" factors) (Greig and Betts, 1992) and emotional stress is most commonly cited as a seizure trigger by patients with epilepsy (Frucht et al., 2000; Spector et al., 2000). What is more, addressing psychiatric symptoms may not only improve these patients' quality of life but also their seizures (Spector et al., 1999; Ramaratnam et al., 2005).

Does this mean that PNES are a manifestation of undetected epileptic activity in the brain? I do not think so. In the large majority of cases, the symptomatology of PNES is very different from that in any known epileptic seizure type. Although practically all individual aspects of PNES semiology may occur in different epileptic seizures, the combination of symptoms and behaviors seen in PNES could not be explained by epileptic activation or disinhibition of centers or neural networks in the brain. What is more, a number of particular observations are made much more frequently in PNES than in epilepsy: suggestive provocation triggers seizures in the majority of patients with PNES but only in very few patients with epilepsy. The methods which have been used to induce seizures not only include the administration of "placebo" (Cohen and Suter, 1982; Walczak et al., 1994) or physiologically challenging stimuli such as hyperventilation, photic stimulation or tilt table testing (Zaidi *et al.*, 1999; McGonigal *et al.*, 2002), but also a psychiatric interview or hypnosis (Barry *et al.*, 2000; Cohen *et al.*, 1992; Barry *et al.*, 2000; Martinez-Taboas, 2002). PNES may also be terminated by verbal interaction with the patient (Kütemeyer *et al.*, 2005), and patients with longstanding PNES disorders can become seizure when the diagnosis is explained to them (Aboukasm *et al.*, 1998; Betts and Boden, 1992).

Does the increased prevalence of epilepsy in patients with PNES mean that epileptic discharges are a "neurologic" risk factor for PNES? Not necessarily. The likely psychodynamic impact of recurrent epileptic seizures is quite apparent in many cases (Rabe, 1970). This means that one does not need to assume direct "neurologic" links to understand the etiological links between epilepsy and non-epileptic seizures.

Neuropsychological deficits

At first sight, the large number of studies describing neuropsychological performance deficits in patients with PNES supports the idea of PNES as a neurologic disorder. In our own study, 60.6% of all patients with PNES (and no additional epilepsy) performed at least 1.5 standard deviations below norm populations in at least one domain of testing (Reuber *et al.*, 2002b). In another study, 63% of patients had a Halstead Impairment Index in the impaired range (Kalogjera-Sackellares and Sackellares, 1999). Many studies comparing performance in patients with PNES with those with epilepsy (i.e., a neurologic brain disorder) found no differences between the two groups when corrections were made for education and age (Hermann, 1993; Wilkus and Dodrill, 1989; Binder *et al.*, 1998; Risse *et al.*, 2000; Swanson *et al.*, 2000). In some studies PNES patients actually did worse than patients with epilepsy (Binder *et al.*, 1994). Although most failed to reveal a characteristic pattern of deficits in patients with PNES (Dodrill and Holmes, 2000), one study comparing PNES patients and healthy controls found motor deficits as an indication of frontal lobe dysfunction (Kalogjera-Sackellares and Sackellares, 2001).

Of course, poor performance in neuropsychological tests cannot prove "neurologic" brain dysfunction. Psychiatric disorders such as depression and posttraumatic stress disorder (PTSD) are also associated with performance deficits (Swanson *et al.*, 2000). What is more, two studies revealed that a qualitative analysis of test results can yield important additional findings in patients with PNES. One study showed that PNES patients had a characteristic tendency to false negative answers (they failed to recognize words which they had seen several times but rarely made false positive identification errors) (Bortz *et al.*, 1995). In the other study, patients were often able to recall words from a list on delayed testing which they had initially failed to remember (Brown *et al.*, 1991). Two studies showed that symptom validity test results (Portland Digit Recognition Test, Word Memory Test) are abnormal in a higher proportion of patients with PNES than with epilepsy, suggesting problems with sustaining effort during neuropsychological tests (Binder *et al.*, 1998; Drane *et al.*, 2006). In fact, one of these studies showed that PNES patients do better in neuropsychological performance tests than patients with epilepsy if effort test results are taken into account (Drane *et al.*, 2006).

Other evidence of brain abnormalities

Many patients with PNES have neuroimaging or electroencephalographic abnormalities. In line with an older study (Lelliott and Fenwick, 1991), we found magnetic resonance imaging (MRI) abnormalities in 27.0% and epileptiform EEG changes in 8.7% of 206 consecutive patients with PNES and no clinical evidence of additional epilepsy (Reuber *et al.*, 2002b). Another study reported MRI abnormalities in as many as 76% and epileptiform EEG changes in 78% of "conversion non-epileptic seizures" (Devinsky *et al.*, 2001). MRI abnormalities were found to be associated with a worse prognosis (Kanner *et al.*, 1999). We found non-specific EEG changes in 18% of patients with PNES (and no epilepsy or other identifiable reason to explain the EEG abnormalities) but only 10% of age-matched healthy controls (Reuber *et al.*, 2002a).

A link between markers of brain abnormalities and PNES would be more likely if a particular pattern of abnormality had emerged in these studies. Indeed, one study suggested that changes were seen more frequently in the right hemisphere (Devinsky *et al.*, 2001). However, we could not confirm this finding in our own studies (Reuber *et al.*, 2002b).

Head injury

Head injuries prior to PNES onset have been reported in 24–65% of patients (Barry *et al.*, 1998; Westbrook *et al.*, 1998; Kalogjera-Sackellares and Sackellares, 2001). It should be noted that all of these studies relied on self-report, although it has been suggested that self-reports are particularly unreliable in this patient group (Schrag *et al.*, 2004). What is more, although PNES can also occur after more severe traumatic brain injuries (Hudak *et al.*, 2004), the head injuries reported by most patients were mild.

In any case, PNES would not have to be a consequence of "neurologic" injury even if patients also performed poorly in neuropsychological tests (Kalogjera-Sackellares and Sackellares, 2001). For instance, Kalogjera-Sackellares (2004) has proposed a psychodynamic formulation which attempts to explain how a relatively minor head injury could lead to chronically disabling PNES. Some reports of head injuries may also have been motivated by other factors: in one series almost all of the 32% of patients who described a head injury before the onset of PNES had gained financially from the injury (Westbrook *et al.*, 1998).

Brain surgery

A number of case series suggest that PNES can also develop after more major brain trauma, for instance, that associated with epilepsy surgery (Krahn et al., 1995; Ney et al., 1998; Glosser et al., 1999; Davies et al., 2000). The risk of PNES after epilepsy surgery ranges from 3.5% to 8.8% in these studies. We reported a series of 17 patients who developed PNES after brain surgery for indications other than epilepsy. They represented 4.6% of all patients with PNES we had seen at the time (Reuber et al., 2002d). Whilst most authors did not identify any neurologic risk factors for postoperative PNES, one study found an increased frequency of low IQ (Ney et al., 1998), another of right-sided procedures (Glosser et al., 1999).

Of course the manifestation of PNES after brain surgery does not prove the link to a "neurologic" cause. Almost all studies noted high rates of preoperative psychopathology. Two studies identified high rates of postoperative complications such as wound infections, which were probably more likely to provoke reactive ("psychogenic") symptoms than cause "neurologic" damage (Ney et al., 1998; Reuber et al., 2002d).

Startle

Blumer suggested that PNES represent a neurally preformed startle or surprise response to a range of possible triggers (Blumer, 2000; Blumer and Adamolekun, 2006). This idea is based on similarities between PNES in humans and startle phenomena in animal models (Matsumoto and Hallet, 1994; Lang et al., 2000). A startle response is usually defined as a sudden, brief, involuntary movement in response to surprise, fright or alarm. Startle responses can be physiologic (as in the Moro Reflex) or exaggerated (as in Hyperekplexia). The startle response is coordinated in the reticular formation on a brainstem level but this area receives inputs from a range of higher centers including the amygdala. This may explain why the startle response in humans (usually measured as an eye blink response to auditory stimuli) is increased when the proband is exposed to fear-generating imagery and reduced by equally arousing pleasant pictures (Hamm et al., 1997). Variations of the startle response can even be seen if pictures are presented so briefly that they are not consciously seen (Ruiz-Padial and Vila, 2007).

Whilst startle could therefore explain some of the phenomenology encountered in patients with PNES, including premonitory symptoms of fear (Vein and Vorob'eva, 1998; Goldstein and Mellers, 2006), there are many important aspects of the semiology and course of PNES which startle cannot explain. To begin with, no other human startle response involves loss of consciousness. In fact, there is a linear correlation between the degree of (very conscious) unpleasantness with which the stimulus is perceived and the reflex startle response (Hamm et al., 1991). This means that, at best, startle could be involved in the start of a seizure, but would not explain one core element of PNES – the dissociation or apparent loss of consciousness. In fact, startle may also not be such a good explanation for the onset of PNES. Startle responses develop and increase as patients are exposed to noxious stimuli. Although the majority of patients with PNES has experienced exceptional trauma and an important minority has been exposed to recurrent trauma or abuse (Reuber *et al.*, 2007), there is usually a significant delay (of hours to several years) between traumatic experience and seizure manifestation. What is more, I would say that it is characteristic that patients report having at least some PNES when they feel relaxed. In fact, any exaggeration of the startle response in patients with PNES may well be an effect of having frightening seizures rather than the cause.

Sleep disorder

Although nearly two thirds of patients with PNES report seizures from sleep (Duncan *et al.*, 2004), closer examination of night-time attacks usually reveals that they start after the patient has woken up (they occur from "pseudosleep") (Thacker *et al.*, 1993; Benbadis *et al.*, 1996). However, PNES arising directly from EEG-documented stage II or III sleep have also been reported in a small number of patients (Orbach *et al.*, 2003). Orbach *et al.* argue that the diagnosis of PNES was correct because the semiology of the observed attacks was not typical of epilepsy (including frontal lobe epilepsy) and that a diagnosis of parasomnia was inappropriate because all patients had similar attacks from a waking state. However, it should perhaps be pointed out that disorders of sleep control can be associated with complex symptoms during the waking state. In narcolepsy for instance, patients may experience daytime sleep attacks, hypnagogic hallucinations and cataplexy. Although the condition can be "medically explained" (Dauvilliers *et al.*, 2007), the onset is often preceded by traumatic life events (Orellana *et al.*, 1994). What is more, cataplexy often occurs in situations which are emotionally charged.

Does this mean that PNES are likely to be a disorder of sleep control? I do not think so. All sleep disorders are characterized by the fact that most of the associated symptoms are related to sleep. Many patients with PNES never have seizures from sleep and the abnormalities of sleep architecture which have been reported in this patient group are quite subtle (Bazil *et al.*, 2003). What is more, sleep regulation and dream content are affected by mental activity in the waking state. Sleep disorders are therefore not clearly "neurologic" in the sense discussed above.

Drug effects in the central nervous system (CNS)

A number of authors have reported that PNES can arise as patients wake up from a general anesthetic (Parry and Hirsch, 1992; Reuber *et al.*, 2000; Lichter *et al.*,

2004). Most of these patients had a previous history of PNES, some had several episodes of PNES after general anesthesia. PNES have also been described in the context of more chronic antiepileptic drug (AED) toxicity (Trimble, 1982). This observation has not been associated with a particular class of AED. The unwanted CNS effects of AED therapy may be exacerbated by painkillers and other drugs with CNS effects often taken by patients with PNES (Hantke *et al.*, 2007). It has been shown that stopping AEDs is not only safe in patients with proven PNES (Oto *et al.*, 2005b), but can also improve attacks, even without additional psychological treatment (Blumer and Adamolekun, 2006).

Of course, an immediate postoperative or chronically toxic state is not only characterized by pharmacologically explicable drug effects. It is easy to imagine that the process of regaining consciousness after surgery could leave a patient in a very vulnerable state with limited access to their normal coping resources. The use of antiepileptic drugs in the outpatient setting not only represents a pharmacologic intervention but also communicates a particular understanding of the problem and raises expectations. Stopping AEDs is an important part of starting a psychotherapeutic process for PNES (Reuber *et al.*, 2005).

"PSYCHOGENIC" FACTORS

Childhood trauma

The original psychoanalytic concepts of "dissociation" and "conversion" were closely related to the idea of the pathogenic role of traumatic childhood experience and its processing (Freud and Breuer, 1892). Sexual abuse in particular was considered etiologically relevant by the younger Freud (Freud, 1896) and many other authors since. More recently the significance of other childhood traumas such as physical abuse and loss (Eisendrath and Valan, 1994) or psychological abuse (Salmon *et al.*, 2003) has also been stressed. In larger patient series, between 32.4% and 88.0% of patients with PNES but only 8.6–37% of patients with epilepsy gave a history of sexual or physical childhood abuse (Betts and Boden, 1992; Alper *et al.*, 1993; Bowman, 1993; Snyder *et al.*, 1994; Arnold and Privitera, 1996; Berkhoff *et al.*, 1998; Rosenberg *et al.*, 2000). In the biggest study, significant differences between the two patient groups were found both for sexual (24.0% vs. 7.1%) and physical abuse (15.5% vs. 2.9%) (Alper *et al.*, 1993). However, other studies showed smaller differences which were not significant for sexual abuse alone (Arnold and Privitera, 1996; Berkhoff *et al.*, 1998). A review and meta-analysis of 34 comparative studies confirmed the increased prevalence of child sexual abuse in patients with PNES compared to patients with epilepsy, but it found variable effect sizes and criticized the methodology employed in many of the original studies (Sharpe and Faye, 2006).

Although there are many studies suggesting an etiologic link between child-hood abuse and of psychopathology in adulthood (Sharpe and Faye, 2006), child-hood abuse is only identifiable in a minority of patients with PNES. What is more, only one quarter of children exposed to severe trauma in childhood develop psy-chopathology later (Paris, 1998). Indeed, it has been suggested that dissociative tendencies may encourage self-reported traumatic experience rather than vice versa (Merckelbach and Muris, 2001). Twin studies support the idea that consti-tutional mechanisms affect perception, processing and recall of traumatic child-hood experience. In one study more than 30% of the variance of self-reported traumatic experiences was explained by inherited factors (Thapar and McGuffin, 1996). In any case, the association of childhood abuse or neglect with PNES would not necessarily prove a "psychogenic" etiology. There are not only psycho-analytical models for a link between childhood trauma and PNES in later life (Bjornaes, 1993). Akin to psychobiological models developed for personality dis-orders (Meares *et al.*, 1999), adverse childhood experience could modulate brain development in early life in a way that would reduce resilience to stressors in later life and make symptoms such as PNES more likely.

Life events

PNES commonly seem to start in the context of adverse life events or irresolv-able dilemmas (Griffith *et al.,* 1998; Bowman, 2000). The list of reported precip-itating factors includes rape (Cartmill and Betts, 1992; Bowman, 1993; Reuber *et al.,* 2007), physical abuse (Bowman and Markand, 1996), surgical procedures and childbirth (Miller, 1983; Ward *et al.,* 1988; Parry and Hirsch, 1992; Reuber *et al.,* 2000), death of or separation from family members or friends (Gardner and Goldberg, 1982), job loss (Bowman and Markand, 1999), road traffic or other accidents (Bowman and Markand, 1999), earthquakes (Watson *et al.,* 2002), rela-tionship discord or family break-up (Griffith *et al.,* 1998; Bowman and Markand, 1999; Reuber *et al.,* 2007), legal disputes (Guberman, 1982) or physical health issues (Reuber *et al.,* 2007). In our own study describing potential etiologic fac-tors we identified at least one predisposing or precipitating factor of this nature in all but 3.4% of patients (Reuber *et al.,* 2007). Another study found potentially stressful "life events" in 91% of patients (Bowman and Markand, 1999). Two stud-ies comparing recent negative life events in patients with PNES and patients with epilepsy found that such events were reported more commonly by patients with PNES (Tojek *et al.,* 2000; Binzer *et al.,* 2004). Reported recent negative life events were also more frequent in patients with PNES than patients with motor conver-sion symptoms (Stone *et al.,* 2004b). The relevance of traumatic life events is fur-ther supported by the common combination of PNES with other symptoms of PTSD (Fleisher *et al.,* 2002; Fiszman *et al.,* 2004).

Although these studies suggest that adverse life events are an important etiologic factor, it is often far from clear how much of the patient's symptomatology they can explain. Whilst the differences in the frequency of life events reported by patients with PNES and epilepsy were statistically significant, they were small. What is more, the authors of one study found that, in 76% of the patients, the traumatic significance of these events only became clear if the patient's current circumstances or previous experiences were taken into account (Bowman and Markand, 1999). In some of these cases the "life events" were of "symbolic" character, but seemed to reactivate mental distress attributed to childhood abuse (Bowman, 1993; Griffith et al., 1998). This suggests that patients' coping resources and strategies may be relevant. This view is supported by a study comparing patients with PNES, epilepsy and healthy controls. Whilst it found similar other-perceived levels of stress in families of patients with PNES and patients with epilepsy, PNES patients themselves experienced their lives as more stressful, were significantly more likely to use escape-avoidant coping strategy and less likely to use more effective "planful" problem solving approaches than patients with epilepsy or healthy controls (Frances et al., 1999).

Illness representations

Illness representations are an essential part of the self-regulation model which proposes that behavior in relation to illness depends on a person's own perception or representation of the illness. Illness representations consist of five elements: identity (symptoms or label), cause, consequences (effects on life or life-style), time-line (time to develop and duration) and controllability or cure (Leventhal et al., 1992). A recent review concluded that there was a clear relationship between illness representations, coping behavior and outcomes (Hagger and Orbell, 2003), and one study demonstrated that (with some minor modification) the self-regulation model was applicable to how patients think about PNES (Green et al., 2004). In fact, in conditions such as PNES, illness representations may actually act as an etiologic factor. One study found that, although patients with PNES had experienced more negative life events, they were less willing than patients with epilepsy to accept that these events could be linked to their seizures (Stone et al., 2004a). It is possible that the tendency not to link life events with symptoms is related to the difficulties many PNES patients have with perceiving and understanding emotions and their physical manifestations. One study demonstrated that 90.5% scored in the "alexithymic range" (Bewley et al., 2005). Another study assessing illness representations and coping styles in PNES patients and healthy controls found that PNES patients experienced their health as more strongly controlled by external factors than the healthy control group and had a greater tendency to use escape-avoidance coping strategies (Goldstein et al., 2000).

Unfortunately, it is not clear whether the patient's illness representations were the same before they developed their seizures as after. It is therefore uncertain whether illness cognitions can be considered a predisposing factor or whether they merely play a perpetuating role. It is also unclear to what extent illness representations are determined by comorbid conditions, especially by depression.

Home environment

Several studies have shown that many PNES patients are exposed to irresolvable conflicts or dilemmas in their home environment that are often (but by no means exclusively) related to sexual or physical abuse (Berkhoff *et al.*, 1998; Griffith *et al.*, 1998; Krawetz *et al.*, 2001; Salmon *et al.*, 2003). Two comparative studies used different measures to demonstrate that patients with PNES experienced their families as less supportive, communicative or helpful than patients with epilepsy (Moore *et al.*, 1994; Krawetz *et al.*, 2001). Another study showed that families of PNES patients had a greater tendency to complain of somatic symptoms and were more likely to criticize each other (Wood *et al.*, 1998). These studies suggest that many patients with PNES live in a home environment in which problems are often not "resolved" through successful communication but by strategies such as somatization.

Of course, the home environment may not only contribute to the etiology of PNES by being a source of stress and trauma. It has been argued that model learning is an important factor in the development of PNES (Sirven and Glosser, 1998). In support of this idea, several authors have found that patients with PNES report a family history of epilepsy more commonly than patients with epilepsy (Stewart *et al.*, 1982; Griffith *et al.*, 1998; Tojek *et al.*, 2000).

Other psychiatric disorders

Most patients with PNES have symptoms of additional axis-I psychiatric disorders. The commonest diagnoses are other somatoform (22–89%), dissociative (22–91%), depressive (57–85%), posttraumatic (35–49%) or anxiety disorders (11–50%) (Griffith, 1990; Bowman, 1993; Bowman and Markand, 1996; Kanner *et al.*, 1999; Krishnamoorthy *et al.*, 2001; Mökleby *et al.*, 2002; Galimberti *et al.*, 2003; Kuyk *et al.*, 2003). Several comparative studies have demonstrated that the prevalence of these psychiatric disorders is higher in patients with PNES than patients with epilepsy (Stewart *et al.*, 1982; Kristensen and Alving, 1992; Roy and Barris, 1993). Quantitative instruments such as measures of general psychopathology (Prueter *et al.*, 2002; Reuber *et al.*, 2003a), anxiety (Donofrio *et al.*, 2000; Owczarek, 2003a), dissociative tendencies (Alper *et al.*, 1997; Bowman and Coons, 2000; Reuber *et al.*, 2003a), and depression (Roy and Barris, 1993; Donofrio *et al.*, 2000) yield higher scores in PNES than epilepsy patient groups or healthy controls (Mökleby *et al.*, 2002).

The strongest argument for the inclusion of other psychiatric disorders in the list of etiologic factors is that in some patients with anxiety, depressive or psychotic disorders, PNES seem to be intimately linked to symptoms such as panic, guilt or hallucinations. Another reason is that the prognosis of PNES is associated with the psychopathological context in which they occur (Bowman, 2001). However, clinical experience suggests that the link between PNES and other psychiatric disorders is not quite as close in some patients, and there is an important subgroup in which even psychiatrists familiar with patients who are reluctant to volunteer emotional symptoms are unable to find any evidence of psychopathology (Bowman and Markand, 1996; Galimberti et al., 2003). In any case, the etiology of other "functional" psychiatric disorders is also controversial, so it is not clear that they should really be discussed under the heading of "psychogenic" factors.

Social/financial illness gains

Although I consider PNES an involuntary expression of distress in the overwhelming majority of patients, it is possible – especially in chronic PNES disorders – that financial and social gains play a perpetuating role. Some patients seem to adopt a "sick role" within their family (Ford, 1993b). The observation that 69% of 84 patients had a job at the time of manifestation of PNES, but only 20% were still working at the time of diagnosis by video-EEG may not only indicate that PNES are disabling but also that avoiding work and obtaining benefit payments can be a pathogenic factor (Martin et al., 2003). Our own long-term follow-up study showed that after a mean of 4.1 years from the time of diagnosis, 41.4% of patients had retired on grounds of ill health and a further 12.4% were unemployed. The mean age of these patients was 38.6 (Reuber et al., 2003c). Two studies suggested that patients with PNES are more likely to receive state benefit payments than patients with similarly severe epilepsy (Kristensen and Alving, 1992; Binder et al., 1994).

Whilst I excluded malingered or factitious seizures from my definition of psychogenic disorders, I agree with others that there is probably no categorical distinction between feigned and "involuntary" symptoms (Fink, 1992). It is conceivable that social or financial illness benefits act as subconscious rewards in some patients.

CONSTITUTIONAL AND DEVELOPMENTAL FACTORS

Female gender

Lesser looked at 21 studies describing consecutive PNES patient series and found reports of 734 women and 230 men (Lesser, 1996b). In line with the gender distribution seen in other somatoform disorders (Chodoff, 1982), women are therefore three times as likely to develop PNES as men. Several studies show that there are some important differences between male and female PNES patients. One study revealed that men tended to have more marked changes on the axes hysteria, depression, hypochondriasis, psychasthenia and schizophrenia of the Minnesota Multiphasic Personality Inventory (MMPI) than women (Holmes et al., 2001). Two studies found that PNES in women were more commonly related to sexual or physical abuse and PNES in men to employment problems (Bowman and Markand, 1999; Oto et al., 2005a).

The higher risk in women is not fully explained by higher rates of sexual or physical abuse (Alper et al., 1993). It has been suggested that PNES (and other forms of somatization) could be a learnt or culturally determined form of expression of female anger, fear or helplessness (Ford, 1993a; Rosenbaum, 2000). On the other hand, genetic mechanisms for gender-specific complex patterns of behavior have been recognized (Breedlove, 1994) and it has been shown that sex hormones have very specific effects on the cerebral processing of cognitive tasks (Fernandéz et al., 2003). It is therefore possible that the asymmetric gender distribution of PNES is in part due to "biologic" gene–environment interactions.

Personality traits

Personality traits predict a person's typical manner of thinking, feeling, perceiving and relating to others in a probabilistic fashion. Genetic influences typically account for 40–50% of the variance of measured traits (Jang et al., 1998). The influence of environmental factors on personality is almost exclusively explained by individual experience (Plomin and Daniels, 1987).

Self-report questionnaires reveal accentuated personality traits in the majority of patients (Kuyk et al., 2003; Reuber et al., 2003d; Cragar et al., 2005). 25–62.0% of patients with PNES fulfill the DSM criteria for the diagnosis of a personality disorder (Stewart et al., 1982; Pakalnis et al., 1991; Ramchandani and Schindler, 1993; Eisendrath and Valan, 1994; Bowman and Markand, 1996; Kanner et al., 1999; Galimberti et al., 2003; Binzer et al., 2004). The association of particular personality profiles with measures of seizure severity or long-term outcome supports the idea that personality traits act as an etiologic factor (Reuber et al., 2003c).

However, the studies which have focused on personality pathology have not consistently revealed one particular profile predisposing patients to PNES. In one study, the largest group of PNES patients had a borderline-like personality profile. The second largest group consisted of a relatively "normal" group that only exhibited exaggerated controlling tendencies (Reuber et al., 2003d). A second study produced similar results using a different psychometric instrument. Whilst the majority of patients were characterized by high levels of neuroticism there was also a subgroup with trait scores in the normal ranges (Cragar et al., 2005). However, in another study, cluster A personality profiles (paranoid, schizoid, schizotypal) were found most commonly (Kuyk et al., 2003). Many other studies have used the MMPI in this patient group (Dodrill and Holmes, 2000). These studies confirm that there is no uniform personality pattern that characterizes patients with PNES, although many patients show elevations of the axes hysteria, hypochondriasis and schizophrenia.

Somatization

Somatization has been defined as the tendency to experience and communicate somatic distress in response to psychosocial stress and to seek medical help for it (Lipowski, 1988). It is discussed here because it behaves like a personality trait in several important ways: it seems to be normally distributed in the population, manifests in late childhood or adolescence, remains relatively stable throughout adult life, and interacts with many aspects of life, aging or other illnesses (Bass and Murphy, 1995). PNES are often associated with other forms of somatization (Ford, 1993a; Devinsky, 1998; Ettinger et al., 1999). PNES patients have more unexplained other symptoms than patients with epilepsy (Kuyk et al., 1999; Owczarek, 2003b; Reuber et al., 2003a). The possibility of an independent etiologic link with PNES is strengthened by the observation of a correlation between PNES severity and outcome and the number of unexplained symptoms listed by patients (Reuber et al., 2003b). This correlation remained significant when measures of dissociation and general psychopathology were introduced as co-variables.

However, it is quite unclear that a "somatization tendency" truly exists as an independent constitutional or developmental factor. For instance, two thirds of the patients with somatization disorder also fulfill the diagnostic criteria for a personality disorder and many factors contribute to both disorders (for instance childhood sexual abuse) (Bass and Murphy, 1995).

Intellectual disability

Several authors have reported that PNES are particularly common in individuals with intellectual disability (Silver, 1982; Krumholz and Niedermeyer, 1983; Ramchandani and Schindler, 1993). The idea of intellectual disability (ID) as an independent etiologic factor is supported by a study which revealed significant differences between PNES patient groups with and without ID. Patients with ID had lower rates of sexual abuse and higher rates of co-existing epilepsy. Immediate eruptional or situational triggers for PNES were more commonly identified (Duncan and Oto, 2008).

A MULTIFACTORIAL ETIOLOGIC MODEL FOR PNES

Somatoform and dissociative disorders including PNES straddle the interface between neurology and psychiatry, body and mind. Our thinking about these disorders has been enriched by the contributions of many experts in the field, as the psychiatric community has geared itself up for the fifth edition of DSM (Kendell, 2001; Sharpe and Mayou, 2004; Mayou *et al.*, 2005; Kroenke and Rosmalen, 2006). This overview can only make a very small contribution to the discussion of the mind–body problem. Focusing on the current knowledge about etiologic factors, I have sought to explore whether PNES can be considered a manifestation of "neurologic" pathology. My examination leads me to draw three conclusions.

Firstly, I have not been able to identify a single factor which is necessary and sufficient to explain the manifestations of all PNES disorders. Patients with PNES could comprise a highly heterogeneous population in whom similar symptoms have very different causes. However, it is more likely that PNES do not occur without the collusion of a range of "neurologic," "psychogenic," constitutional or developmental factors.

Secondly, my limited understanding of the working of brain and the interaction of constitutional, developmental, environmental and acquired factors does not enable me to make neat distinctions between the categories "neurologic," "psychogenic" and constitutional or developmental. In fact, it is practically impossible to use the terms "neurologic" and "psychogenic" as if they represented separate etiologic domains, if the neurobiological underpinnings of CNS dysfunction are considered. It is becoming increasingly clear that the reason why early childhood experiences have behavioral effects in adult life is that the experience modulates brain development (Bremner *et al.*, 2003). Even traumatic experience in adulthood is associated with structural changes in the brain (Yamasue *et al.*, 2003). The same is true of learning (Castro-Caldas *et al.*, 1999; Draganski *et al.*, 2004; Draganski *et al.*, 2006). The situation is even more complex in variable conditions such as epilepsy in which behavioral or symptomatic manifestations are the result of stable, long-term structural changes (such as destruction of neuronal tissue), more variable structural changes (such as an increase in certain neuron populations related to epileptic seizure activity), submicroscopic structural changes (such as reversible changes to transmitter binding sites, ion channels or mitochondrial function), functional changes (such as those caused by epileptic discharges) and changes (perhaps on all of these levels) related to learning or biological responses to environmental stimuli.

Thirdly, I conclude that the etiology of PNES, like that of other functional ("medically unexplained") somatic syndromes (Wessely *et al.*, 1999; Sharpe, 2002), is best described by a multidimensional model that encompasses a range of different predisposing, precipitating, perpetuating and triggering factors (Reuber *et al.*, 2007). Predisposing factors confer an increased vulnerability to PNES, precipitating factors determine the timing of the manifestation of seizures, perpetuating factors contribute to a chronically recurrent course, and triggering factors start off individual seizures. It is a particular advantage of this model that it can be used to produce an etiology-based formulation in an individual patient which can inform treatment decisions. All of the factors discussed above, other axis-I pathology, coping resources and trigger factors can be considered in such a formulation (Reuber *et al.*, 2005). Factors can interact with each other (a history of childhood neglect may become particularly relevant after a distressing life event in adult life), and one factor can act in different ways. Additional epileptic seizures, for instance, can act as a predisposing factor by reducing a patient's confidence to be able to remain in control in a moment of particular distress, an epileptic aura can precipitate or trigger a dissociative reaction, and certain emotional reactions or environmental responses that patients experience when they have epileptic seizures could make a chronically recurrent course of PNES more likely.

In conclusion, I cannot really answer the question whether PNES are an expression of "neurologic" pathology. The current evidence suggests that PNES do not occur without "psychogenic", constitutional or developmental factors, however, "neurologic" factors can play an important etiologic role.

REFERENCES

Aboukasm, A., Mahr, G., Gabry, B. R., Thomas, A., Barkley, G. L. (1988). Retrospective analysis of the effects of psychotherapeutic interventions on outcomes of psychogenic nonepileptic seizures. *Epilesia* **39**, 470–473.

Alper, K., Devinsky, O., Perrine, K., Vazquez, B., Luciano, D. (1993). Nonepileptic seizures and childhood sexual and physical abuse. *Neurology* **43**, 1950–1953.

Alper, K., Devinsky, O., Perrine, K., Luciano, D., Vazquez, B., Pacia, S., Rhee, E. (1997). Dissociation in epilepsy and conversion nonepileptic seizures. *Epilepsia* **38**, 991–997.

Arnold, L. M., Privitera, M. D. (1996). Psychopathology and trauma in epileptic and psychogenic seizure patients. *Psychosomatics* **37**, 438–443.

Bare, M. A., Burnstine, T. H., Fisher, R. S., Lesser, R. P. (1994). Electroencephalographic changes during simple partial seizures. *Epilepsia* **35**, 715–720.

Barry, E., Krumholz, A., Bergey, G. K., Chatha, H., Alemayehu, S., Grattan, L. (1998). Nonepileptic posttraumatic seizures. *Epilepsia* **39**, 427–431.

Barry, J. J., Atzman, O., Morrell, M. J. (2000). Discriminating between epileptic and nonepileptic events: the utility of hypnotic seizure induction. *Epilepsia* **41**, 81–84.

Bass, C., Murphy, M. (1995). Somatoform and personality disorder: syndromal comorbidity and overlapping developmental pathways. *J Psychosom Res* **39**, 403–427.

Bazil, C. W., Legros, B., Kenny, E. (2003). Sleep structure in patients with psychogenic nonepileptic seizures. *Epilepsy Behav* **4**, 395–398.

Benbadis, S. R., Lancman, M. E., King, L. M., Swanson, S. J. (1996). Preictal pseudosleep: a new finding in psychogenic seizures. *Neurology* **47**, 63–67.

Berkhoff, M., Briellmann, R. S., Radanov, B. P., Donati, F., Hess, C. W. (1998). Developmental background and outcome in patients with nonepileptic versus epileptic seizures: a controlled study. *Epilepsia* **39**, 463–469.

Betts, T., Boden, S. (1992). Diagnosis, management and prognosis of a group of 128 patients with nonepileptic attack disorder. Part II. Previous childhood sexual abuse in the aetiology of these disorders. *Seizure* **1**, 27–32.

Bewley, J., Murphy, P. N., Mallows, J., Baker, G. A. (2005). Does alexithymia differentiate between patients with nonepileptic seizures, patients with epilepsy and nonpatient controls? *Epilepsy Behav* **7**, 1165–1173.

Binder, L. M., Salinsky, M. C., Smith, S. P. (1994). Psychological correlates of psychogenic seizures. *J Clin Exp Neuropsychol* **16**, 524–530.

Binder, L. M., Kindermann, S., Heaton, R., Salinsky, M. (1998). Neuropsychological impairment in patients with nonepileptic seizures. *Arch Neuropsychol* **13**, 513–522.

Binzer, M., Stone, J., Sharpe, M. (2004). Recent onset pseudoseizures – clues to aetiology. *Seizure* **13**, 146–155.

Bjornaes, H. (1993). Aetiological models as a basis for individualized treatment of pseudo-epileptic seizures. In *Pseudo-epileptic seizures* (L. Gram, S. I. Johnnessen, P. O. Osterman, M. Sillanpää, eds), pp. 81–98. Petersfield and Bristol, PA: Wrightson Biomedical Publishing Ltd..

Blumer, D. (2000). On the psychobiology of non-epileptic seizures. In *Nonepileptic seizures* (J. R. Gates, A. J. Rowan, eds), pp. 305–310. Boston, MA: Butterworth-Heinemann.

Blumer, D., Adamolekun, B. (2006). Treatment of patients with coexisting epileptic and nonepileptic seizures. *Epilepsy Behav* **9**, 498–502.

Bortz, J. J., Prigatano, G. P., Blum, D., Fisher, R. S. (1995). Differential response characteristics in nonepileptic and epileptic seizure patients on a test of verbal learning and memory. *Neurology* **45**, 2029–2034.

Bowman, E. S. (1993). Etiology and clinical course of pseudoseizures. Relationship to trauma, depression, and dissociation. *Psychosomatics* **34**, 333–342.

Bowman, E. S. (2000). Relationship of remote and recent life events to the onset and course of nonepileptic seizures. In *Non-epileptic seizures* (J. R. Gates, A. J. Rowan, eds), pp. 269–283. Boston, MA: Butterworth-Heinemann.

Bowman, E. S. (2001). Psychopathology and outcome in pseudoseizures. In *Psychiatric issues in epilepsy – a practical guide to diagnosis and treatment* (A. B. Ettinger, A. Kanner, eds), pp. 355–378. Philadelphia, PA: Lippincott Williams & Wilkins.

Bowman, E. S., Markand, O. N. (1996). Psychodynamics and psychiatric diagnoses of pseudoseizure subjects. *Am J Psychiatry* **153**, 57–63.

Bowman, E. S., Markand, O. N. (1999). The contribution of life events to pseudoseizure occurrence in adults. *Bull Menninger Clin* **63**, 70–88.

Bowman, E. S., Coons, P. M. (2000). The differential diagnosis of epilepsy, pseudoseizures, dissociative identity disorder, and dissociative disorder not otherwise specified. *Bull Menninger Clin* **64**, 164–180.

Breedlove, S. M. (1994). Sexual differentiation of the human nervous system. *Ann Rev Psychol* **45**, 389–418.

Bremner, J. D., Vythilingam, M., Vermetten, S. M., Southwick, S. M., McGlashan, T., Nazeer, A., Khan, S., Vaccarino, L. V., Soufer, R., Garg, P. K. (2003). MRI and PET study of deficits in hippocampal structure and function in women with childhood sexual abuse and posttraumatic stress disorder. *Am J Psychiatry* **160**, 924–932.

Brown, M. C., Levin, B. E., Ramsay, R. E., Katz, D. A., Duchowny, M. S. (1991). Characteristics of patients with nonepileptic seizures. *J Epilepsy* **4**, 225–229.

Cartmill, A., Betts, T. (1992). Seizure behaviour in a patient with post-traumatic stress disorder following rape. Notes on the aetiology of "pseudoseizures". *Seizure* **1**, 33–36.

Castro-Caldas, A., Caleiro Miranda, P., Carmo, I., Reis, A., Leote, F., Ribeiro, C., Ducla-Soares, E. (1999). Influence of learning to read and write on the morphology of the corpus callosum. *Eur J Neurol* **6**, 23–28.

Thomas Saville, ed. (1889). *Clinical Lectures on Diseases of the Nervous System.* London: The New Sydenham Society.

Chodoff, P. (1982). Hysteria and women. *Am J Psychiatry* **139**, 545–551.

Cohen, L. M., Howard, G. F., III., Bongar, B. (1992). Provocation of pseudoseizures by psychiatric interview during EEG and video monitoring. *Int J Psychiatry Med* **22**, 131–140.

Cohen, R. J., Suter, C. (1982). Hysterical seizures: suggestion as a provocative EEG test. *Ann Neurol* **11**, 391–395.

Cragar, D. E., Berry, D. T. R., Schmitt, F. A., Fakhoury, T. A. (2005). Cluster analysis of normal personality traits in patients with psychogenic nonepileptic seizures. *Epilepsy Behav* **6**, 593–600.

Dauvilliers, Y., Arnulf, I., Mignot, E. (2007). Narcolepsy with cataplexy. *Lancet* **369**, 499–511.

Davies, K. G., Blumer, D. P., Lobo, S., Hermann, B. P., Phillips, B. L. B., Montouris, G. D. (2000). *De novo* nonepileptic seizures after intracranial surgery for epilepsy: incidence and risk factors. *Epilepsy Behav* **1**, 436–443.

Devinsky, O. (1998). Nonepileptic psychogenic seizures: quagmires of pathophysiology, diagnosis, and treatment. *Epilepsia* **39**, 458–462.

Devinsky, O., Gordon, E. (1998). Epileptic seizures progressing into nonepileptic conversion seizures. *Neurology* **51**, 1293–1296.

Devinsky, O., Sanchez-Villasenor, F., Vazquez, B., Kothari, M., Alper, K., Luciano, D. (1996). Clinical profile of patients with epileptic and nonepileptic seizures. *Neurology* **46**, 1530–1533.

Devinsky, O., Kelley, K., Porter, R. J., Theodore, W. R. (1988). Clinical and electrographic features of simple partial seizures. *Neurology* **18**, 1347–1352.

Devinsky, O., Mesad, S., Alper, K. (2001). Nondominant hemisphere lesions and conversion nonepileptic seizures. *J Neuropsychiatry Clin Neurosci* **13**, 367–373.

Dodrill, C. B., Holmes, M. D. (2000). Part summary: psychological and neuropsychological evaluation of the patient with non-epileptic seizures. In *Non-epileptic seizures* (J. R. Gates, A. J. Rowan, eds), pp. 169–181. Boston, MA: Butterworth-Heinemann.

Donofrio, N., Perrine, K., Alper, K., Devinsky, O. (2000). Depression and anxiety in patients with non-epileptic versus epileptic seizures. In *Non-epileptic seizures* (J. R. Gates, A. J. Rowan, eds), pp. 151–158. Boston, MA: Butterworth-Heinemann.

Draganski, B., Gaser, C., Busch, V., Schuierer, G., Bogdahn, U., May, A. (2004). Neuroplasticity: changes in gray matter induced by training. *Nature* **427**, 311–312.

Draganski, B., Gaser, C., Kempermann, G., Kuhn, H. G., Winkler, J., Buchel, C., May, A. (2006). Temporal and spatial dynamics of brain structure changes during extensive learning. *J Neurosci* **26**, 6314–6317.

Drane, D. L., Williamson, D. J., Stroup, E. S., Holmes, M. D., Jung, M., Koerner, E., Chaytor, N., Wilensky, A. J., Miller, J. W. (2006). Cognitive impairment is not equal in patients with epileptic and psychogenic nonepileptic seizures. *Epilepsia* **47**, 1879–1886.

Duncan, R., Oto, M., Russel, A. J., Conway, P. (2004). Pseudosleep events in patients with psychogenic non-epileptic seizures: prevalence and associations. *J Neurol Neurosurg Psychiatry* **75**, 1009–1012.

Duncan, R., Oto, M. (2008). Psychogenic non-epileptic seizures in patients with learning difficulties: comparison with patients with no learning difficulties. *Epilepsy Behav* **12**, 183–186.

Ebner, A., Dinner, D. S., Noachtar, S., Lüders, H. (1995). Automatisms with preserved responsiveness: a lateralizing sign in psychomotor seizures. *Neurology* **45**, 61–64.

Eisendrath, S. J., Valan, M. N. (1994). Psychiatric predictors of pseudoepileptic seizures in patients with refractory seizures. *J Neuropsych Clin Neurosci* **6**, 257–260.

Ettinger, A. B., Devinsky, O., Weisbrot, D. M., Goyal, A., Shashikumar, S. (1999). Headaches and other pain symptoms among patients with psychogenic non-epileptic seizures. *Seizure* **8**, 424–426.

Fernandéz, G., Weis, S., Stoffel-Wagner, B., Tendolkar, I., Reuber, M., Beyenburg, S., Klaver, P., Fell, J., de Greiff, A., Ruhlmann, J., Reul, J., Elger, C. E. (2003). Menstrual cycle-dependent neural plasticity in the adult human brain is hormone, task and region specific. *J Neurosci* **23**, 3790–3795.

Finerman, R., Bennet, L. A. (1995). Guilt, blame and shame: responsibility in health and sickness. *Soc Sci Med* **40**, 1–3.

Fink, P. (1992). Physical complaints and symptoms of somatizing patients. *J Psychosom Res* **36**, 125–136.

Fiszman, A., Alves-Leon, S. V., Nunes, R. G., D'Andrea, I., Figuera, I. (2004). Traumatic events and posttraumatic stress disorder in patients with psychogenic nonepileptic seizures: a critical review. *Epilepsy Behav* **5**, 818–825.

Fleisher, W., Staley, D., Krawetz, P., Pillay, N., Arnett, J. L., Maher, J. (2002). Comparative study of trauma-related phenomena in subjects with pseudoseizures and subjects with epilepsy. *Am J Psychiatry* **159**, 660–663.

Ford, C. (1993a). Somatization and non-epileptic seizures. In *Non-epileptic seizures* (J. R. Gates, A. J. Rowan, eds), pp. 153–164. Boston, MA: Butterworth-Heinemann.

Frances, P. L., Baker, G. A., Appleton, P. L. (1999). Stress and avoidance in pseudoseizures: testing the assumptions. *Epilepsy Res* **34**, 241–249.

Freud, S. (1896). Zur Ätiologie der Hysterie. In *Gesammelte Werke* (A. Freud, E. Bibring, W. Hoffer, E. Kris, O. Isakower, eds), pp. 425–459. London: Imago.

Freud, S., Breuer, J. (1892). Studien über Hysterie. In *Gesammelte Werke* (A. Freud, E. Bibring, W. Hoffer, E. Kris, O. Isakower, eds), pp. 75–98. London: Imago.

Frucht, M. M., Quigg, M., Schwaner, C., Fountain, N. B. (2000). Distribution of seizure precipitants among epilepsy syndromes. *Epilepsia* **41**, 1534–1539.

Galimberti, C. A., Ratti, M. T., Murelli, R., Marchioni, E., Manni, R., Tartara, A. (2003). Patients with psychogenic nonepileptic seizures, alone or epilepsy associated, share a psychological profile distinct of epilepsy patients. *J Neurol* **250**, 338–346.

Gardner, D. L., Goldberg, R. L. (1982). Psychogenic seizures and loss. *Int J Psychiatry Med* **12**, 121–128.

Glosser, G., Roberts, D., Glosser, D. S. (1999). Nonepileptic seizures after resective epilepsy surgery. *Epilepsia* **40**, 1750–1754.

Goldstein, L. H., Mellers, J. D. (2006). Ictal symptoms of anxiety, avoidance behaviour, and dissociation in patients with dissociative seizures. *J Neurol Neurosurg Psychiatry* **77**, 616–621.

Goldstein, L. H., Drew, C., Mellers, J., Mitchell-O'Malley, S., Oakley, D. A. (2000). Dissociation, hypnotizability, coping styles and health locus of control: characteristics of pseudoseizure patients. *Seizure* **9**, 314–322.

Green, A., Payne, S., Barnitt, R. (2004). Illness representations among people with non-epileptic seizures attending a neuropsychiatry clinic: a qualitative study based on the self-regulation model. *Seizure* **13**, 331–339.

Greig, E., Betts, T. (1992). Epileptic seizures induced by sexual abuse. Pathogenic and pathoplastic factors. *Seizure* **1**, 269–274.

Griffith, J. L. (1990). The mind/body problem revisited: pseudoseizure patients. *Fam Syst Med* **8**, 71–89.

Griffith, J. L., Polles, A., Griffith, M. E. (1998). Pseudoseizures, families, and unspeakable dilemmas. *Psychosomatics* **39**, 144–153.

Guberman, A. (1982). Psychogenic pseudoseizures in non-epileptic patients. *Can J Psychiatry* **27**, 401–404.

Hagger, M. S., Orbell, S. (2003). A meta-analytic review of the common-sense model of illness representations. *Psychol Health* **18**, 141–184.

Hamm, A. O., Greenwald, M. K., Bradley, M. M., Cuthbert, B. N., Lang, P. J. (1991). The fear potentiated startle effect. Blink reflex modulation as a result of classical aversive conditioning. *Integr Physiol Behav Sci* **26**, 119–126.

Hamm, A. O., Cuthbert, B. N., Globisch, J., Vaitl, D. (1997). Fear and the startle reflex: blink modulation and autonomic response patterns in animal and mutilation fearful subjects. *Psychophysiol* **34**, 97–107.

Hantke, N. C., Doherty, M. J., Haltiner, A. M. (2007). Medication use profiles in patients with psychogenic nonepileptic seizures. *Epilepsy Behav* **10**, 333–335.

Henry, T. R., Drury, I. (1997). Non-epileptic seizures in temporal lobectomy candidates with medically refractory seizures. *Neurology* **48**, 1374–1382.

Hermann, B. P. (1993). Neuropsychological assessment in the diagnosis of non-epileptic seizures. In *Non-epileptic seizures* (A. J. Rowan, J. R. Gates, eds), pp. 221–232. Stoneham, MA: Butterworth-Heinemann.

Holmes, M. D., Dodrill, C. B., Bachtler, S., Wilensky, A. J., Ojemann, L. M., Miller, J. W. (2001). Evidence that emotional maladjustment is worse in men than in women with psychogenic nonepileptic seizures. *Epilepsy Behav* **2**, 568–573.

Horne, R. (1999). Patients' beliefs about treatment: the hidden determinant of treatment outcome? *J Psychosom Res* **47**, 491–495.

Hudak, A. M., Trivedi, K., Harper, C. R., Booker, K., Caesar, R. R., Agostini, M., Van Ness, P. C., Diaz-Arrastia, R. (2004). Evaluation of seizure-like episodes in survivors of moderate and severe traumatic brain injury. *J Head Trauma Rehabil* **19**, 290–295.

Jang, K. L., McCrae, R. R., Angleitner, A., Riemann, R., Livesley, W. J. (1998). Heritability of facet-level traits in a cross-cultural twin sample: support for a hierarchical model of personality. *J Pers Soc Psychol* **74**, 1556–1565.

Jung, C. G. (1939). On the psychogenesis of schizophrenia. *J Ment Sci* **85**, 999–1011.

Kalogjera-Sackellares, D. (2004). *Psychodynamics and Psychotherapy of Pseudoseizures.* Camarthen and Williston, VT: Crown House Publishing, Camarthen, Wales and Williston.

Kalogjera-Sackellares, D., Sackellares, J. C. (1999). Intellectual and neuropsychological features of patients with psychogenic pseudoseizures. *Psychiatry Res* **86**, 73–84.

Kalogjera-Sackellares, D., Sackellares, J. C. (2001). Impaired motor function in patients with psychogenic pseudoseizures. *Epilepsia* **42**, 1600–1606.

Kanner, A. M., Parra, J., Frey, M., Stebbins, G., Pierre-Louis, S., Iriarte, J. (1999). Psychiatric and neurologic predictors of psychogenic pseudoseizure outcome. *Neurology* **53**, 933–938.

Kendell, R. E. (2001). The distinction between mental and physical illness. *Br J Psychiatry* **178**, 490–493.

Krahn, L. E., Rummans, T. A., Sharbrough, F. W., Jowsey, S. G., Cascino, G. D. (1995). Pseudoseizures after epilepsy surgery. *Psychosomatics* **36**, 487–493.

Krawetz, P., Fleisher, W., Pillay, N., Staley, D., Arnett, J., Maher, J. (2001). Family functioning in subjects with pseudoseizures and epilepsy. *J Nerv Ment Dis* **189**, 38–43.

Krishnamoorthy, E. S., Brown, R. J., Trimble, M. (2001). Personality and psychopathology in nonepileptic attack disorder: a prospective study. *Epilepsy Behav* **2**, 418–422.

Kristensen, O., Alving, J. (1992). Pseudoseizures – risk factors and prognosis. A case-control study. *Acta Neurol Scand* **85**, 177–180.

Kroenke, K., Rosmalen, J. G. (2006). Symptoms, syndromes, and the value of psychiatric diagnostics in patients who have functional somatic disorders. *Med Clin North Am* **90**, 603–626.

Krumholz, A., Niedermeyer, E. (1983). Psychogenic seizures: a clinical study with follow-up data. *Neurology* **33**, 498–502.

Kütemeyer, M., Masuhr, K. M., Schultz-Venrath, U. (2005). Kommunikative Anfallsunterbrechung – zum ärztlichen Umgang mit Patienten im Status pseudoepilepticus. *Z Epileptol* **18**, 71–77.

Kuyk, J., Spinhoven, P., van Emde, B. W., van Dyck, R. (1999). Dissociation in temporal lobe epilepsy and pseudo-epileptic seizure patients. *J Nerv Ment Dis* **187**, 713–720.

Kuyk, J., Swinkels, W. A. M., Spinhoven, P. (2003). Psychopathologies in patients with and without comorbid epilepsy: how different are they? *Epilepsy Behav* **4**, 13–18.

Lang, P. J., Davis, M., Ohman, A. (2000). Fear and anxiety: animal models and human cognitive psychophysiology. *J Affect Disord* **61**, 137–159.

Lelliott, P. T., Fenwick, P. (1991). Cerebral pathology in pseudoseizures. *Acta Neurol Scand* **83**, 129–132.

Lesser, R. P. (1996). Psychogenic seizures. *Neurology* **46**, 1499–1507.

Lesser, R. P., Lueders, H., Dinner, D. S. (1983). Evidence for epilepsy is rare in patients with psychogenic seizures. *Neurology* **33**, 502–504.

Leventhal, H., Diefenbach, M., Leventhal, E. A. (1992). Illness cognitions: using common sense to understand treatment adherence and affect cognition interactions. *Cogn Ther Res* **16**, 143–163.

Lichter, I., Goldstein, L. H., Toone, B. K., Mellers, J. D. (2004). Nonepileptic seizures following general anaesthetics: a report of five cases. *Epilepsy Behav* **5**, 1005–1013.

Lipowski, Z. J. (1988). Somatization: the concept and its clinical application. *Am J Psychiatry* **145**, 1358–1368.

Martin, R., Bell, B., Hermann, B., Mennemeyer, S. (2003). Nonepileptic seizures and their costs: the role of neuropsychology. In *Clinical neuropsychology and cost outcome research: a beginning* (G. P. Prigatano, N. H. Pliskin, eds), pp. 235–258. New York: Psychology Press.

Martinez-Taboas, A. (2002). The role of hypnosis in the detection of psychogenic seizures. *Am J Clin Hypnosis* **45**, 11–20.

Matsumoto, J., Hallett, M. (1994). Startle syndromes. In *Movement Disorders* (C. D. Marsden, S. Fahn, eds) 3rd edition, pp. 418–433. Oxford: Butterworth-Heinemann.

Mayou, R., Kirmayer, L. J., Simon, G., Kroenke, K., Sharpe, M. (2005). Somatoform disorders: time for a new approach in DSM-V. *Am J Psychiatry* **162**, 847–855.

McGonigal, A., Oto, M., Russel, A. J., Greene, J., Duncan, R. (2002). Outpatient video-EEG recording in the diagnosis of non-epileptic seizures: a randomized controlled trial of simple suggestion techniques. *J Neurol Neurosurg Psychiatry* **72**, 549–551.

Meares, R., Stevenson, J., Gordon, E. (1999). A Jacksonian and biopsychosocial hypothesis concerning boderline and related phenomena. *Aust N Z J Psychiatry* **33**, 831–840.

Merckelbach, H., Muris, P. (2001). The causal link between self-reported trauma and dissociation: a critical review. *Behav Res Ther* **39**, 245–254.

Miller, H. R. (1983). Psychogenic seizures treated by hypnosis. *Am J Clin Hypn* **25**, 248–252.

Mökleby, K., Blomhoff, S., Malt, U. F., Dahlström, A., Taubøll, E., Gjerstad, L. (2002). Psychiatric comorbidity and hostility in patients with psychogenic nonepileptic seizures compared with somatoform disorders and healthy controls. *Epilepsia* **43**, 193–198.

Moore, P. M., Baker, G. A., McDade, G., Chadwick, D., Brown, S. (1994). Epilepsy, pseudoseizures and perceived family characteristics: a controlled study. *Epilepsy Res* **18**, 75–83.

Ney, G. C., Barr, W. B., Napolitano, C., Decker, R., Schaul, N. (1998). New-onset psychogenic seizures after surgery for epilepsy. *Arch Neurol* **55**, 726–730.

Orbach, D., Ritaccio, A., Devinsky, O. (2003). Psychogenic nonepileptic seizures associated with video-EEG confirmed sleep. *Epilepsia* **44**, 64–68.

Orellana, C., Villemin, E., Tafti, M., Carlander, B., Besset, A., Billiard, M. (1994). Life events in the year preceding the onset of narcolepsy. *Sleep* **17**, S50–S53.

Oto, M., Conway, P., McGonigal, A., Russel, A. J., Duncan, R. (2005a). Gender differences in psychogenic non-epileptic seizures. *Seizure* **14**, 33–39.

Oto, M., Espie, C., Pelosi, A., Selkirk, M., Duncan, R. (2005b). The safety of antiepileptic drug withdrawal in patients with non-epileptic seizures. *J Neurol Neurosurg Psychiatry* **76**, 1682–1685.

Owczarek, K. (2003a). Anxiety as a differential factor in epileptic versus psychogenic pseudoepileptic seizures. *Epilepsy Res* **52**, 227–232.

Owczarek, K. (2003b). Somatisation indexes as differential factors in psychogenic pseudoepileptic and epileptic seizures. *Seizure* **12**, 178–181.

Pakalnis, A., Drake, M. E., Jr., Phillips, B. (1991). Neuropsychiatric aspects of psychogenic status epilepticus. *Neurology* **41**, 1104–1106.

Papacostas, S. S., Myrianthopoulou, P., Papathanasiou, E. (2006). Epileptic seizures followed by nonepileptic manifestations: a video-EEG diagnosis. *Electromyogr Clin Neurophysiol* **46**, 323–327.

Paris, J. (1998). Does childhood trauma cause personality disorder in adults? *Can J Psychiatry* **43**, 148–153.

Parry, T., Hirsch, N. (1992). Psychogenic seizures after general anaesthesia. *Anaesthesia* **47**, 534.

Peters, S., Stanley, I., Rose, M., Salmon, P. (1998). Patients with medically unexplained symptoms: sources of patients' authority and implications for demand on medical care. *Soc Sci Med* **46**, 559–565.

Plomin, R., Daniels, D. (1987). Why are children in the same family so different from one another? *Behav Brain Sci* **10**, 1–16.

Prueter, C., Schultz-Venrath, U., Rimpau, W. (2002). Dissociative and associated psychopathological symptoms in patients with epilepsy, pseudoseizures and both seizure forms. *Epilepsia* **43**, 188–192.

Rabe, F. (1970). *Die Kombination hysterischer and epileptischer Anfälle-das Problem der "Hysteroepilepsie" in neuer Sicht*. Berlin: Springer Verlag.

Ramaratnam, S., Baker, G. A., Goldstein, L. H. (2005). Psychological treatments for epilepsy. *Cochrane Database Syst Rev* . CD002029

Ramchandani, D., Schindler, B. (1993). Evaluation of pseudoseizures. A psychiatric perspective. *Psychosomatics* **34**, 70–79.

Reuber, M., Elger, C. E. (2003). Psychogenic nonepileptic seizures: review and update. *Epilepsy Behav* **4**, 205–216.

Reuber, M., Enright, S. M., Goulding, P. J. (2000). Postoperative pseudostatus: not everything that shakes is epilepsy. *Anaesthesia* **55**, 74–78.

Reuber, M., Fernández, G., Bauer, J., Singh, D. D., Elger, C. E. (2002a). Interictal EEG abnormalities in patients with psychogenic non-epileptic seizures. *Epilepsia* **43**, 1013–1020.

Reuber, M., Fernández, G., Helmstaedter, C., Qurishi, A., Elger, C. E. (2002b). Evidence of brain abnormality in patients with psychogenic nonepileptic seizures. *Epilepsy Behav* **3**, 246–248.

Reuber, M., Kurthen, M., Fernández, G., Schramm, J., Elger, C. E. (2002c). Epilepsy surgery in patients with additional psychogenic seizures. *Arch Neurol* **59**, 82–86.

Reuber, M., Kurthen, M., Kral, T., Elger, C. E. (2002d). New-onset psychogenic seizures after intracranial neurosurgery. *Acta Neurochir* **144**, 901–908.

Reuber, M., Fernández, G., Helmstaedter, C., Bauer, J., Qurishi, A., Elger, C. E. (2003a). Are there physical risk factors for psychogenic nonepileptic seizures in patients with epilepsy? *Seizure* **12**, 561–567.

Reuber, M., House, A. O., Pukrop, R., Bauer, J., Elger, C. E. (2003b). Somatization, dissociation and psychopathology in patients with psychogenic nonepileptic seizures. *Epilepsy Res* **57**, 159–167.

Reuber, M., Pukrop, R., Bauer, J., Helmstaedter, C., Tessendorf, N., Elger, C. E. (2003c). Outcome in psychogenic nonepileptic seizures: 1 to 10-year follow-up in 164 patients. *Ann Neurol* **53**, 305–311.

Reuber, M., Pukrop, R., Derfuss, R., Bauer, J., Elger, C. E. (2003d). Multidimensional assessment of personality in patients with psychogenic nonepileptic seizures. *J Neurol Neurosurg Psychiatry* **75**, 743–748.

Reuber, M., Howlett, S., Kemp, S. (2005). Psychologic treatment for patients with psychogenic nonepileptic seizures. *Expert Opin Neurotherapeut* **5**, 737–752.

Reuber, M., Howlett, S., Khan, A., Grünewald, R. (2007). Nonepileptic seizures and other functional neurological symptoms: predisposing, precipitating and perpetuating factors. *Psychosomatics* **48**, 230–238.

Risse, G. L., Mason, S. L., Kent Mercer, D. (2000). Neuropsychological performance and cognitive complaints in epileptic and nonepileptic seizure patients. In *Non-epileptic seizures* (J. R. Gates, A. J. Rowan, eds), pp. 139–150. Boston, MA: Butterworth-Heinemann.

Rosenbaum, M. (2000). Psychogenic seizures – why women? *Psychosomatics* **41**, 147–149.

Rosenberg, H. J., Rosenberg, S. D., Williamson, P. D., Wolford, G. L. (2000). A comparative study of trauma and posttraumatic stress disorder prevalence in epilepsy patients and psychogenic nonepileptic seizure patients. *Epilepsia* **41**, 447–452.

Roy, A., Barris, M. (1993). Psychiatric concepts in psychogenic nonepileptic seizures. In *Non-epileptic seizures* (A. J. Rowan, J. R. Gates, eds), pp. 143–152. Boston, MA: Butterworth-Heinemann.

Ruiz-Padial, E., Vila, J. (2007). Fearful and sexual pictures not consciously seen modulate the startle reflex in human beings. *Biol Psychiatry* **61**, 996–1001.

Salanova, V., Morris, H. H., Van Ness, P., Kotagal, P., Wyllie, E., Lüders, H. (1995). Frontal lobe seizures: electroclinical syndromes. *Epilepsia* **36**, 16–24.

Salmon, P., Al-Marzooqi, S. M., Baker, G., Reilly, J. (2003). Childhood family dysfunction and associated abuse in patients with nonepileptic seizures: towards a causal model. *Psychosom Med* **65**, 695–700.

Scambler, G. (1994). Patient perceptions of epilepsy and of doctors who manage epilepsy. *Seizure* **3**, 287–293.

Schrag, A., Brown, R. J., Trimble, M. R. (2004). Reliability of self-reported diagnosis in patients with neurologically unexplained symptoms. *J Neurol Neurosurg Psychiatry* **75**, 608–611.

Sharpe, D., Faye, C. (2006). Non-epileptic seizures and child sexual abuse: a critical review of the literature. *Clin Psychol Rev* **26**, 1020–1040.

Sharpe, M. (2002). Medically unexplained symptoms and syndromes. *Clin Med* **2**, 501–504.

Sharpe, M., Mayou, R. (2004). Somatoform disorders: a help or a hindrance to good patient care? *Br J Psychiatry* **184**, 470–476.

Silver, L. B. (1982). Conversion disorder with pseudoseizures in adolescence: a stress reaction to unrecognized and untreated learning disabilities. *J Am Acad Child Psychiatry* **21**, 508–512.

Sirven, J. I., Glosser, D. S. (1998). Psychogenic nonepileptic seizures. *Neuropsych Neuropsychol Behav Neurol* **11**, 225–235.

Smith, D., Defalla, B. A., Chadwick, D. W. (1999). The misdiagnosis of epilepsy and the management of refractory epilepsy in a specialist clinic. *QJM* **92**, 15–23.

Snyder, S. L., Rosenbaum, D. H., Rowan, A. J., Strain, J. J. (1994). SCID diagnosis of panic disorder in psychogenic seizure patients. *J Neuropsychiatry Clin Neurosci* **6**, 261–266.

Spector, S., Cull, C., Goldstein, L. H. (2000). Seizure precipitants and perceived self-control of seizures in adults with poorly controlled epilepsy. *Epilepsy Res* **38**, 207–216.

Spector, S., Tranah, A., Cull, C., Goldstein, L. H. (1999). Reduction in seizure frequency following a short-term group intervention for adults with epilepsy. *Seizure* **8**, 297–303.

Stewart, R. S., Lovitt, R., Stewart, R. M. (1982). Are hysterical seizures more than hysteria? A research diagnostic criteria, DMS-III, and psychometric analysis. *Am J Psychiatry* **139**, 926–929.

Stone, J., Binzer, M., Sharpe, M. (2004a). Illness beliefs and locus of control: a comparison of patients with pseudoseizures and epilepsy. *J Psychosom Res* **57**, 541–547.

Stone, J., Sharpe, M., Binzer, M. (2004b). Motor conversion symptoms and pseudoseizures: a comparison of clinical characteristics. *Psychosomatics* **45**, 492–499.

Swanson, S. J., Springer, J. A., Benbadis, S. R., Morris, G. L. (2000). Cognitive and psychological functioning in patients with nonepileptic seizures. In *Non-epileptic seizures* (J. R. Gates, A. J. Rowan, eds), pp. 124–137. Boston, MA: Butterworth-Heinemann.

Thacker, K., Devinsky, O., Perrine, K., Alper, K., Luciano, D. (1993). Nonepileptic seizures during apparent sleep. *Ann Neurol* **33**, 414–418.

Thapar, A., McGuffin, P. (1996). Genetic influences on life events in childhood. *Psychol Med* **26**, 813–820.

Tojek, T. M., Lumley, M., Barkley, G., Mahr, G., Thomas, A. (2000). Stress and other psychosocial characteristics of patients with psychogenic nonepileptic seizures. *Psychosomatics* **41**, 221–226.

Trimble, M. R. (1982). Anticonvulsant drugs and hysterical seizures. In *Pseudoseizures* (T. L. R. A. Riley, ed.), pp. 148–158. Baltimore, MA and London: Williams and Wilkins.

Vein, A. M., Vorob'eva, O. V. (1998). [A neurophysiological model of the "paroxysmal brain" (cerebral mechanisms in the genesis of paroxysmal states)]. *Vestn Ross Akad Med Nauk* **8**, 32–36.

Walczak, T. S., Williams, D. T., Berten, W. (1994). Utility and reliability of placebo infusion in the evaluation of patients with seizures. *Neurology* **44**, 394–399.

Ward, P. E., McCarthy, D. J., Nyman, G. W. (1988). Podiatric implications of psychogenic seizures. *J Foot Surg* **27**, 222–225.

Watson, N. F., Doherty, M. J., Dodrill, C. B., Farrel, D., Miller, J. W. (2002). The experience of earthquakes by patients with epileptic and psychogenic nonepileptic seizures. *Epilepsia* **43**, 317–320.

Wessely, S., Nimnuan, C., Sharpe, M. (1999). Functional somatic syndromes: one or many. *Lancet* **354**, 936–939.

Westbrook, L. E., Devinsky, O., Geocadin, R. (1998). Nonepileptic seizures after head injury. *Epilepsia* **39**, 978–982.

Wilkus, R. J., Dodrill, C. B. (1989). Factors affecting the outcome of MMPI and neuropsychological assessments of psychogenic and epileptic seizure patients. *Epilepsia* **30**, 339–347.

Wilner, A. N., Bream, P. (1993). Status epilepticus and pseudostatus epilepticus. *Seizure* **2**, 257–260.

Wood, B. L., McDaniel, S., Burchfiel, K., Erba, G. (1998). Factors distinguishing families of patients with psychogenic seizures from families of patients with epilepsy. *Epilepsia* **39**, 432–437.

Yamasue, H., Kasai, K., Iwanami, A., Ohtani, T., Yamada, H., Abe, O., Kuroki, N., Fukuda, R., Tochigi, M., Furukawa, S., Sadamatsu, M., Sasaki, T., Aoki, S., Ohtomo, K., Asukai, N., Kato, N. (2003). Voxel-based analysis of MRI reveals anterior cingulate gray-matter volume reduction in posttraumatic stress disorder due to terrorism. *Proc Natl Acad Sci* **100**, 9039–9043.

Zaidi, A., Clough, P., Cooper, P., Scheepers, B., Fitzpatrick, A. P. (2000). Misdiagnosis of epilepsy: many seizure-like attacks have a cardiovascular cause. *J Am Coll Cardiol* **36**, 181–184.

Zaidi, A., Crampton, S., Clough, P., et al. (1999). Head-up tilting is a useful test for psychogenic nonepileptic seizures. *Seizure* **8**, 353–355.

CHAPTER 12

Are Psychiatric Disorders a Risk for the Development of Neurological Disorders?

Andres M. Kanner

INTRODUCTION

"Melancholics ordinarily become epileptics, and epileptics melancholics: what determines the preference is the direction the malady takes; if it bears upon the body, epilepsy, if upon the intelligence, melancholy" (Hippocrates, 400BC):

The implication of Hippocrates' statement is that patients suffering from a depressive disorder are likely to develop epilepsy and vice versa (Lewis, 1934). While most clinicians take the latter implied association between the two conditions for granted, such is not the case with the former. Yet, as shown in this chapter, there is a growing body of literature that suggests that psychiatric disorders, particularly mood disorders, appear to be risk factors for the development of a variety of neurological disorders. Do these bidirectional relations imply causality, or rather, the existence of common pathogenic mechanisms that facilitate the development of a neurologic disorder in the presence of a psychiatric disorder? The purpose of this chapter is to discuss these questions in some detail. The second section of this chapter is devoted to a review of the common pathogenic mechanisms that facilitate the existence of bidirectional relations between psychiatric and neurologic disorders, while the final section analyzes the impact that a psychiatric disorder, particularly depression, has on the course of a neurologic entity.

DO PSYCHIATRIC DISORDERS PRECEDE THE DEVELOPMENT OF NEUROLOGIC DISORDERS?

Depression preceding the onset of epilepsy

The first modern study to support Hippocrates' observation consisted of a case series of 51 inpatients with late-onset epilepsy, 16% of whom had a history of depression before the initial seizure (Dominian *et al.*, 1963). This study had several methodologic problems, however. First, it had no control group; second, it included a biased patient sample as the inpatients had probably been suffering from more severe forms of epilepsy. The second study was a Swedish population-based case-control study in which depression was found to be seven times more frequent among patients with new-onset epilepsy, preceding the seizure disorder, than among age- and sex-matched controls (Forsgren and Nystrom, 1990). When analyses were restricted to cases with partial epilepsy, depression was found to be 17 times more common among cases than among controls. These data were confirmed by two recent population-based, case-control studies of patients with newly diagnosed adult-onset epilepsy. The first study included all adults aged 55 years and older at the time of the onset of their epilepsy living in Olmstead County, MN (Hesdorffer *et al.*, 2000). In this study, the investigators found that a diagnosis of depression preceding the time of the first seizure was 3.7 times more frequent among cases than among controls after adjusting for medical therapies for depression. As in the Swedish study (Forsgren and Nystrom, 1990), this increased risk was greater among cases with partial-onset seizures. An interesting finding of this study was that among people with epileptic seizures, an episode of major depression had taken place closer to the time of the first seizure than for controls. The second study, carried out in Iceland, investigated the role of specific symptoms of depression in predicting the development of unprovoked seizures or epilepsy in a population-based study of 324 children and adults, aged 10 years and older with a first unprovoked seizure or newly diagnosed epilepsy and 647 controls (Hesdorffer *et al.*, 2006). Major depression was associated with a 1.7-fold increased risk for developing epilepsy while a history of attempted suicide was 5.1-fold more common among cases than among controls.

Attention deficit disorder preceding epilepsy

The average annual incidence of seizures in children aged 5–16 years is approximately 0.0470 per year (Dunn *et al.*, 1997). Studies that investigated the development of new-onset seizures in selected populations of children with attention deficit hyperactivity disorder (ADHD) found a significantly higher incidence, ranging between 0.2% and 2% (Hughes *et al.*, 2000; Hemmer *et al.*, 2001; Williams *et al.*, 2001; Holtmann *et al.*, 2003). By the same token, three case-control studies of children with new-onset seizures or epilepsy have found a higher prevalence of behavioral problems in children and adolescents before the first seizure ever occurred (Austin *et al.*, 2001; Hesdorffer *et al.*, 2004; Jones *et al.*, 2007).

The first report consisted of a case-control study of 148 children with a first unprovoked seizure and 89 seizure-free sibling controls (Austin *et al.*, 2001). Attention problems were 2.4-fold more common in children who went on to develop seizures, prior to the first seizure, than in controls (8.1% vs. 3.4%). The second study consisted of a population-based case-control study conducted among Icelandic children (Hesdorffer *et al.*, 2004); a history of

ADHD-predominantly inattentive type according to DSM-IV diagnostic criteria was identified almost four times more frequently (OR = 3.7; 95% CI = 1.1–13) in children with an unprovoked seizure *prior* to seizure onset than in age- and gender-matched controls. Jones *et al.* (2007) reported similar findings in a study of 103 children aged 8–18 years, 53 with recent onset epilepsy (<1 year in duration) of idiopathic etiology and 50 healthy children matched for age. Each child underwent a structured psychiatric diagnostic interview to characterize the spectrum of lifetime-to-date history of comorbid psychiatric disorder. Compared with the control group, children with epilepsy exhibited an elevated rate of lifetime-to-date Diagnostic and Statistical Manual of Mental Disorders, fourth edition (DSM-IV) Axis I disorders, including significantly higher rates of depressive disorders (22.6% vs. 4%), anxiety disorders (35.8% vs. 22%), and attention deficit–hyperactivity disorder (26.4% vs. 10%). Of note, 45% of children with epilepsy exhibited DSM-IV Axis I disorders before the first recognized seizure.

In a review of the literature, Hesdorffer and Hauser (2007) concluded: "there was an increased risk for developing unprovoked seizures in children with ADHD, and the reported increased risk is smaller in case-control studies than in cohort studies, which were limited by small numbers of ensuing unprovoked seizures during short follow-up periods in selected populations". A cautionary note is in order: some of these children may have an unrecognized seizure disorder that is associated with the development of ADHD. This is illustrated in a study of 234 children (179 males, 9.1 ± 3.6 years of age; 55 females, 9.6 ± 3.9 years of age) with uncomplicated ADHD who underwent an electroencephalogram (EEG); epileptiform abnormalities were found in the recordings of 36 (15.4%) children. Benign focal rolandic epileptiform discharges of childhood accounted for 40% of the abnormal EEGs (Hemmer *et al.*, 2001).

Data from another study suggested that psychiatric pathology could be a "risk factor" for the development of unprovoked non-febrile seizures and epilepsy in children. Thus McAfee *et al.* (2007) conducted a retrospective cohort study of 133,440 pediatric patients (age 6–17 years) without history of seizures or prior use of anticonvulsant medications. The data source for this study was a research database containing pharmacy and medical claims for members of a large US-based managed care organization. The incidence rate of seizures among children without psychiatric diagnoses was 149/100,000 person-years (95% CI = 122–180), while that among children with psychiatric diagnoses other than ADHD was 513/100,000 person-years (95% CI = 273–878).

Depression preceding the onset of stroke

Five studies have investigated the impact of depression on the risk of stroke in large cohorts ranging from 1,703 to 6,675 subjects (Colantonio *et al.*, 1992; Everson *et al.*, 1998; Jonas and Mussolino, 2000; Larson *et al.*, 2001; May *et al.*,

2002). Four found that depression increased the risk of developing stroke, after controlling for other risk factors of stroke (Everson *et al.,* 1998; Jonas and Mussolino, 2000; Larson *et al.,* 2001; May *et al.,* 2002). For example, Larson *et al.* (2001) followed 1,703 subjects for a period of 13 years; patients with a depressive disorder or depressive symptoms had a 2.67 (CI = 1.08–6.63) relative risk of developing a stroke, after controlling for vascular risk factors (hypertension, diabetes, hyperlipidemia, heart disease and use of tobacco). May *et al.* (2002) followed 2,201 men aged 45–59 years for 14 years. Patients with significant symptoms of depression had a 3.36 (CI = 1.29–8.71) relative risk of developing a fatal stroke.

As shown below, the increased risk of stroke in patients with depression may be mediated through a direct impact on coagulation and central nervous system (CNS) vascular parameters, and indirectly by increasing risks of cardiovascular disease, hypertension, cardiac arrhythmias and diabetes.

A history of depression preceding dementia

A prior history of depressive episodes and family history of depression have been identified as risk factors for developing depression in Alzheimer's dementia (AD) (Pearlson *et al.,* 1990; Migliorelli *et al.,* 1995; Lyketsos *et al.* 1996; Judd 1997; Garre-Olmo *et al.,* 2003). Furthermore, multiple studies have demonstrated an increased history of depression in patients who go on to develop dementia, compared with those who do not (Lee and Lyketsos, 2003). In a study by Lyketsos *et al.* (1996), 27% of patients with AD had experienced a prior history of major depression. Similar conclusions can be derived from population-based studies. For example, in a study of 1,003 elderly subjects (all with a Mini Mental State score of more than 26), the presence of significant depressive symptoms at baseline predicted a higher risk of cognitive decline 4 years later (Paterniti *et al.,* 2002). The severity of a mood disorder was also associated with the risk of developing dementia. Data from a case-register study of almost 23,000 patients with an affective disorder suggested that increasing severity, expressed as the number of major depressive episodes leading to an inpatient admission, increased the risk of developing dementia (Kessing and Andersen, 2004). Thus, patients with three admissions had close to a 3-fold increased risk of dementia (95% CI = 0.64–13.2), compared to patients with only one admission.

In a recent study, Dal Forno *et al.* (2005) determined the relative hazard of premorbid depressive symptomatology for the development of dementia and AD. They investigated the risk for the incidence of dementia and AD over a 14-year period in 1,357 community-dwelling men and women, and prospectively carried out comprehensive medical and neuropsychological evaluations every 2 years to identify depressive symptoms. Premorbid depressive symptoms significantly increased the risk for dementia, particularly the risk of AD in men (but not in women), with hazard ratios approximately two times greater than for individuals without a history of depressive symptoms. This finding was independent of vascular disease.

Similar data were found in one study of 265 depressed individuals and 138 healthy, non-depressed controls age 60 and older who were followed for at least 1 year, during which time they underwent periodic clinical evaluation by a geriatric psychiatrist. Among the depressed subjects, 27 were found to develop dementia (11 with AD, 8 with vascular dementia, and 8 with dementia of undetermined etiology) and 25 additional subjects displayed other forms of cognitive impairment (Steffens *et al.*, 2004). In contrast, only two of the non-depressed controls developed substantial cognitive impairment with clinical diagnoses of dementia. Clearly, these data revealed significantly higher incidence rates of dementia for this age among depressed subjects than would be expected.

The presence of depression in individuals with mild cognitive impairment (MCI) is predictive of a higher likelihood of developing AD. Indeed, of 114 patients with amnesic mild cognitive impairment that were followed for a 3-year period, 41 (36%) displayed a depressive disorder at baseline. After 3 years, 35 (85%) of these patients had developed AD, in comparison with 32% of the non-depressed subjects, yielding a relative risk of developing AD of 2.6 (95% CI = 1.8–3.6) (Modrego and Ferrandez, 2004).

Depressive disorders preceding the development of Parkinson's disease

Recent studies have demonstrated a variety of symptoms other than motor symptoms preceding the typical manifestations of Parkinson's disease (PD), including constipation, loss of smell, sleep disturbances such as rapid eye movement (REM) sleep behavior disorder (RBD) and depression. Three population-based studies are worth reviewing in greater detail (Nilsson *et al.*, 2001; Leentgens *et al.*, 2003; Reijnders *et al.*, 2007). In the first study all subjects diagnosed with depression between 1975 and 1990 were included and matched with subjects with the same birth year who were never diagnosed with depression (Leentgens *et al.*, 2003). Follow-up went up to April 30, 2000. Among the 1,358 depressed subjects, 19 developed PD, and among the 67,570 non-depressed subjects, 259 developed PD. Thus, people with depression were three times more likely to develop PD than non-depressed people [hazard ratio of 3.13 (95% CI = 1.95–5.01)] in multivariable analysis.

In the second study, investigators compared the incidence of depression in patients preceding the onset of PD with that of a matched control population (Nilsson *et al.*, 2001). To that end, data from an ongoing general practice-based register study that included a population of 105,416 people was used. Among patients who went on to develop PD, 9.2% had a history of depression, compared with 4.0% of the control population, yielding an odds ratio for a history of depression for these patients of 2.4 (95% CI = 2.1–2.7) (Leentgens *et al.*, 2003).

A third population-based study compared the risk of developing PD between patients with affective disorders and two groups of medically ill patients, one with osteoarthritis and the second with diabetes using linkage of public hospital registers from 1977 to 1993 (Reijnders *et al.*, 2007). A total of 164,385 patients entered the study database. The risk of being given a diagnosis of PD was significantly increased for patients with affective disorder, when compared to patients with osteoarthritis [odds ratio 2.2 (CI = 95% 1.7–2.8)] or diabetes [odds ratio: 2.2 (CI = 95% 1.7–2.9)] (Nilsson *et al.*, 2001).

COMMON PATHOGENIC MECHANISMS BETWEEN PSYCHIATRIC AND NEUROLOGIC DISORDERS?

The data reviewed above do not necessarily imply that psychiatric disorders are a cause of the various neurologic entities. The existence of common pathogenic mechanisms operant in psychiatric and neurologic disorders offers an attractive and more plausible explanation for the higher incidence of neurologic disorders in patients with a prior history of depression, and vice versa, the well-established (though not reviewed in this chapter) higher incidence of depressive, anxiety, ADHD and psychotic disorders in patients with epilepsy (Tellez-Zenteno *et al.*, 2007), and the higher incidence of depressive disorders in patients with stroke, PD, and dementias (Kanner, 2005). The data supporting this theory will be reviewed below.

Common pathogenic mechanisms between mood disorders and epilepsy?

Common pathogenic mechanisms between mood disorders and epilepsy can be inferred from experimental studies carried out in animal models of epilepsy and clinical studies done in patients with primary mood disorders and in patients with epilepsy.

Neurotransmitter abnormalities

Neurotransmitter abnormalities are one set of potential common pathogenic mechanisms shared by the two disorders. The neurotransmitters in question include: (1) serotonin (5-HT), specifically 5-HT_{1A} and (2) norepinephrine (NE), dopamine (DA), γ-aminobutyric acid (GABA) and glutamate.

Data from experimental animal studies

In some animal models of epilepsy, decreased 5-HT and NE activity has been shown to facilitate the kindling of seizures, to exacerbate seizure severity and intensify seizures (Jobe et al., 1999). For example, the two strains of the genetic epilepsy-prone rats (GEPR), GEPR-3 and GEPR-9, have a genetically determined predisposition to sound-induced generalized tonic/clonic seizures (GTCS) (Jobe et al., 1999). Both strains of rats have innate pre- and postsynaptic NE and 5-HT transmission deficits, the former resulting from deficient arborization of neurons arising from the locus coeruleus coupled with excessive presynaptic suppression of stimulated NE release in the terminal fields and lack of postsynaptic compensatory upregulation (Jobe et al., 1999). GEPR-9 rats have a more pronounced NE transmission deficit and, in turn, exhibit more severe seizures than GEPR-3 rats (Jobe et al., 1994). Abnormal 5-HT arborization has also been identified in the GEPR's brain coupled with deficient postsynaptic $5HT_{1A}$-receptor density in the hippocampus (Dailey et al., 1992). Of note, GEPRs display similar endocrine abnormalities to those identified in patients with major depressive disorder (MDD), such as increased corticosterone serum levels, deficient secretion of growth hormone and hypothyroidism (Jobe, 2006). Abnormal 5-HT and NE transmission in the brain has been recognized as some of the pivotal pathogenic mechanisms of mood disorders in humans.

Increments of either NE and/or 5-HT transmission with the selective serotonin reuptake inhibitor (SSRI) sertraline resulted in a dose-dependent seizure-frequency reduction in the GEPR, which correlated with the extracellular thalamic serotonergic concentration (Yan et al. 1993; Yan et al., 1995). In addition, the 5-HT precursor 5-HTP has been shown to have anticonvulsant effects in GEPRs when combined with a mono-amino-oxidase inhibitor (MAOI) (Jobe et al., 1999), while SSRIs and MAOIs have been found to exert anticonvulsant effects in genetically prone epilepsy mice (Jobe et al., 1999; Jobe, 2006) and baboons (Meldrum et al., 1982) as well as in non-genetically prone cats (Polc et al., 1979), rabbits (Piette et al., 1963) and rhesus monkeys (Yanagita et al., 1980). Conversely, drugs that interfere with the release or synthesis of NE or 5-HT exacerbate seizures in the GEPRs (Jobe et al., 1999; Jobe, 2006).

An anticonvulsant effect of 5-HT activity has been reported in other animal models of epilepsy. Lopez-Meraz et al. (2005) studied the impact of two 5-HT_{1A} receptor agonists, 8-OH-DPAT and Indorenate, in three animal models of epileptic seizures (clonic–tonic induced by pentylenetetrazol (PTZ), status epilepticus of limbic seizures induced by kainic acid (KA) and tonic–clonic seizures induced by amygdala kindling) in Wistar rats. They found that 8-OH-DPAT lowered the incidence of seizures and the mortality induced by PTZ, increased the latency and reduced the frequency of wet-dog shake and generalized seizures induced by KA and at high doses diminished the occurrence and delayed the establishment of status epilepticus. Indorenate increased the latency to the PTZ-induced seizures and decreased the percentage of rats that showed tonic extension and death, augmented the latency to wet-dog shake and generalized seizures and diminished the number of generalized seizures.

Furthermore, antiepileptic drugs (AEDs) with established psychotropic effects (carbamazepine – CBZ, valproic acid – VPA and lamotrigine – LTG) have been found to cause an increase in 5-HT (Whitton and Fowler, 1991; Yan *et al.*, 1992; Dailey *et al.*, 1998; Dailey *et al.*, 1997a, b; Southam *et al.*, 1998). In fact, the anti-convulsant protection of CBZ can be blocked with 5-HT depleting drugs in GEPRs (Yan *et al.*, 1992). Likewise, in a recent study, Clinckers *et al.* (2005a, b) investigated the impact of oxcarbazepine (OXC) infusion on the extracellular hippocampal concentrations of 5-HT and DA in the focal pilocarpine model for limbic seizures. When OXC was administered together with verapamil or proben-ecid (so as to ensure its passage through the blood-brain barrier), complete seizure remission was obtained, which was associated with an increase in 5-HT and DA extracellular concentrations (Clinckers *et al.*, 2005a, b).

In addition, it has been suggested that the anticonvulsant effect of vagus nerve stimulation (VNS) in the rat could be mediated by NE and 5-HT mechanisms, as depletion of NE and 5-HT neurons in the rat prevents or reduces significantly the anticonvulsant effect of VNS against electroshock or PTZ-induced seizures (Naritoku *et al.*, 1995; Browning *et al.*, 1997). Furthermore, the effect of VNS on the locus coeruleus and raphe may be responsible for the reported antidepressant effects identified in humans (Nahas *et al.*, 2005).

Neurotransmitter changes in humans

Abnormal binding of 5-HT_{1A} receptors has been found in patients with temporal lobe epilepsy (TLE) and others with primary mood disorders, as demonstrated with studies carried out with positron emission tomography (PET). For example, in a PET study of patients with TLE using the 5-HT_{1A} receptor antagonist ([18F] *trans*-4-fluro-*N*-2-[4-(2-methoxyphenyl)piperazin-1-yl]ethyl-*N*-(2-pyridyl) cyclohexan-ecarboxamide), reduced 5-HT_{1A} binding was found in mesial temporal structures ipsilateral to the seizure focus in patients with and without hippocampal atrophy (Toczek *et al.*, 2003). In addition, a 20% binding reduction was found in the raphe and 34% lower binding in the ipsilateral thalamic region to the seizure focus (this difference yielded a statistical trend). In a separate PET study aimed at quantifying 5-HT_{1A} receptor binding in 14 patients with TLE, decreased binding was identified in the hippocampus, amygdala, anterior cingulate and lateral temporal neocortex ipsilateral to the seizure focus, as well as in the contralateral hippocampus, but to a lesser degree, and in the raphe nuclei (Savic *et al.*, 2004).

These changes of 5-HT_{1A} receptor binding are quite similar to those identi-fied in PET studies of patients with primary major depressive disorders (MDDs) Deficits in 5-HT transmission in human depression is thought to be partially related to a paucity of serotonergic innervation of its terminal areas, suggested by a reduction of 5-HT levels in brain tissue, plasma and platelets and by a deficit in 5-HT transporter binding sites in postmortem human brain. For example, Sargent *et al.* (2000) demonstrated reduced 5-HT_{1A} receptor binding potential values in frontal, temporal, and limbic cortex with PET studies using [11C]WAY-100635 in both unmedicated and medicated depressed patients compared with healthy volunteers. Of note, binding potential values in medicated patients were similar to those in unmedicated patients. Drevets *et al.* (1999) using the same radioligand reported a decreased binding potential of $5\text{-}HT_{1A}$ receptors in mesial temporal cortex and in the raphe in 12 patients with familial recurrent major depressive episodes, compared to controls.

A prominent pathogenic role of 5-HT has been identified as well in the brains of patients who committed a successful suicide. To cite one example, a deficit in the density or affinity of postsynaptic $5\text{-}HT_{1A}$ receptors has been reported in the hippocampus and amygdala of untreated depressed patients who committed suicide (Oguendo *et al.*, 2003).

Other investigators using the $5\text{-}HT_{1A}$ tracer, 4,2-(methoxyphenyl)-1-[2-(*N*-2-pyridinyl)-*p*-fluorobenzamido]ethylpiperazine ([(18)F]MPPF), found that the decrease in binding of $5\text{-}HT_{1A}$ was significantly greater in the areas of seizure onset and propagation identified with intracranial electrode recordings. As in the other studies, reduction in $5\text{-}HT_{1A}$ binding was present even when quantitative and qualitative magnetic resonance imaging (MRI) studies were normal (Merlet *et al.*, 2004).

Hasler *et al.* (2007) compared $5\text{-}HT_{1A}$ receptor binding between 37 TLE patients with and without MDD with interictal PET using the $5\text{-}HT_{1A}$ antagonist [(18)F]FCWAY. The MDD was diagnosed by clinical and structured psychiatric interviews. They found that in addition to decreased $5\text{-}HT_{1A}$ receptor binding in the epileptic focus, patients with TLE and MDD exhibited a significantly more pronounced reduction in $5\text{-}HT_{1A}$ receptor binding extending into non-lesional limbic brain areas outside the epileptic focus. The side of the ictal focus and the presence of mesial temporal sclerosis were not associated with the presence of comorbid depression. In a second study of 45 patients with TLE, Theodore *et al.* (2006) demonstrated an inverse correlation between increased severity of symptoms of depression identified on the Beck Depression Inventory and $5\text{-}HT_{1A}$ receptor binding at the hippocampus ipsilateral to the seizure focus and to a lesser degree at the contralateral hippocampus and midbrain raphe. These changes of $5\text{-}HT_{1A}$ receptor binding are quite similar to those identified in PET studies of patients with primary MDDs (Meschaks *et al.*, 2005).

Common pathogenic mechanisms between mood disorders and dementia

The presence of hippocampal atrophy is one of the characteristic findings of minimal cognitive impairment that worsens in magnitude as the patient develops an overt AD. By the same token, atrophy of temporal and frontal lobe structures are prominent in other types of dementia such as fronto-temporal dementia, to name one.

Patients with primary mood disorders such as recurrent MDDs and bipolar disorders have been found to have structural changes presenting as atrophy of temporal lobe structures (identified by high resolution MRI and volumetric measurements) that include amygdala, hippocampus, entorhinal cortex and temporal lateral neocortex; frontal lobe structures involving prefrontal, orbitofrontal and mesial frontal cortex; and to a lesser degree, thalamic nuclei and basal ganglia (Sheline, 2006).

Neurotransmitter changes: Serotonin has been implicated in the pathogenic mechanisms identified in various dementing processes. For example, Truchot *et al.* (2007) investigated the binding potential of 5-HT$_{1A}$ receptors in the hippocampal formation of 10 patients with mild AD, 11 patients with MCI and 21 aged paired control subjects. To measure the 5-HT$_{1A}$ receptor density they used PET with a selective 5-HT(1A) antagonist, 2′-methoxyphenyl-(*N*-2′-pyridinyl)-*p*-[(18)F]fluorobenzamidoethylpiperazine ([(18)F]MPPF). They found significant differences in binding potential of 5-HT$_{1A}$ receptors among the three groups; compared to controls, patients with MCI had 59% higher binding potential, while it was 35% lower in patients with AD. These changes were not related to the magnitude of hippocampal atrophy. The authors interpreted these findings as reflecting a compensatory mechanism illustrated by an upregulation of 5-HT metabolism at the stage of amnesic mild cognitive impairment in contrast with a dramatic decrease at later stages of AD.

Other studies have supported the pathogenic role played by 5-HT$_{1A}$ receptors in AD and MCI, but have not supported an upregulation of 5-HT metabolism suggested by Truchot *et al.* In a PET study of eight patients with AD, six with MCI and five controls, Kepe *et al.* (2006) found that patients with AD had significantly lower receptor densities in the hippocampi and in raphe nuclei compared to controls. The average mean decrease of 5-HT$_{1A}$ receptor density was 24% in patients with mild cognitive impairment and 49% in patients with AD. In addition there was significant correlation of 5-HT$_{1A}$ receptor density decreases in hippocampus with worsening of clinical symptoms, measured with the Mini Mental State Exam scores.

The pathogenic role of 5-HT$_{1A}$ has also been suggested in fronto-temporal dementias. In a pilot study using PET, Lanctôt *et al.* (2007) demonstrated significantly decreased serotonin 5-HT$_{1A}$ binding potential in four patients with fronto-temporal dementia compared with controls in all 10 brain regions examined.

Common pathogenic mechanisms in depression and cerebrovascular disease?

Potential common pathogenic mechanisms that can be operant in mood and cerebrovascular disorders may be divided into those that facilitate the development of vascular disease in a *direct fashion*, either through a change in coagulation parameters that cause hypercoagulable states, or the development of vasculitic processes. Other potential pathogenic mechanisms include those that increase the risk of stroke by *indirect means*, that is, by facilitating the development of comorbid disorders that are well-known risk factors of cerebrovascular disorder such as hypertension, diabetes and cardiac disease.

Depression can contribute to the development of vascular disease through three direct mechanisms: (1) activation of the hypothalamic-pituitary-adrenal axis, (2) sympatho-adrenal hyperactivity and (3) vascular inflammation and facilitation of hypercoagulability.

The activation of the hypothalamic-pituitary-adrenal axis causes elevation of cortisol, which promotes the development of atherosclerosis and causes sympatho-adrenal hyperactivity. This hyperactivity results in increased secretion of catecholamines that in turn leads to vasoconstriction, platelet activation and elevated heart rate (Troxler et al., 1977).

The vascular inflammatory reaction has been attributed to elevated plasma levels of inflammatory markers such as C-reactive protein and pro-inflammatory cytokines (including interleukin-1 and interleukin-6). All of these markers have been found to be elevated in patients with depression with and without vascular disease (Thompson et al., 1995; Leonard, 2000; Koenig, 2001; Leonard, 2001; Mulvihill and Foley, 2002). These cytokines are also known to be secreted in the presence of increased stress (Musselman et al., 1996; Kuijpers et al., 2002).

The suspected role of depression in the development of hypercoagulable states has been related to hypothalamic-pituitary-adrenal axis and sympatho-adrenal hyperactivity, as both are known to stimulate the coagulation process. This is mediated by the occurrence of hypercortisolism, which can cause elevation of Factor VIII and von Willebrand factor. Furthermore, higher secretion of catecholamines can lead to an increase in coagulation and fibrinolysis (Mendelson, 2002). In addition, abnormal platelet function in untreated depressed patients can mediate a hypercoagulable state, as depressed patients have been found to have platelet reactivity of up to 40% greater than controls (Serebruany et al., 2001b). Abnormal platelet function has been associated with abnormalities in platelets' 5-HT receptors (Serebruany et al., 2001a), while treatment with an SSRI has been shown to normalize platelet activation (Musselman et al., 2000; Serebruany et al., 2001a; Serebruany et al., 2001b).

Impact of depression on hypertension

The impact of depression on the risk of hypertension is not clear. One prospective study failed to find any association between a history of depression and an increased risk of developing hypertension (Jonas and Lando, 2000), while two other studies did find that the presence of depression at baseline was associated with an increased likelihood of hypertension at follow-up (Carney et al., 1993; Davidson et al., 2000; Carney et al., 2001).

Impact of depression on the risk of cardiac disease and arrhythmias

In an epidemiological longitudinal study of 93,676 postmenopausal women followed for an average of 4.1 years, depression was significantly related to risk factors of cardiovascular disease and a history of cardiovascular morbidity. Among women with no history of cardiovascular disease, depression was an independent predictor of cardiovascular disease and all-cause mortality; while among women with a prior history of cardiovascular disease, depression was significantly associated with the development of a stroke. In addition, five other studies found that depression predicted a worse prognosis of coronary artery disease and/or cardiac mortality (Colantonio *et al.*, 1992; Migliorelli *et al.*, 1995; Everson *et al.*, 1998; Jonas and Mussolino, 2000; Larson *et al.*, 2001).

Decreased heart-rate variability has been blamed for the development of cardiac arrhythmias and sudden death, and has been attributed to increased sympathetic or decreased parasympathetic activities. In a study of 700 postmyocardial infarction patients, those with depression had significantly decreased heart rate variability relative to controls (Carney *et al.*, 2001). In a separate study, the presence of depressive symptoms was associated with a 30% decrease of baroreflex sensitivity (Watkins and Grossman, 1999), while other studies revealed higher QT variability among depressed patients than controls (Nahshoni *et al.*, 2000; Yeragani *et al.*, 2000). Furthermore, patients with depression have been found to face a higher risk of ventricular tachycardia (Carney *et al.*, 1993).

Impact of depression on the risk and/or severity of diabetes

Depression may be an independent risk factor for the development of Type 2 diabetes after controlling for confounding factors such as race, age, gender, socio-economic status, level of education and body weight (Eaton *et al.*, 1996). The depression-related biological mechanisms responsible for the greater risk of developing diabetes include an increased release of counter-regulatory hormones (glucocorticoids, growth hormone and glucagon) and catecholamines, which counteract the hypoglycemic action of insulin by raising levels of glucagon. These hormonal changes may be responsible, in turn, for the development of insulin resistance. Furthermore, once patients develop diabetes, the presence of depression may be associated with worse glycemic control, related to an increased difficulty of adhering to a proper diet, and worse compliance with hypoglycemic agents.

Common pathogenic mechanisms of depression and PD?

A review of the neuroanatomic structural and functional abnormalities of PD, and the resulting monoaminergic changes involving DA, 5-HT and NE reveals some parallels with abnormalities identified in primary depressive disorders.

Using PET studies, patients with PD and major depression were found to have decreased metabolic activity in the caudate, anterior temporal cortex and orbital-inferior frontal cortex compared with non-depressed PD patients (Mayberg et al., 1990). The metabolic activity in the inferior frontal cortex was inversely proportional to the degree of depression. In a follow-up study, depressed PD patients successfully treated with fluoxetine exhibited increased and normalized metabolic activity in the dorsal frontocortical areas, with a decrease in the ventral paralimbic areas (Mayberg et al., 1997). It has been concluded from PET studies that the medial prefrontal cortex is a common area of dysfunction in depressed PD patients and in those with a primary depression. Furthermore, decreased metabolism in the frontal cortex–basal ganglia–thalamic loop of PD patients is similar to the pattern shown in metabolic studies of non-PD patients with major depression (McDonald and Krishnan, 1992; Ring et al., 1994). Other neuroanatomical structural abnormalities that are commonly identified in patients with PD and patients with primary major depression include a smaller than normal caudate nucleus, putamen and thalamus (Lisanby et al., 1993).

From a pathological standpoint, PD is characterized by neuronal cell loss and the formation of Lewy bodies in several brainstem nuclei, including the substantia nigra, ventral tegmental area, locus coeruleus, dorsal raphe nucleus and nucleus basalis. In addition, several cortical structures are also affected, albeit to a lesser degree, including the temporal cortex (involving the entorhinal cortex) and the frontal cortex (involving the anterior cingulate cortex) (Cummings and Mega, 2003).

Degeneration of the ventral tegmental area and substantia nigra, which project to the mesocortical and mesolimbic areas and striatum, can result in depletion of DA, 5-HT and NE (Cummings and Mega, 2003). The role of these monoamines in the pathogenesis of primary depression is well known and is the basis for the pharmacotherapy of depressive disorders. Furthermore, patients with PD who suffered from depression have been found to show greater loss of ventral tegmental area dopaminergic neurons at autopsy than non-depressed PD patients (Torack and Morris, 1988; Kuzis et al., 1997).

While disturbances in DA are the hallmark of PD, comorbid depressive episodes may not be solely explained by an abnormal DA secretion pattern. Since DA and 5-HT are inter-related, DA terminal destruction may result in depletion of 5-HT systems. In fact, some authors have suggested that dopaminergic cell loss in the ventral tegmental area may not be directly causative of depression, but may be so indirectly, by facilitating a loss of serotonergic cells in the raphe, which then more directly causes depression. In fact, Becker et al. (1997) have used transcranial sonography to show significantly reduced echogenicity in the mesencephalic brainstem dorsal raphe nuclei of depressed PD patients.

Other evidence supporting the role of 5-HT in depressive disorders of PD patients includes the finding of decreased levels of the major 5-HT metabolite, 5-hydroxyindoleacetic acid (5-HIAA) in the cerebrospinal fluid, and an inverse correlation with the severity of depressive symptoms (Mayeux et al., 1984; Mayeux, 1990).

IMPACT OF MOOD DISORDERS ON THE COURSE OF THE NEUROLOGIC DISORDER AND THE PATIENT'S QUALITY OF LIFE

Epilepsy

Several studies have demonstrated the negative impact of depressive disorders on the quality of life of patients with epilepsy (Perrine *et al.*, 1995; Gilliam *et al.*, 1997; Lehrner *et al.*, 1999). For example, in a study of 56 patients with epilepsy carried out in Germany by Lehrner *et al.* (1999), depression was the single strongest predictor for each domain of health-related quality of life (HRQOL). The significant association of depression with HRQOL persisted after controlling for seizure frequency, seizure severity and other psychosocial variables. In another study of 257 patients with epilepsy by Perrine *et al.* (1995), the "mood factor" had the highest correlations with scales of the QOLIE-89 (Quality of Life in Epilepsy Inventory-89) and was the strongest predictor of poor quality of life in regression analyses. In a separate study, Gilliam *et al.* (1997) investigated the variables responsible for poor quality of life measured with the QOLIE-89 in 194 adult patients with refractory partial epilepsy. Patients averaged 9.7 seizures/month (range: 0.3–51), but there was no correlation between the type or the frequency of seizures and the QOLIE-89 scores. The presence of symptoms of depression and neurotoxicity from AEDs were the only independent variables significantly associated with poor quality-of-life scores on the QOLIE-89 summary score.

A negative impact of depressive disorders on the responses to pharmacologic and surgical treatments has also been identified (Anhoury *et al.*, 2000; Kanner *et al.*, 2006; Hitiris *et al.*, 2007). In a study of 780 patients with new-onset epilepsy, Hitiris *et al.* (2007) found that individuals with a history of psychiatric disorders, and particularly depression, were almost three times less likely to be seizure-free with AED treatment (median follow-up period was 79 months) than patients without a history of psychiatric disorders. Similarly, among 121 patients who underwent a temporal lobectomy, Anhoury *et al.* (2000) reported a worse postsurgical seizure outcome for patients with a psychiatric history compared with those without a psychiatric history. In a study of 100 patients who had a temporal lobectomy and were followed for a mean period of 8.8 ± 3.3 years, Kanner *et al.* (2006) investigated the role of a lifetime history of depression as a predictor of postsurgical seizure outcome. Using a multivariate logistic regression model, the investigators evaluated the covariates of a lifetime history of depression, cause of TLE (i.e., mesial temporal sclerosis, lesional or idiopathic), duration of seizure disorder, occurrence of GTCS and extent of resection of mesial temporal structures. A lifetime history of depression and a smaller resection of mesial temporal structures were the only independent predictors of persistent auras in the absence of disabling seizures in multivariate analyses. A lifetime history of depression was also an independent predictor of failure to reach freedom from disabling seizures in univariate analyses. The data in these three studies raise the question of whether a history of depression may be a marker of a more severe form of epilepsy.

Impact of poststroke depression on the course after the stroke

The presence of poststroke depression (PSD) has been found to have a negative impact on recovery of cognitive function, recovery of ability to perform activities of daily living and mortality risks. With respect to the impact of PSD on cognitive functions, Starkstein et al. (1988) demonstrated that patients with major PSD had significantly more cognitive deficits than non-depressed patients who experienced a similar location and size of left-hemisphere stroke. However, this was not the case for strokes in the right hemisphere. In a follow-up study of 140 patients, Robinson et al. (1983) also found that the presence of major PSD was associated with greater cognitive impairment 2 years after a stroke.

Regarding the impact of PSD on the recovery of the ability to perform activities of daily living (ADL), Parikh et al. (1990) found that in-hospital PSD was the most important variable predicting poor recovery in ADL over a 2-year period. In fact, the score of in-hospital ADL was not associated with the 2-year recovery.

Likewise, the negative impact of PSD on the course of strokes is reflected in the associated higher mortality risk. Indeed, in a study of 976 stroke patients followed for 1 year, those with PSD had 50% higher mortality than those without (Wade et al., 1987).

Impact of depression on the course and quality of life in PD

The presence of depression in PD patients has been associated with a more rapid deterioration of motor and cognitive functions, especially executive function, and a greater likelihood of displaying psychotic symptoms and physical disability (Starkstein et al., 1990; Starkstein et al., 1992; Kuzis et al., 1997; Kuopio et al., 2000; The Global Parkinson's Disease Survey Steering Committee, 2002; Weintraub et al., 2004; Ravina et al., 2007). Such impact is appreciated even when the depressive disorder occurs in the early stage of the disease. For example, in a study by Ravina et al. (2007) a total of 114 (27.6%) patients were identified among a group of 413 patients with a depressive disorder during the average 14.6 months of follow-up. About 40% of these subjects were neither treated with antidepressants nor referred for further psychiatric evaluation. Depression was a significant predictor of more impairment in ADLs and increased need for symptomatic therapy of PD (hazard ratio = 1.86; 95% CI = 1.29, 2.68).

The cognitive disturbances of depressed PD patients include poor insight and judgment, and problems with planning. However, memory impairment is seen less frequently. Neuropsychological tests that evaluate executive functioning have demonstrated significant deficits in PD patients, indicating frontal subcortical impairment – an impairment that may also be operant in the development of mood disorders. In a study that compared performance in neuropsychological testing between PD patients with major depression, non-depressed PD patients, patients with primary major depression, and healthy controls, depressed PD patients exhibited impairments in set shifting and concept formation. These abnormalities were unique to this group of patients (Kuzis et al., 1997). However, cognitive deterioration in depressed PD patients can be mitigated with treatment of the depressive disorder. In a longitudinal study that compared cognitive performance in depressed and non-depressed PD patients that were followed for a 3- to 4-year period, cognitive functions deteriorated more quickly in the depressed PD group. However, in the depressed PD patients that were treated, there was an attenuated decrement in cognitive scores (11%), compared with untreated patients (Starkstein et al., 1990).

Depression also has been found to have a negative impact on the quality of life in PD patients, just as in stroke and epilepsy. For example, in a multicenter study conducted by the The Global Parkinson's Disease Survey Steering Committee (2002) and involving six countries, data were obtained from 2,020 PD patients and 687 caregivers, and depression was found to be the most significant predictor variable in poor HRQOL. It is worth noting that patients often failed to recognize their depressive disorder, as only 1% of the patients reported feeling depressed, while 50% were considered depressed by study criteria of a score of >10 in the Beck Depression Inventory. Furthermore, in a community-based study of 228 people with PD, depression was the factor most closely related to a poor quality of life, while the stage of PD, duration and cognitive impairment had a lesser impact (Kuopio et al., 2000). Others have confirmed these findings (Schrag et al., 2000).

Impact of depression on quality of life in AD

Depressive disorders have a negative impact at various levels on the quality of life of patients with AD. They are associated with greater impairment in ADL (Bassuk et al., 1998) and have been found to be a predictor of cognitive decline (Kopetz et al., 2000). They are associated with an earlier need for placement in a nursing home and for discharge from an assisted-living facility to a higher level facility (Gonzalez-Salvador et al., 2000). Finally, the presence of depressive disorders in AD has been found to increase the caregiver's depression and burden.

CONCLUDING REMARKS

In this chapter, data were presented that suggest a bidirectional relationship between several psychiatric disorders, particularly mood disorders, and various neurologic disorders. These data do not presuppose a causative relation between neurologic and psychiatric disorders; rather, these data suggest the presence of common pathogenic mechanisms mediating both types of disorders. This hypothesis was supported by an abundance of evidence in this chapter. Clearly, the studies reviewed in this chapter point to the need to change the way we view psychiatric comorbid disorders in neurologic disease. They cannot solely be considered as the expression of a reactive process to a variety of obstacles posed by the neurologic disorders. Thus, in the evaluation of a neurologic disorder, it is of the essence to investigate the presence of psychiatric comorbidity, perhaps even antedating the initial neurologic symptoms, and conversely, in the evaluation of any psychiatric disorder, we must investigate the presence of any risk factors for the development of neurologic disorders. Finally, the presence of comorbid psychiatric disorders, particularly depression, has a direct negative impact on the course of the neurologic disorder and patients' quality of life.

REFERENCES

Anhoury, S., Brown, R. J., Krishnamoorthy, E. S., Trimble, M. R. (2000). Psychiatric outcome after temporal lobectomy: a predictive study. *Epilepsia* **41**, 1608–1615.

Austin, J. K., Harezlak, J., Dunn, D. W., *et al.* (2001). Behavior problems in children before first recognized seizure. *Pediatrics* **107**, 115–122.

Bassuk, S. S., Berkman, L. F., Wypip, D. (1998). Depressive symptomatology and incident cognitive decline in an elderly community sample. *Arch Gen Psychiatry* **55**, 1073–1081.

Becker, T., Becker, G., Seufert, J., *et al.* (1997). Parkinson's disease and depression: evidence for an alteration of the basal limbic system detected by transcranial sonography. *J Neurol Neurosurg Psychiatry* **63**, 590–596.

Browning, R. A., Clark, K. B., Naritoku., D. K., Smith, D. C., Jensen, R. A. (1997). Loss of anticonvulsant effect of vagus nerve stimulation in the pentylenetetrazol seizure model following treatment with 6-hydroxydopamine or 5,7-dihydroxy-tryptamine. *Soc Neurosci* **23**, 2424.

Carney, R. M., Freedland, K. E., Rich, M. W., Smith, L. J., Jaffe, A. S. (1993). Ventricular tachycardia and psychiatric depression in patients with coronary artery disease. *Am J Med* **95**, 23–28.

Carney, R. M., Blumenthal, J. A., Stein, P. K., Watkins, L., Catellier, D., Berkman, L. F., *et al.* (2001). Depression, heart rate variability, and acute myocardial infarction. *Circulation* **104**, 2024–2028.

Clinckers, R., Smolders, I., Meurs, A., Ebinger, G., Michotte, Y. (2005a). Hippocampal dopamine and serotonin elevations as pharmacodynamic markers for the anticonvulsant efficacy of oxcarbazepine and 10,11-dihydro-10-hydroxycarbamazepine. *Neurosci Lett* **16**(390), 48–53.

Clinckers, R., Smolders, I., Meurs, A., Ebinger, G., Michotte, Y. (2005b). Quantitative *in vivo* microdialysis study on the influence of multidrug transporters on the blood-brain barrier passage of oxcarbazepine: concomitant use of hippocampal monoamines as pharmacodynamic markers for the anticonvulsant activity. *J Pharmacol Exp Ther* **314**, 725–731.

Colantonio, A., Kasi, S. V., Ostfeld, A. M. (1992). Depressive symptoms and other psychosocial factors as predictors of stroke in the elderly. *Am J Epidemiol* **136**, 884–894.

Cummings, J. L., Mega, M. S. (2003). Parkinson's disease. *Neuropsychiatry and Behavioral Neuroscience*, pp. 256–257. New York: Oxford University Press.

Dailey, J. W., Mishra, P. K., Ko, K. H., Penny, J. E., Jobe, P. C. (1992). Serotonergic abnormalities in the central nervous system of seizure-naive genetically epilepsy-prone rats. *Life Sci* **50**, 319–326.

Dailey, J. W., Reith, M. E., Yan, Q. S., Li, M. Y., Jobe, P. C. (1997a). Carbamazepine increases extracellular serotonin concentration: lack of antagonism by tetrodotoxin or zero Ca^{2+}. *Eur J Pharmacol* **328**(2–3), 153–162.

Dailey, J. W., Reith, M. E. A., Yan, Q. S., Li, M. Y., Jobe, P. C. (1997b). Anticonvulsant doses of car-bamazepine increase hippocampal extracellular serotonin in genetically epilepsy-prone rats: dose response relationships. *Neurosci Lett* **227**(1), 13–16.

Dailey, J. W., Reith, M. E., Steidley, K. R., Milbrandt, J. C., Jobe, P. C. (1998). Carbamazepine-induced release of serotonin from rat hippocampus *in vitro. Epilepsia* **39**(10), 1054–1063.

Dal Forno, G., Palermo, M. T., Donohue, J. E., Karagiozis, H., Zonderman, A. B., Kawas, C. H. (2005). Depressive symptoms, sex, and risk for Alzheimer's disease. *Ann Neurol* **57**(3), 381–387.

Davidson, K., Jonas, B. S., Dixon, K. E., Markovitz, J. H. (2000). Do depression symptoms predict early hypertension incidence in young adults in the CARDIA study? Coronary Artery Risk Development in Young Adults. *Arch Intern Med* **160**, 1495–1500.

Dominian, M. A., Serafetidines, E. A., Dewhurst, M. (1963). A follow-up study of late-onset epilepsy: II. Psychiatric and social findings. *Br Med J* **1**, 431–435.

Drevets, W. C., Frank, E., Price, J. C., Kupfer, D. J., Holt, D., Greer, P. J. N., Huang, Y., Gautier, C., Mathis, C. (1999). PET imaging of serotonin 1A receptor binding in depression. *Biol Psychiatry* **46**, 1375–1387.

Dunn, D. W., Austin, J. K., Huster, G. A. (1997). Behavior problems in children with new-onset epilepsy. *Seizure* **6**, 283–287.

Eaton, W. W., Armenian, H., Gallo, J., Pratt, L., Ford, E. (1996). Depression and risk for onset of type II diabetes: a prospective population-based study. *Diabetes Care* **22**, 1097–1102.

Everson, S. A., Roberts, R. E., Goldberg, D. E., Kaplan, G. A. (1998). Depressive symptoms and increased risk of stroke mortality over a 29-year period. *Arch Intern Med* **158**, 1133–1138.

Forsgren, L., Nystrom, L. (1990). An incident case referent study of epileptic seizures in adults. *Epilepsy Res* **6**, 66–81.

Garre-Olmo, J., Lopez-Pousa, S., Vilalta-Franch, J., Turon-Estrada, A., Hernandez-Ferrandiz, M., Lozano-Gallego, M., Fajardo-Tibau, C., Puig-Vidal, O., Morante-Munoz, V., Cruz-Reina, M. M. (2003). Evolution of depressive symptoms in Alzheimer disease: one-year follow-up. *Alzheimer Dis Assoc Disord* **17**, 77–85.

Gilliam, F., Kuzniecky, R., Faught, E., *et al.* (1997). Patient-validated content of epilepsy-specific qual-ity-of-life measurement. *Epilepsia* **38**(2), 233–236.

Gonzalez-Salvador, T., Arango, C., Lyketsos, C. G., Barba, A. C. (2000). The stress and psychological morbidity of the Alzheimer's patient caregiver. *Int J Geriatr Psychiatry* **14**, 701–710.

Hasler, G., Bonwetsch, R., Giovacchini, G., Toczek, M. T., Bagic, A., Luckenbaugh, D. A., Drevets, W. C., Theodore, W. H. (2007). 5-HT(1A) receptor binding in temporal lobe epilepsy patients with and without major depression. *Biol Psychiatry* **62**,1258–1264.

Hemmer, S. A., Pasternak, J. F., Zecker, S. G., Trommer, B. L. (2001). Stimulant therapy and seizure risk in children with ADHD. *Pediatr Neurol* **24**, 99–102.

Hesdorffer, D. C., Hauser, A. W. (2007). Epidemiologic considerations. In *Psychiatric Issues in Epilepsy: A Practical Guide to Diagnosis and Treatment* (A. B. Ettinger, A. M. Kanner, eds) 2nd edition, pp. 1–16. Philadelphia, PA: Lipincott, Williams and Wilkins.

Hesdorffer, D. C., Hauser, W. A., Annegers, J. F., *et al.* (2000). Major depression is a risk factor for sei-zures in older adults. *Ann Neurol* **47**, 246–249.

Hesdorffer, D. C., Ludvigsson, P., Olafsson, E., Gudmundsson, G., Kjartansson, O., Hauser, W. A. (2004). ADHD as a risk factor for incident unprovoked seizures and epilepsy in children. *Arch Gen Psychiatry* **61**, 731–736.

Hesdorffer, D. C., Hauser, W. A., Olafsson, E., Ludvigsson, P., Kjartansson, O. (2006). Depression and suicidal attempt as risk factor for incidental unprovoked seizures. *Ann Neurol* **59**(1), 35–41.

Hitiris, N., Mohanraj, R., Norrie, J., Sills, G. J., Brodie, M. J. (2007). Predictors of pharmacoresistant epilepsy. *Epilepsy Res* **5**(2–3), 192–196.

Holtmann, M., Becker, K., Kentner-Figura, B., Schmidt, M. H. (2003). Increased frequency of rolandic spikes in ADHD children. *Epilepsia* **44**, 1241–1244.

Hughes, J. R., DeLeo, A. J., Melyn, M. A. (2000). The electroencephalogram in attention deficit-hyperactivity disorder: Emphasis on epileptiform discharges. *Epilepsy Behav* **1**, 271–277.

Jobe, P. C. (2006). Affective disorder and epilepsy comorbidity in the genetically epilepsy prone-rat (GEPR). In *Depression and Brain Dysfunction* (F. Gilliam, A. M. Kanner, Y. I. Sheline, eds). London: Taylor & Francis, pp. 121–157.

Jobe, P. C., Mishra, P. K., Browning, R. A., *et al.* (1994). Noradrenergic abnormalities in the genetically epilepsy-prone rat. *Brain Res Bull* **35**, 493–504.

Jobe, P. C., Dailey, J. W., Wernicke, J. F. (1999). A noradrenergic and serotonergic hypothesis of the linkage between epilepsy and affective disorders. *Crit Rev Neurobiol* **13**, 317–356.

Jonas, B. S., Mussolino, M. E. (2000). Symptoms of depression as a prospective risk factor for stroke. *Psychosom Med* **62**, 463–471.

Jonas, B. S., Lando, J. F. (2000). Negative affect as a prospective risk factor for hypertension. *Psychosom Med* **62**, 188–196.

Jones, J. E., Watson, R., Sheth, J., Caplan, R., Koehn, M., Seidenberg, M., Hermann, B. (2007). Psychiatric comorbidity in patients with epilepsy. *Dev Med Child Neurol* **49**, 493–497.

Judd, L. L. (1997). The clinical course of major depressive disorders. *Arch Gen Psychiatry* **54**, 989–991.

Kanner, A. M. (2005). *Depression in Neurologic Disorders.* Cambridgeshire: Cambridge Medical Communications.

Kanner, A. M., Byrne, R., Smith, M. C., Balabanov, A., Frey, M. (2006). Does a lifetime history of depression predict a worse postsurgical seizure outcome following a temporal lobectomy? *Ann Neurol* **60**(Suppl 10), 19. (Abstract)

Kepe, V., Barrio, J. R., Huang, S. C., Ercoli, L., Siddarth, P., Shoghi-Jadid, K., Cole, G. M., Satyamurthy, N., Cummings, J. L., Small, G. W., Phelps, M. E. (2006). Serotonin 1A receptors in the living brain of Alzheimer's disease patients. *Proc Natl Acad Sci USA* **103**(3), 702–707.

Kessing, L. V., Andersen, P. K. (2004). Does the risk of developing dementia increase with the number of episodes in patients with depressive disorder and in patients with bipolar disorder? *J Neurol Neurosurg Psychiatry* **75**, 1662–1666.

Koenig, W. (2001). Inflammation and coronary heart disease: an overview. *Cardiol Rev* **9**, 31–35.

Kopetz, E. S., Steele, C. D., Brandt, J., *et al.* (2000). Characteristics and outcome of dementia residents in assisted living facility. *Int J Geriatr Psychiatry* **15**, 586–593.

Kuijpers, P. M., Hamulyak, K., Strik, J. J., Wellens, H. J., Honig, A. (2002). Beta-thromboglobulin and platelet factor 4 levels in post-myocardial infarction patients with major depression. *Psychiatry Res* **109**, 207–210.

Kuopio, A. M., Marttila, R. J., Helenius, H., *et al.* (2000). The quality of life in Parkinson's disease. *Mov Disord* **15**, 216–223.

Kuzis, G., Sabe, L., Tiberti, C., *et al.* (1997). Cognitive function in major depression and Parkinson's disease. *Arch Neurol* **54**, 982–986.

Lanctôt, K. L., Herrmann, N., Ganjavi, H., Black, S. E., Rusjan, P. M., Houle, S., Wilson, A. A. (2007). Serotonin-1A receptors in frontotemporal dementia compared with controls. *Psychiatry Res* **156**(3), 247–250. Epub 2007

Larson, S. L., Owens, P. L., Ford, D., Eaton, W. (2001). Depressive disorder, dysthymia, and risk of stroke. Thirteen-year follow-up from the Baltimore Epidemiological Catchment Area Study. *Stroke* **32**, 1979–1983.

Lee, H. B., Lyketsos, C. G. (2003). Depression in Alzheimer's disease: heterogeneity and related issues. *Biol Psychiatry* **54**, 353–362.

Leentgens, A. F. G., Van Der Akker, M., Metsemakers, J. F. M., *et al.* (2003). Higher incidence of depression preceding the onset of Parkinson's disease: a register study. *Mov Disord* **18**, 414–418.

Lehrner, J., Kalchmayr, R., Serles, W., *et al.* (1999). Health-related quality of life (HRQOL), activity of daily living (ADL) and depressive mood disorder in temporal lobe epilepsy patients. *Seizure* **8**(2), 88–92.

Leonard, B. E. (2000). Evidence for a biochemical lesion in depression. *J Clin Psychiatry* **61**(Suppl 6), 12–17.

Leonard, B. E. (2001). The immune system, depression and the action of antidepressants. *Prog Neuropsychopharmacol Biol Psychiatry* **25**, 767–780.

Lewis, A. (1934). Melancholia: a historical review. *J Ment Sci* **80**, 1–42.

Lisanby, S. H., McDonald, W. M., Massey, E. W., *et al.* (1993). Diminished subcortical nuclei volumes in Parkinson's disease by MR imaging. *J Neural Transm Suppl* **40**, 13–21.

Lopez-Meraz, M. L., Gonzalez-Trujano, M. E., Neri-Bazan, L., Hong, E., Rocha, L. L. (2005). 5-HT1A receptor agonists modify seizures in three experimental models in rats. *Neuropharmacology* **49**, 367–375.

Lyketsos, C. G., Tune, L. E., Pearlson, G., Steele, C. (1996). Major depression in Alzheimer's disease. An interaction between gender and family history. *Psychosomatics* **37**, 380–384.

May, M., McCarron, P., Stansfeld, S., Ben-Shlomo, Y., Gallacher, J., Yarnell, J., *et al.* (2002). Does psychological distress predict the risk of ischemic stroke and transient ischemic attack? The Caerphilly Study. *Stroke* **33**, 7–12.

Mayberg, H. S., Starkstein, S. E., Sadzot, B., *et al.* (1990). Selective hypometabolism in the inferior frontal lobe in depressed patients with Parkinson's disease. *Ann Neurol* **28**, 57–65.

Mayberg, H. S., Brannan, S. K., Mahurin, R. K., *et al.* (1997). Cingulate function in depression: a potential predictor of treatment response. *Neuroreport* **8**, 1057–1061.

Mayeux, R., Stern, Y., Cote, L., Williams, J. B. (1984). Altered serotonin metabolism in depressed patients with Parkinson's disease. *Neurology* **34**, 642–646.

Mayeux, R. (1990). The serotonin hypothesis for depression in Parkinson's disease. *Adv Neurol* **53**, 163–166.

McAfee, A. T., Chilcott, K. E., Johannes, C. B., Hornbuckle, K., Hauser, W. A., Walker, A. M. (2007). The incidence of first unprovoked seizure in pediatric patients with and without psychiatric diagnsoses. *Epilepsia* **48**, 1075–1082.

McDonald, W. M., Krishnan, K. R. (1992). Magnetic resonance in patients with affective illness. *Eur Arch Psychiatry Clin Neurosci* **241**, 283–290.

Meldrum, B. S., Anlezark, G. M., Adam, H. K., Greenwod, D. T. (1982). Anticonvulsant and proconvulsant properties of viloxazine hydrochloride: pharmacological and pharmacokinetic studies in rodents and epileptic baboon. *Psychopharmacology (Berlin)* **76**, 212.

Mendelson, S. D. (2002). The current status of the platelet 5-HT(2A) receptor in depression. *J Affect Disord* **57**, 13–24.

Merlet, I., Ostrowsky, K., Costes, N., Ryvlin, P., Isnard, J., Faillenot, I., Lavenne, F., Dufournel, D., Le Bars, D., Mauguiere, F. (2004). 5-HT1A receptor binding and intracerebral activity in temporal lobe epilepsy: an [18F]MPPF-PET study. *Brain* **127**, 900–913.

Meschaks, A., Lindstrom, P., Halldin, C., Farde, L., Savic, I. (2005). Regional reductions in serotonin 1A receptor binding in juvenile myoclonic epilepsy. *Arch Neurol* **62**, 946–960.

Migliorelli, R., Teson, A., Sabe, L., Petracchi, M., Leiguarda, R., Starkstein, S. E. (1995). Prevalence and correlates of dysthymia and major depression among patients with Alzheimer's disease. *Am J Psychiatry* **152**, 37–44.

Modrego, P. J., Ferrandez, J. (2004). Depression in patients with mild cognitive impairment increases the risk of developing dementia of Alzheimer type: a prospective cohort study. *Arch Neurol* **61**, 1290–1293.

Mulvihill, N. T., Foley, J. B. (2002). Inflammation in acute coronary syndromes. *Heart* **87**, 201–204.

Musselman, D. L., Marzec, U. M., Manatunga, A., Penna, S., Reemsnyder, A., Knight, B. T., *et al.* (2000). Platelet reactivity in depressed patients treated with paroxetine: preliminary findings. *Arch Gen Psychiatry* **57**, 875–882.

Musselman, D. L., Tomer, A., Manatunga, A. K., Knight, B. T., Porter, M. R., Kasey, S., *et al.* (1996). Exaggerated platelet reactivity in major depression. *Am J Psychiatry* **153**, 1313–1317.

Nahas, Z., Marangell, L. B., Husain, M. M., Rush, A. J., Sackeim, H. A., Lisanby, S. H., Martinez, J. M., George, M. S. (2005). Two-year outcome of vagus nerve stimulation (VNS) for treatment of major depressive episodes. *J Clin Psychiatry* **66**(9), 1097–1104.

Nahshoni, E., Aizenberg, D., Strasberg, B., Dorfman, P., Sigler, M., Imbar, S., *et al.* (2000). QT dispersion in the surface electrocardiogram in elderly patients with major depression. *J Affect Disord* **60**, 197–200.

Naritoku, D. K., Terry, W. J., Helfert, R. H. (1995). Regional induction of Fos immunoreactivity in the brain by anticonvulsant stimulation of the vagus nerve. *Epilepsy Res* **22**, 53.

Nilsson, F. M., Kessing, L. V., Bowlig, T. G. (2001). Increased risk of developing Parkinson's disease for patients with major affective disorder: a register study. *Acta Psychiatr Scand* **104**, 380–386.

Oguendo, M. A., Placidi, G. P., Malone, K. M., et al. (2003). Positron emission tomography of regional brain metabolic responses to a serotonergic challenge and lethality of suicide attempts in major depression. *Arch Gen Psychiatry* **60**, 14–22.

Parikh, R. M., Robinson, R. G., Lipsey, J. R., Starkstein, S. E., Fedoroff, J. P., Price, T. R. (1990). The impact of post-stroke depression on recovery in activities of daily living over two year follow-up. *Arch Neurol* **47**, 785–789.

Paterniti, S., Verdier-Taillefer, M. H., Dufouil, C., et al. (2002). Depressive symptoms and cognitive decline in elderly people. Longitudinal study. *Br J Psychiatry* **181**, 406–410.

Pearlson, G. D., Ross, C. A., Lohr, W. D., Rovner, B. W., Chase, G. A., Folstein, M. F. (1990). Association between family history of affective disorder and the depressive syndrome of Alzheimer's disease. *Am J Psychiatry* **147**, 452–456.

Perrine, K., Hermann, B. P., Meador, K. J., et al. (1995). The relationship of neuropsychological functioning to quality of life in epilepsy [see comments]. *Arch Neurol* **52**(10), 997–1003.

Piette, Y., Delaunois, A. L., De Shaepdryver, A. F., Heymans, C. (1963). Imipramine and electroshock threshold. *Arch Int Pharmacodyn Ther* **144**, 293–297.

Polc, P., Schneeberger, J., Haefely, W. (1979). Effects of several centrally active drugs on the sleep wakefulness cycle of cats. *Neuropharmacology* **8**, 259–267.

Ravina, B., Camicioli, R., Como, P. G., Marsh, L., Jankovic, J., Weintraub, D., Elm, J. (2007). The impact of depressive symptoms in early Parkinson disease. *Neurology* **69**(4), E2–E3.

Reijnders, J. S., Ehrt, U., Weber, W. E., Aarsland, D., Leentjens, A. F. (2008). A systematic review of prevalence studies of depression in Parkinson's disease. *Mov Disord* **23**(2), 183–189.

Ring, H. A., Bench, C. J., Trimble, M. R., et al. (1994). Depression in Parkinson's disease. A positron emission study. *Br J Psychiatry* **165**, 333–339.

Robinson, R. G., Starr, L. B., Kubos, K. L., Price, T. R. (1983). A two year longitudinal study of post-stroke mood disorders: findings during the initial evaluation. *Stroke* **14**, 736–744.

Sargent, P. A., Kjaer, K. H., Bench, C. J., Rabiner, E. A., Messa, C., Meyer, J., Gunn, R. N., Grasby, P. M., Cowen, P. J. (2000). Brain serotonin 1A receptor binding measured by positron emission tomography with [11C]WAY-100635: effects of depression and antidepressant treatment. *Arch Gen Psychiatry* **57**, 174–180.

Savic, I., Lindstrom, P., Gulyas, B., Halldin, C., Andree, B., Farde, L. (2004). Limbic reductions of 5-HT1A receptor binding in human temporal lobe epilepsy. *Neurology* **62**, 1343–1351.

Schrag, A., Jahanshahi, M., Quinn, N. (2000). What contributes to quality of life in patients with Parkinson's disease? *J Neurol Neurosurg Psychiatry* **69**, 308–312.

Serebruany, V. L., Gurbel, P. A., O'Connor, C. M. (2001a). Platelet inhibition by sertraline and N-demethylsertraline: a possible missing link between depression, coronary events, and mortality benefits of selective serotonin reuptake inhibitors. *Pharmacol Res* **43**, 453–462.

Serebruany, V. L., O'Connor, C. M., Gurbel, P. A. (2001b). Effect of selective serotonin reuptake inhibitors on platelets in patients with coronary artery disease. *Am J Cardiol* **87**, 1398–1400.

Sheline, Y. I. (2006). Brain structural changes associated with depression. In *Depression and Brain Dysfunction* (F. Gilliam, A. M. Kanner, Y. I. Sheline, eds). London: Taylor & Francis, pp. 85–104.

Southam, E., Kirkby, D., Higgins, G. A., Hagan, R. M. (1998). Lamotrigine inhibits monoamine uptake *in vitro* and modulates 5-hydroxytryptamine uptake in rats. *Eur J Pharmacol* **358**(1), 19–24.

Starkstein, S. E., Robinson, R. G., Price, T. R. (1988). Comparison of patients with and without post-stroke major depression matched for age and location of lesion. *Arch Gen Psychiatry* **45**, 247–252.

Starkstein, S. E., Bolduc, P. L., Mayberg, H. S., et al. (1990). Cognitive impairments and depression in Parkinson's disease: a follow up study. *J Neurol Neurosurg Psychiatry* **53**, 597–602.

Starkstein, S. E., Mayberg, H. S., Leiguarda, R., *et al.* (1992). A prospective longitudinal study of depression, cognitive decline, and physical impairments in patients with Parkinson's disease. *J Neurol Neurosurg Psychiatry* **55**, 377–382.

Steffens, D. C., Welsh-Bohmer, K. A., Burke, J. R., Plassman, B. L., Beyer, J. L., Gersing, K. R., Potter, G. G. (2004). Methodology and preliminary results from the neurocognitive outcomes of depression in the elderly study. *J Geriatr Psychiatry Neurol* **17**(4), 202–211.

Tellez-Zenteno, J. F., Patten, S. B., Jetté, N., Williams, J., Wiebe, S. (2007). Psychiatric comorbidity in epilepsy: a population-based analysis. *Epilepsia* **48**, 2336–2344.

The Global Parkinson's Disease Survey Steering Committee (2002). Factors impacting on quality of life in Parkinson's disease: results from an international survey. *Mov Disord* **17**, 60–67.

Theodore, W. H., Giovacchini, G., Bonwetsch, R., Bagic, A., Reeves-Tyer, P., Herscovitch, P., Carson, R. E. (2006). The effect of antiepileptic drugs on 5-HT-receptor binding measured by positron emission tomography. *Epilepsia* **47**(3), 499–503.

Thompson, S. G., Kienast, J., Pyke, S. D., Haverkate, F., van de Loo, J. C. (1995). Hemostatic factors and the risk of myocardial infarction or sudden death in patients with angina pectoris. European Concerted Action on Thrombosis and Disabilities Angina Pectoris Study Group. *New Engl J Med* **332**, 635–641.

Toczek, M. T., Carson, R. E., Lang, L., Ma, Y., Spanaki, M. V., Der, M. G., Fazilat, S., Fazilat, S., Kopylev, L., Herscovitch, P., Eckelman, W. C., Theodore, W. H. (2003). PET imaging of 5-HT1A receptor binding in patients with temporal lobe epilepsy. *Neurology* **60**, 749–756.

Torack, R. M., Morris, J. C. (1988). The association of ventral tegmental area histopathology with adult dementia. *Arch Neurol* **45**, 497–501.

Troxler, R. G., Sprague, E. A., Albanese, R. A., Fuchs, R., Thompson, A. J. (1977). The association of elevated plasma cortisol and early atherosclerosis as demonstrated by coronary angiography. *Atherosclerosis* **26**, 151–162.

Truchot, L., Costes, S. N., Zimmer, L., Laurent, B., Le Bars, D., Thomas-Antérion, C., Croisile, B., Mercier, B., Hermier, M., Vighetto, A., Krolak-Salmon, P. (2007). Up-regulation of hippocampal serotonin metabolism in mild cognitive impairment. *Neurology* **69**(10), 1012–1017.

Wade, D. T., Legh-Smith, J., Hewer, R. A. (1987). Depressed mood after stroke, a community study of its frequency. *Br J Psychiatry* **151**, 200–205.

Watkins, L. L., Grossman, P. (1999). Association of depressive symptoms with reduced baroreflex cardiac control in coronary artery disease. *Am Heart J* **137**, 453–457.

Weintraub, D., Moberg, P. J., Duda, J. E., Katz, I. R., Stern, M. B. (2004). Effect of psychiatric and other nonmotor symptoms on disability in Parkinson's disease. *J Am Geriatr Soc* **52**, 784–788.

Whitton, P. S., Fowler, L. J. (1991). The effect of valproic acid on 5-hydroxytryptamine and 5- hydroxyindoleacetic acid concentration in hippocampal dialysates *in vivo. Eur J Pharmacol* **200**, 167–169.

Williams, J., Schultz, E. G., Griebel, M. L. (2001). Seizure occurrence in children diagnosed with ADHD. *Clin Pediatr* **40**, 221–224.

Yan, Q. S., Mishra, P. K., Burger, R. L., Bettendorf, A. F., Jobe, P. C., Dailey, J. W. (1992). Evidence that carbamazepine and antiepilepsirine may produce a component of their anticonvulsant effects by activating serotonergic neurons in genetically epilepsy-prone rats. *J Pharmacol Exp Ther* **261**, 652–659.

Yan, Q. S., Jobe, P. C., Dailey, J. W. (1993). Evidence that a serotonergic mechanism is involved in the anticonvulsant effect of fluoxetine in genetically epilepsy-prone rats. *Eur J Pharmacol* **252**(1), 105–112.

Yan, Q. S., Jobe, P. C., Dailey, J. W. (1995). Further evidence of anticonvulsant role for 5-hydroxytryptamine in genetically epilepsy prone rats. *Br J Pharmacol* **115**, 1314–1318.

Yanagita, T., Wakasa, Y., Kiyohara, H. (1980). Drug-dependance potential of viloxazine hydrochloride tested in rhesus monkeys. *Pharmacol Biochem Behav* **12**, 155.

Yeragani, V. K., Pohl, R., Jampala, V. C., Balon, R., Ramesh, C., Srinivasan, K. (2000). Increased QT variability in patients with panic disorder and depression. *Psychiatry Res* **93**, 225–235.

CHAPTER 13

Do Peri-ictal Psychiatric Symptoms Account for the Differences between Depressive Disorders in Patients with and without Epilepsy?

Andres M. Kanner

INTRODUCTION

Patients with epilepsy have been found to experience mood and anxiety disorders, attention deficit–hyperactivity disorders (ADHD) and psychotic disorders with a higher frequency than the general population (Kanner and Weisbrot, 2007). However, controversy has arisen concerning whether these disorders in epilepsy patients differ in their phenomenology from that of people without epilepsy, particularly as it relates to mood and psychotic disorders (Kanner and Barry, 2001). In fact, new terminology has emerged to describe these disorders based on special clinical semiology, for example, interictal dysphoric disorder (IDD) (Blumer and Altshuler, 1998) or "dysthymic-like disorder of epilepsy"(DLDE) (Kanner et al., 2000).

Psychiatric episodes that are intimately related to the seizure occurrence, also referred to as peri-ictal episodes, are obviously unique to patients with epilepsy (Kanner, 2007). These include episodes preceding (preictal), occurring as an expression of the seizure activity (ictal) or following the seizure (postictal), which can begin either immediately after the seizure (immediate postictal period) or more characteristically, 12–120 hours (h) after a seizure. It is reasonable to assume that peri-ictal symptoms and episodes may account for the clinical differences of psychiatric disorders between patients with and without epilepsy, although this question has not been addressed in a systematic manner. The purpose of this chapter is to answer the following two questions: (1) Do mood disorders differ between patients with and without epilepsy? and (2) Are peri-ictal symptoms and episodes responsible for these differences? We will review first the available data of peri-ictal depressive symptoms and episodes of depression.

Ictal depressive symptoms or episodes

Ictal depressive symptoms or episodes are the clinical expression of a simple partial seizure in which the symptoms of depression consists of its sole (or predominant) semiology. It has been estimated that psychiatric symptoms occur in 25% of "auras;" 15% of these involve affect or mood changes (Weil, 1955; Williams, 1956; Daly, 1958). For example, ictal depression ranked second after anxiety/fear as the most common type of ictal affect in one study (Weil, 1955; Williams, 1956). At times, mood changes represent the only expression of simple partial seizures, and consequently, it may be difficult to recognize them as epileptic phenomena. They typically are brief, stereotypical, occur out of context, and are associated with other ictal phenomena. The most frequent symptoms include feelings of anhedonia, guilt and suicidal ideation. More typically, however, ictal symptoms of depression are followed by alteration of consciousness as the ictus evolves from a simple to a complex partial seizure.

Preictal depressive symptoms or episodes

There are very few reports in the literature of this form of depressive episodes, which typically presents as a dysphoric mood preceding a seizure. At times, prodromal symptoms may extend for hours or even 1–2 days prior to the onset of a seizure. The best example of this phenomenon was illustrated in the study of Blanchet and Frommer (1986) who assessed mood changes during 56 days in 27 patients with epilepsy who were asked to rate their mood on a daily basis. Mood ratings pointed to a dysphoric state 3 days prior to a seizure in 22 patients. This change in mood was more accentuated during the 24 h preceding the seizure. It has been our experience that in children, these dysphoric moods often take the form of irritability, poor frustration tolerance and aggressive behavior. It is not rare that patients or parents of children with epilepsy report the occurrence of dysphoric symptoms, including increased irritability, mood lability and overt symptoms of depression, preceding their seizures, just to disappear the day after the ictus.

Postictal symptoms of depression

Postictal symptoms of depression have been recognized for a very long time (Gowers 1881; Hughlings, 1931). Their prevalence in large populations of patients with epilepsy has yet to be established, however. In a study carried out at the Rush Epilepsy Center, we investigated the presence of postictal symptoms of depression in 100 consecutive patients with poorly controlled partial seizure disorders (Kanner et al., 2004a). Patients were given a 42-item questionnaire designed to identify psychiatric symptoms occurring in the postictal period, defined as the 72 h that followed recovery of consciousness from the last seizure. Twelve of the 42 questions were directed toward identifying symptoms of depression, which included: anhedonia, irritability, poor frustration tolerance, feelings of hopelessness and helplessness, suicidal ideation, feelings of guilt and self-deprecation, and crying bouts. Six questions were devoted to identify neurovegetative symptoms, including changes in sleep and appetite patterns and sexual drive. We estimated the average duration of each symptom, investigated the presence of these symptoms during the interictal period, and when present, we compared their severity during the two periods. We only included a symptom in our analysis if it occurred following more than 50% of seizures. Symptoms identified during interictal and postictal periods were included only if they were significantly more severe during the postictal period. Among the 100 patients, 79 patients had temporal lobe epilepsy (TLE) and 21 had seizures of extratemporal origin. Half of the patients had only complex partial seizures and the other half had complex partial and secondarily generalized tonic–clonic seizures. Seventy-eight patients had more than one seizure per month. Fifty-two patients had a past psychiatric history, consisting of depression, anxiety disorders and attention deficit disorders, but none had a history of a psychotic disorder.

We identified a mean of 4.8 ± 2.4 postictal symptoms of depression (range $= 2$–9; median $= 5$) in 43 (43%) of patients. A median duration of 24 h was reported in all but one of the symptoms (range $= 0.1$–240 h). Thirteen of these 43 patients reported a cluster of at least seven symptoms and 18 patients, six symptoms that lasted 24 h or longer and mimicked symptoms of major depressive episode, with exception of the time-frame criterion. Of note, symptoms of depression always occurred in combination with other postictal psychiatric symptoms, particularly anxiety, irritability and neurovegetative symptoms. Seven of the 100 patients reported postictal psychotic symptoms in addition to the postictal symptoms of depression and anxiety. Clearly, postictal depressive episodes have a pleomorphic presentation.

Thirteen patients reported habitual postictal suicidal ideation; eight experienced passive and active suicidal thoughts, while five only reported passive suicidal ideation. No patient ever acted on these symptoms. Ten of these 13 patients (77%) had a history of either major depression or bipolar disorder and this association was highly significant.

Among the 43 patients with postictal symptoms of depression, 25 had a history of an interictal mood disorder and 11 of an anxiety disorder. In fact, there was a significantly greater number of postictal symptoms of depression in the presence of an interictal history of depression and anxiety disorders. Furthermore, there was a significant association between the occurrence of postictal symptoms of depression and prior psychiatric hospitalizations, primarily with the occurrence of symptoms related to postictal suicidal ideation. It is worth noting that some reports have suggested that adverse life events may be contributory to the occurrence of postictal depression, but this point is yet to be confirmed. Also, symptoms of depression can outlast the ictus for up to 2 weeks, and at times, have led patients to commit suicide.

Postictal symptoms of anxiety

Postictal symptoms of anxiety were reported by 34 patients with postictal symptoms of depression. Accordingly, it is fitting to include our findings of postictal symptoms of anxiety when discussing postictal depressive episodes. We identified a mean of 2 ± 1 postictal symptoms of anxiety (range = 1–5; median = 2) in 45 patients. Fifteen patients (33%) experienced a cluster of four symptoms that lasted at least 24 h. A prior history of anxiety disorder was identified in 15 patients (33%). There was a significant association between a history of anxiety disorder and the occurrence of two postictal symptoms of anxiety: constant worrying and panicky feelings. In addition, there were a significantly greater number of postictal symptoms of anxiety in the presence of a history of anxiety and depressive disorders.

In addition to these postictal psychiatric symptoms, we identified interictal symptoms of depression and anxiety that worsened in severity during the postictal period. We referred to these symptoms as *interictal symptoms with a postictal exacerbation* (ISPE). Symptoms in which the severity did not differ during interictal and postictal periods were coded as interictal symptoms. We identified interictal symptoms of depression, anxiety and neurovegetative symptoms with postictal exacerbation in 36 patients; in 19, all recorded symptoms were coded as ISPE, while the other 17 patients reported symptoms coded as both ISPE and interictal symptoms.

Among the 36 patients with ISPE, 30 (83%) also experienced separate postictal symptoms of depression or anxiety. In fact, there was a significant association between the occurrence of ISPE and postictal symptoms of depression and anxiety. Among these 30 patients, the number of postictal symptoms of depression and anxiety were significantly greater than those of ISPE of depression (4.7 ± 2.1 vs. 2.4 ± 1.7) and anxiety (2.0 ± 0.9 vs. 1.5 ± 1.1), respectively.

In 13 patients, antidepressant medication was started for the sole treatment of postictal symptoms of depression and anxiety. Yet, these psychotropic drugs did not prevent the occurrence of postictal symptoms. Others have also observed the failure of postictal depressive episodes to respond to antidepressant medication (Barry, personal communication).

Clearly, postictal symptoms of depression are relatively frequent among patients with poorly controlled epilepsy and must account in great part for the poor quality of life of these patients and for the patient's perceived chronic dysphoric states. When occurring by themselves, these symptoms account for the "atypical" presentation of the semiology of depression in patients with epilepsy. Furthermore, failure of postictal symptoms of depression to respond to psychotropic drugs suggests a different pathogenic mechanism than that mediating symptomatology of idiopathic depressive disorders. When occurring in conjunction with interictal depressive disorders, they may add a "pleomorphic" picture, if the interictal disorder is identical to a primary mood disorder. If the interictal depressive disorder is an IDD, postictal symptoms of depression further contribute to its pleomorphic clinical manifestation (see below).

The relatively high prevalence of pre- and postictal symptoms of depression identified in patients with persistent seizures raises the question of their role in "shaping"
the psychiatric symptomatology in intractable partial epilepsy. To the best of our
knowledge, in none of the published studies aimed at identifying psychiatric disorders according to the Diagnostic and Statistical Manual of Mental Disorders (DSM)
criteria did investigators ever discriminate among psychiatric symptoms with an
interictal, preictal or postictal occurrence. Had we attempted to formulate a psychiatric diagnosis according to DSM-IV criteria, most of our patients would have
ended with a diagnosis of "atypical depression (or anxiety) not otherwise specified."

ATYPICAL MANIFESTATIONS OF INTERICTAL DEPRESSION IN EPILEPSY

In 1923, Kraepelin described the changes in mood of patients with epilepsy
(Kraepelin, 1903), which he called "Periodic Dysphorias" (Verstimmungszustände).
In his opinion, they represent the most common form of psychiatric disorders in
these patients. He identified irritability, with or without outbursts of rage, as a cardinal symptom. In addition patients presented symptoms of depression and anxiety, as
well as pain, particularly headaches and insomnia. Of note, patients could "paradoxically" present intermixed euphoric moods. Clearly, this clinical picture had a pleomorphic characteristic. They begin without triggers and in the midst of a clear state
of consciousness and end just as abruptly as they started, having a duration ranging
from several hours to 2 days and recurring on a regular basis with variable frequency,
ranging from a few days to every few months. These periodic dysphorias were interictal events, that is, they occurred independently of seizures, but he identified the
same symptomatology in peri-ictal periods, either preictally or postictally. In addition,
Kraepelin reported the occurrence of interictal hallucinatory or delusional episodes
usually lasting a few days but at times persisting for weeks or even months. In his
opinion, these psychotic episodes were mere expansions of the dysphoric moods.

In 1949, Bleuler confirmed Kraepelin's observations when describing the pleomorphic dysphoric moods of epilepsy according to Kraepelin's concept (Bleuler,
1949). Gastaut expanded on Kraepelin's initial observations (Gastaut *et al.*, 1953;
Gastaut *et al.*, 1955) and Blumer coined the term of IDD and published several
case series of patients with these characteristics (Blumer and Altshuler, 1998).
According to Blumer, this form of depression presents as a pleomorphic condition with predominance of symptoms of irritability, which often overshadow other
symptoms of depression. In addition, patients present symptoms of anxiety, pain
and intermixed symptoms of euphoria. He concurred with Kraepelin's observation
with respect to the intermittent recurrence of this cluster of symptoms lasting from
a few hours to several days, interrupted by symptom-free periods of days to weeks.

I do not pretend to suggest that *all* forms of depression in patients with epilepsy
differ from that of nonepileptic groups. For example, in a study of 193 consecutive

outpatients from five epilepsy centers, psychiatric comorbid disorders were identified with two structured interview instruments developed for patients without
epilepsy (PWE)-The Structured Clinical Interview for Axis I DSM-IV Disorders
(SCID) and the Mini International Neuropsychiatric Interview (MINI) (Jones et al.,
2005; Kanner et al., submitted). Patients were also asked to complete self-rating
screening instruments of symptoms of depression, such as the Beck Depression
Inventory-II (BDI-II) and the Center for Epidemiologic Studies-Depression
(CES-D). A diagnosis of subsyndromic depression was made in patients who failed
to meet any axis I diagnosis according to DSM-IV-TR criteria with the SCID
and MINI, but whose score on the BDI-II and/or CES-D were greater than 12
and 16, respectively. Among the 193 patients, 101 were totally asymptomatic, 10
met a DSM-IV criterion of major depressive disorder, 30 of an anxiety disorder,
24 of mixed major depressive and anxiety disorders, four of dysthymia and 22 met
our diagnostic criterion of subsyndromic depressive episode.

In this study, we were also able to confirm the pleomorphic nature of interictal
depressive episodes suggested by Blumer in mood and anxiety disorders that met
DSM-IV criteria (Jones et al., 2005; Kanner et al., submitted). We used a 46-item
instrument, developed in large part on Blumer's observations. The instrument, called
Mood and Anxiety Symptoms in Epilepsy Inventory (MASEI), is a self-rating
questionnaire that included symptoms from the following eight domains: depression, anxiety, irritability, self-consciousness, physical symptoms, disturbances in
socialization, suicidal ideation and hypomanic-like symptoms. Each item is rated by
the patient (and a companion) on a 4-point Lickert scale (1–4) and was completed
twice, 2 weeks apart, as the items inquire about the presence of symptoms for the
last 2 weeks. Ratings of 3 (some of the time) and 4 (all the time) were considered
to reflect significant symptomatology over the previous 2 weeks. The MASEI had
a high internal consistency with a Cronbach's α of 0.96. Among the 193 patients,
52 were symptomatic on the MASEI alone. More than 90% of these patients
reported symptoms of depression, anxiety and irritability, 51% reported symptoms of
self-consciousness, 67% hypomanic-like symptoms and 71% physical symptoms.
There was a high correlation in the occurrence of symptoms of depression, anxiety
and irritability, and among symptoms of self-consciousness, depression and anxiety
correlated in a significant manner. We identified a mean of 5.5 $\pm$ 1.1 domains with
positive symptoms in these 52 patients. By the same token, patients who met criteria for major depression displayed symptoms in a mean of 6.6 $\pm$ 1 domains with
the MASEI, patients with anxiety disorder had symptoms of 5.7 $\pm$ 1.6 domains
and patients with mixed major depression and anxiety disorders had symptoms
that belonged to a mean of 7.1 $\pm$ 0.6 domains. Clearly, these data show that even
when psychiatric disorders met DSM-IV diagnostic criteria, they exhibited additional symptoms that conveyed a pleomorphic presentation.

We investigated the impact of the "typical" depressive and anxiety disorders
and atypical depressive episodes on the quality of life of these patients, assessed
with the quality of life in epilepsy inventory-89 (QOLIE-89) (Jones et al., 2005;

Kanner *et al.*, submitted). Patients with major depressive disorders and mixed major depressive and anxiety disorders had the worst scores on the QOLIE-89, while patients with anxiety disorders and patients symptomatic only on the MASEI did not differ in their QOLIE-89 scores but scored significantly worse than asymptomatic patients.

A diagnosis of atypical depression not otherwise specified is based on the failure to meet a DSM criterion. For example, in a study by Mendez *et al.*, almost 50% had to be classified as depressive disorder not otherwise specified (DDNOS), or atypical depression, according to DSM-III-R criteria (Kanner *et al.*, 2004b); likewise Wiegartz *et al.* (1999) found that the depressive disorder of 25% of PWE and depression were also classified as DDNOS (Mendez *et al.*, 1986).

In 2000, we reported on the clinical semiology of 97 consecutive patients with epilepsy with symptoms of depression of sufficient severity to require pharmacotherapy with an antidepressant drug (Kanner *et al.*, 2000). Only 28 of our 97 patients met DSM-IV criteria for major depressive disorder. The remaining 69 patients failed to meet criteria for major depressive disorder, dysthymia, cyclothymia or bipolar disorder. Their semiology, however, most resembled a dysthymic disorder, but the recurrence of symptom-free periods intermittently precluded DSM criteria for this diagnosis. We therefore referred to this form of depression as Dysthymic-like Disorder of Epilepsy (DLDE). These 69 patients presented a pleomorphic clinical picture consisting of anhedonia, with or without hopelessness, fatigue, anxiety, irritability and poor frustration tolerance, and mood liability with bouts of crying. Changes in appetite and sleep patterns, and problems with concentration were also reported by some patients. In 33 of these 69 patients, the predominant and most disabling symptom was anhedonia, while in the remaining 36 patients irritability and poor frustration tolerance were the most disabling symptoms. Of note, patients with DLDE in whom anhedonia was the predominant symptom (vs. irritability) were significantly more likely to have experienced a prior history of major depressive disorder (45.5% vs. 19.5%). Most symptoms presented with a waxing and waning course, with repeated interspersed symptom-free periods of one to several days duration. Nevertheless, symptomatology was severe enough in all patients to disrupt their activities, interpersonal relations and overall quality of life and to make them seek treatment. We did not identify symptoms of euphoria in these patients. Of note, a complete symptom remission was obtained in 54% of patients at relatively low doses of sertraline [108 ± 56.9 mg/day (range = 25–200)]. Such efficacy represents a difference from response to treatment in PWE and dysthymic disorders. Clearly, all of these data suggest that interictal depressive disorders present with atypical manifestations and the differences with depressive disorders in PWE are not necessarily related (only) to the presence of postictal symptoms of depression.

In 2007, a subcommission of the International League Against Epilepsy (ILAE) Commission on Psychobiology of Epilepsy (now the Commission on Neuropsychiatric Aspects) published a classification proposal of various psychiatric disorders that are "specific to epilepsy" (Krishnamoorthy *et al.*, 2007). The aim of this proposal was "to separate disorders comorbid with epilepsy and those that reflect ongoing epileptiform activity from epilepsy-specific disorders and to attempt to subclassify the epilepsy-specific disorders alone."

The authors of the classifications referred to the mood disorders specific to epilepsy as "Affective-somatoform (dysphoric) disorders of epilepsy." This entity is reminiscent of Kraepelin's periodic dysphorias. According to the authors "they manifest in a pleomorphic pattern and include eight symptoms: irritability, depressive moods, anergia, insomnia, atypical pains, anxiety, phobic fears and euphoric moods. These symptoms occur at various intervals and tend to last from hours to 2 or 3 days, although they might, on occasion, last longer. Some of the symptoms may be present continually at a baseline from which intermittent fluctuations occur. The presence of at least three symptoms generally coincides with significant disability. The same affective-somatoform symptoms occur during the prodromal and postictal phases and need to be coded as such if they are of clinical significance."

The authors divided this condition into four separate entities, none of which could meet ICD-10 or DSM-IV criteria for major depression, dysthymia, and cyclothymia: (1) IDD, which includes three of the eight symptoms listed above, each to a troublesome degree. (2) Prodromal dysphoric disorder, which presents with irritability or other dysphoric symptoms preceding a seizure by hours or days and which can cause significant impairment. (3) Postictal dysphoric disorder, in which symptoms of anergia or headaches, as well as depressed mood, irritability, and anxiety, may develop after a seizure and be prolonged or exceptionally severe. (4) Alternative affective-somatoform syndromes. In this entity, depression, anxiety, depersonalization, derealization, and even nonepileptic seizures have been reported as presenting manifestations of forced normalization; that is, with remission of epileptic seizures after a state of persistent seizures, with normal electroencephalogram (EEG) recordings, though the latter is not required.

CONCLUDING REMARKS

This chapter reviewed the role of peri-ictal symptomatology on the "atypical" clinical manifestations of mood disorders. From the data reviewed it can be clearly established that, while peri-ictal semiology may be an important contributor, interictal mood episodes present atypical clinical characteristics by themselves. Recognition of peri-ictal episodes and in particular, of postictal depressive and anxiety episodes, is of the essence, as they have a very negative impact on the quality of life of these patients. Unfortunately, postictal psychiatric symptoms are rarely investigated in the course of an evaluation either in the clinic or in research studies. The atypical characteristics of mood disorders require that screening instruments and structured psychiatric interviews be developed specifically for patients with epilepsy.

REFERENCES

Blanchet, P., Frommer, G. P. (1986). Mood change preceding epileptic seizures. *J Nerv Ment Dis* **174**, 471–476.

Bleuler, E. (1949). *Lehrbuch der Psychiatrie* 8th edition. Berlin: Springer-Verlag.

Blumer, D., Altshuler, L. L. (1998). Affective disorders. Epilepsy: A Comprehensive Textbook, **Vol. II** (J. Engel, T. A. Pedley, eds), pp. 2083–2099. Philadelphia, PA: Lippincott-Raven.

Daly, D. (1958). Ictal affect. *Am J Psych* **115**, 97–108.

Gastaut, H., Roger, J., Lesèvre, N. (1953). Différenciation psychologique des épileptiques en fonction des formes électrocliniques de leur maladie. *Rev Psychol Appl* **3**, 237–249.

Gastaut, H., Morin, G., Lesèvre, N. (1955). Étude du comportement des épileptiques psychomoteurs dans l'intervalle de leurs crises: les troubles de l'activité globale et de la sociabilité. *Ann Med Psychol (Paris)* **113**, 1–27.

Gowers, W. R. (1881). *Epilepsy and other chronic and convulsive diseases*. London: JA Churchill.

Hughlings Jackson, J. (1931). *Selected Writings of John Hughlings Jackson* (J. Taylor, G. Holmes, F. M. R. Walshe, eds). London: Hodder and Stoughton.

Jones, J. E., Hermann, B. P., Barry, J. J., Gilliam, F., Kanner, A. M., Meador, K. J. (2005). Clinical assessment of Axis I psychiatric morbidity in chronic epilepsy: a multicenter investigation. *J Neuropsychiatry Clin Neurosci* **17**(2), 172–179.

Kanner, A. M. (2007). Peri-ictal psychiatric phenomena: clinical characteristics and implications of past and future psychiatric disorders. In *Psychiatric Issues in Epilepsy: A Practical Guide to Diagnosis and Treatment* (A. Ettinger, A. M. Kanner, eds) 2nd edition, pp. 321–345. Philadelphia, PA: Lippincott, Williams and Wilkins.

Kanner, A. M., Barry, J. J. (2001). Depression and psychotic disorders associated with epilepsy – are they unique? *Epilepsy Behav* **2**, 170–186.

Kanner, A. M., Weisbrot, D. (2007). Psychiatric evaluation of the adult and pediatric patient with epilepsy: a practical approach for the "nonpsychiatrist". In *Psychiatric Issues in Epilepsy: A Practical Guide to Diagnosis and Treatment* (A. Ettinger, A. M. Kanner, eds) 2nd edition, pp. 119–132. Philadelphia, PA: Lippincott, Williams and Wilkins.

Kanner, A. M., Kozak, A. M., Frey, M. (2000). The use of sertraline in patients with epilepsy: is it safe? *Epilepsy Behav* **1**(2), 100–105.

Kanner, A. M., Soto, A., Gross-Kanner, H. (2004a). Prevalence and clinical characteristics of postictal psychiatric symptoms in partial epilepsy. *Neurology* **62**, 708–713.

Kanner, A. M., Wuu, J., Barry, J., Hermann, B., Meador, K. J., Gilliam, F. (2004b). Atypical depressive episodes in epilepsy: a study of their clinical characteristics and impact on quality of life. *Neurology* **62**(Suppl 5), A249.

Kraepelin, E. (1903). *Psychiatrie. 7. Auflage*. Liepzig: JA Barth.

Krishnamoorthy, E. S., Trimble, M. R., Blumer, D. (2007). The classification of neuropsychiatric disorders in epilepsy: a proposal by the ILAE Commission on psychobiology of epilepsy. *Epilepsy Behav* **10**(3), 349–353.

Mendez, M. F., Cummings, J., Benson, D., *et al.* (1986). Depression in epilepsy. Significance and phenomenology. *Arch Neurol* **43**, 766–770.

Weil, A. (1955). Depressive reactions associated with temporal lobe uncinate seizures. *J Nerv Ment Dis* **121**, 505–510.

Wiegartz, P., Seidenberg, M., Woodard, A., Gidal, B., Hermann, B. (1999). Co-morbid psychiatric disorder in chronic epilepsy: recognition and etiology of depression. *Neurology* **53**(Suppl 2), S3–S8.

Williams, D. (1956). The structure of emotions reflected in epileptic experiences. *Brain* **79**, 29–67.

CHAPTER 14

Can Psychological Testing Replace Psychiatric Evaluations in Patients with Epilepsy? Or Can Psychiatric Evaluations Replace Psychological Testing in Patients with Epilepsy?

David W. Loring and Bruce P. Hermann

INTRODUCTION

The presence of psychiatric comorbidities in epilepsy, treatment emergent psychiatric effects of anti-epilepsy drugs (AEDs) as well as their possible psychotropic benefits are increasingly appreciated by both epilepsy researchers and clinicians. Although estimates vary, comorbid depression may be present in over half of patients with poorly controlled epilepsy (Gilliam and Kanner, 2002), and epilepsy is associated with increased rates of anxiety (Beyenburg *et al.*, 2005), bipolar disorder (Ettinger *et al.*, 2005), and even comorbid schizophrenia (Gaitatzis *et al.*, 2004). The high incidence of depression in patients with epilepsy led the Epilepsy Foundation to develop their Epilepsy and Mood Disorders Initiative (http://www. epilepsyfoundation.org/about/related/mood/index.cfm), which not only is intended to increase the recognition of depression in epilepsy patients, but also to develop better management and treatment outcomes strategies.

Depression in epilepsy is a clinically significant condition that is associated with increased medical costs (Cramer *et al.*, 2004) and consistently lower reported quality of life (QoL) – more closely associated with lower QoL than seizures themselves (Cramer *et al.*, 2003; Tracy *et al.*, 2007). Despite its high prevalence, depression in epilepsy is under detected and under treated (Harden, 2002; Kanner, 2003). Suicidal ideation and mortality secondary to suicide is significantly increased in epilepsy (Hitiris *et al.*, 2007), and it is critically important to address issues related to mood and other disorders in epilepsy as part of standard clinical care (Kanner, 2005). Tasks include identifying mood and other disorders (anxiety), determining their relationship to seizures (interictal, periictal, postictal), and making pertinent treatment decisions/referrals for intervention. Similar concerns are present in pediatric epilepsy, in which high rates of affective and anxiety disorders exist but in whom appropriate treatment options are infrequently implemented (Caplan *et al.*, 2005). Psychiatric evaluation of the children involves many of the same issues as adults, but proper diagnosis is even more critical in this group in order to establish programs of early intervention when therapeutic benefit can be expected to be the greatest.

The ability to characterize psychiatric features of epilepsy, and to monitor either positive or negative AED effects during treatment, will vary based on the training, interest, and biases of the clinicians providing care to epilepsy patients, the availability of psychiatrists or clinical psychologists with experience and interest in epilepsy, or of the clinical investigators examining behavioral characteristics of epilepsy. Currently, there is often limited availability of appropriate resources and expertise in psychiatric aspects of epilepsy when developing diagnostic formulations and treatment strategies. At many epilepsy centers, a common approach to identify patients in need of more careful psychiatric evaluation and treatment has been to "screen" for psychological symptoms. In addition, given the recent interest in psychiatric comorbidities, psychiatric contributions are often explicitly assessed in many clinical research studies. In the present chapter, we will compare and contrast the contributions of psychiatric interviews and psychological testing in both clinical and research applications of epilepsy care.

DIAGNOSTIC ACCURACY

Correct identification of psychiatric characteristics of epilepsy or explicit psychiatric comorbidities is necessary to maximize the likelihood of successful treatment and intervention. Formal psychiatric diagnoses are made according to criteria presented in the Diagnostic and Statistical Manual (Fourth Edition) published by the American Psychiatric Association, or the International Classification of Diseases (Ninth Edition) published by the World Health Organization. Correct classification depends on the ability to elicit the necessary information to formulate a correct psychiatric diagnosis, as well as to determine subsyndromal symptoms that do not meet criteria for formal diagnostic classification. This is particularly true for depression and mood disorders in epilepsy since patients may not be aware of depressive symptoms. Patients may not volunteer this information spontaneously, and even upon questioning, may tend to minimize or deny symptoms making it important to obtain or verify symptoms from other sources such as the family.

Psychiatric diagnostic accuracy is variable in non-psychiatric settings. Primary care physicians commonly fail to recognize depression in the majority of patients, and when depression is diagnosed, the diagnosis is as likely to be correct as it is to be incorrect (Rogers, 2003). In more specialized practices such as neurology and epilepsy, patients are often asked to complete a medical questionnaire containing items related to depressive symptomatology, and in some practices, patients are given a formal depression questionnaire in order to elicit symptoms of depression. Information obtained through formal or informal screening/questionnaires alerts the clinician to the possible presence and severity of common depressive symptoms.

Screening patients for depression has been referred to as "case-finding" (Pignone *et al.*, 2002), and differs from less structured methods used by mental health providers in which clinical features of possible depression (e.g. feeling fatigued, headaches) are elicited through clinical interview and followed up as appropriate. The case-finding approach is often preferable in general (i.e. non-psychiatric) practice since non-structured interviewing depends on specific training and expertise in psychiatric interviewing, which many clinicians have not received. Without proper questioning, appropriate symptoms will not be elicited.

In the above scenarios, the instruments used to identify psychiatric symptoms become the initial stage of what ultimately should result in proper psychiatric diagnoses. If a patient reports symptoms such as losing interest in pleasurable tasks, or obtains a score on a depression questionnaire that exceeds a specified threshold, then either a more thorough evaluation of psychiatric symptoms is conducted or a referral made to an appropriate mental health provider who will perform an appropriate diagnostic psychiatric interview. Regardless of how the initial index of suspicion is raised, a psychiatric diagnosis has not been made and the goal of the screening is to identify patients in whom further diagnostic evaluation is warranted.

Formal psychiatric diagnoses require evaluation to determine whether diagnostic criteria (DSM or ICD) have been met. Psychiatric interviews include non-structured, semi-structured, and structured approaches, and the reliability of diagnosis varies across interview approaches. A non-structured interview is not scripted, and the order in which content is elicited and the language used for questioning is interviewer dependent. Non-structured interviews may be affected by "confirmatory bias", meaning that after an initial diagnostic impression is made, there is a tendency to seek additional information consistent with that impression. Data consistent with initial impressions tend to be overvalued while information that does not support initial impressions tends to be discounted. An additional risk associated with non-structured interviews is the premature discontinuation of relevant questioning after the initial diagnostic impressions, thereby decreasing the likelihood of diagnosing less frequent conditions.

Because of the above limitations, structured and semi-structured interviews have been developed to increase the diagnostic accuracy for both psychiatric and non-psychiatric use. Although both approaches are more reliable than non-structured interviews, structured interviews that follow standard language with a fixed sequence of questioning are more reliable than semi-structured approaches (Miller *et al.*, 2001).

The gold standard of structured psychiatric interviews is the Structured Clinical Interview for the DSM-IV (SCID), and generally can be administered in 60–90 minutes by appropriately trained clinicians. However, for many purposes, semi-structured psychiatric interviews such as the Mini International Neuropsychiatric Interview (MINI) (Sheehan *et al.*, 1998) or Brief Psychiatric Rating Scale (BPRS) (Overall and Gorham, 1962) are sufficient, with both approaches being significantly shorter than the SCID and requiring approximately 15–30 minutes to complete. The diagnostic concordance between the MINI and SCID in patients with chronic epilepsy is high, particularly for major depressive episodes (0.86) (Jones *et al.*, 2005a). Consequently, in many settings where psychiatric expertise is not readily available, the semi-structured MINI interview provides an acceptable alternative for obtaining psychiatric diagnoses. While highly reliable, even semi-structured psychiatric interviews require proper training, and since they are not the standard of care at many epilepsy centers, the persons performing such assessments generally do so outside what might be described as their standard of clinical care. Approaches such as the MINI also sacrifice information compared to a SCID since the coverage between the two is not identical. For example, the MINI provides information on current diagnoses rather than lifetime to present.

A research form of the SCID has been developed targeting only Axis I (i.e. non-personality) disorders (First *et al.*, 2001), and a self-report questionnaire (e.g. Psychiatric Diagnostic Screening Questionnaire, PDSQ) has been adapted to assess Axis I disorders by including appropriate time frames (Zimmerman and Mattia, 2001). With the PDSQ, the clinician is able to efficiently screen for a variety of Axis I disorders as well as being able to identify subsyndromal symptoms. While the PDSQ by itself is not diagnostic, it serves to ensure that significant disorders are not missed, provides an efficient approach to identify possible disorders with careful patient interview, or serves as a screening device to determine which cases should have appropriate mental health referrals. The PDSQ is well tolerated by patients, and is particularly useful in identifying less common disorders that may otherwise be overlooked (e.g. post-traumatic stress disorder (PTSD), eating disorders).

Summary

Psychiatric interviews are necessary to establish formal psychiatric diagnoses. The best psychiatric interviews are structured, and are valid and reliable approaches to patient diagnoses. However, they require significant personnel training to be used properly, and also require significant administration time, both of which decrease the frequency of use outside of primary psychiatric settings. Less structured interviews may not be diagnostic by themselves, but with judicious use, are able to identify patients in whom more comprehensive evaluation is warranted, and also explore a variety of symptoms beyond mood/depression that may need to be the focus of intervention and treatment.

MONITORING CHANGE

Although the presence of major depression identified with structured psychiatric interviewing may at times be used as a dichotomous outcome variable in formal psychiatry settings (e.g. ECT treatment outcomes (Kellner *et al.*, 2006)), diagnostic classification may not reflect smaller degrees of change, particularly when the presence of mood alteration is not part of a larger psychiatric syndrome. In one report describing the development of psychiatric features surgery, pre- and postoperative SCIDs were obtained in a sample of 70 patients undergoing anterior temporal lobectomy (Pintor *et al.*, 2007). For statistical analysis, however, it was necessary to form single diagnostic groups by combining conditions (i.e. depressive disorders consisting of major depressive episodes, adjustment disorder with depressed mood, etc.; anxiety disorders including generalized anxiety disorders, social phobia, etc.), thereby losing much of the diagnostic precision obtained in the first place by performing the SCID. With greater power associated with larger samples, it is not necessary to collapse across groups in order to obtain statistical significance to demonstrate the beneficial effects of epilepsy surgery on depression and anxiety (Devinsky *et al.*, 2005). However, dichotomous diagnostic classification will always require greater sample sizes to reach statistical significance compared to parametric measures with continuous variables.

There are multiple questionnaires and inventories that are used to assess mood, some of which are completed by the patient (e.g. Beck Depression Inventory-II (BDI-II) (Beck, 1996), Minnesota Multiphasic Personality Inventory-II (MMPI-II) (Butcher *et al.*, 1989), Zung Self-Rating Depression Scale (Zung, 1965)), and others are rating scales that are completed by the clinician (e.g. Hamilton Depression Rating Scale (HAM-D or HDRS) (Hamilton, 1960), Montgomery Åsberg Depression Rating Scale) (Montgomery and Åsberg, 1979), Cornell Dysthymia Scale (Mason *et al.*, 1993), and Center for Epidemiologic Studies Depression Scale (CES-D) (Radloff, 1977)). Although epilepsy and its treatment may be associated with factors such as fatigue and sleep disturbance that are also common components of depression, the presence of independent depressive symptoms can be identified with high reliability with few but highly specific questions (Neurological Disorders Depression Inventory for Epilepsy (NDDI-E) (Gilliam *et al.*, 2006)). Even though the NDDI-E is brief, it has better predicative power for depression in epilepsy than the longer BDI (Gilliam *et al.*, 2006; Jones *et al.*, 2005b), likely reflecting the successful effort to eliminate questions that could also be affected by medications.

There are significant advantages associated with psychological questionnaires and inventories. Many scales are self-report measures that require minimal or no assistance from the clinician. Like all self-report measures, however, these scales are subject to response bias, with the patient having the potential to distort test results.

Some scales require the clinician to rate various psychiatric symptoms. The HDRS (Hamilton, 1960) is perhaps the most commonly used scale in drug trials. The HDRS was developed to quantify the intensity of depression in patients already diagnosed. Nine items are rated on a 5-point scale and eight items are rated on a 3-point scale. Of course, the validity of HDRS is dependent on the skill of the clinical interviewer performing the rating. Clinical rating scales, like self-report inventories, may also be associated with symptom exaggeration or minimization.

Of the various scales presented, only the MMPI-II has explicit scales to measure both response distortion tendencies, and also has scales that indicate whether a patient may simply be responding randomly or carelessly (Butcher et al., 1989). The MMPI-2 has 10 primary clinical scales that may assist in diagnosis, but by themselves are not diagnostic. Although the MMPI has been used in many studies of epilepsy, it is not typically part of multicenter trials since it typically requires the on-site presence of a clinical psychologist at each clinical site. The response burden on the patient is also quite high (567 questions) and results do not always readily conform to current diagnostic terminology, in part because it was developed without directly addressing diagnostic information such as symptom duration.

Because of their brevity, the Beck Depression Inventory (BDI) (Beck 1996) and Beck Anxiety Inventory (BAI) (Beck et al., 1988) are among the most commonly used instruments to characterize psychiatric features of epilepsy in research studies not involving drug treatment. Like all screening measures, the BDI is not diagnostic of depression in epilepsy; it is associated with a high negative predictive value indicating a low probability of depression when low Beck scores are obtained (Jones et al., 2005b). In this context, it serves as a good and convenient screening measure to identify patients requiring more comprehensive evaluation. Similar patterns of low false-positive identifications are present for other depression measures such as the CES-D (Jones et al., 2005b). However, when the BDI or CES-D is given as a screening measure, it should likely be supplemented with an anxiety measure such as the BAI to insure adequate coverage of symptoms to screen for suicide and other risks.

When used as screening measures, psychological tests serve primarily to identify patients in whom further evaluation should be considered. Thus, it is easy to develop a system where measures such as the BDI/BAI are administered initially, and if there is a suggestion of abnormality on either scale, then an evaluation such as the MINI (or SCID) is conducted.

The other benefit of screening using psychological tests is their ability to identify subsyndromal disorders. Even in patients who do not meet diagnostic criteria for major depression, the presence of depression symptoms endorsed on psychological tests is related to overall QoL (Loring *et al.*, 2004; Tracy *et al.*, 2007). This indicates that subsyndromal mood states are important to characterize as part of comprehensive epilepsy evaluation and management.

Summary

Psychological measures are not by themselves diagnostic of depression or other psychiatric conditions. However, unlike psychiatric diagnoses, they are able to easily measure subsyndromal mood states and, in the case of research applications, they are able to effectively monitor change due to treatment or disease progression, and provide greater statistical power as psychiatric outcome variables. When used clinically, psychological tests serve as effective screening measures to identify patients needing further evaluation, and are associated with relatively low false-positive rates of incorrect patient identification.

CONCLUSION

The answer to both questions contained in the title to this chapter is "no." Both psychiatric interviews and psychological testing fill different needs in both clinical and research applications. Because diagnosis guides treatment, psychiatric interviews will be necessary during initial clinical evaluations. Also, because psychiatric interviews provide specific diagnostic classification, they are necessary in epidemiologic studies examining incidence and prevalence in static populations and in specific clinical outcome studies.

Psychological testing, while not in itself diagnostic, is useful in identifying patients for whom a more comprehensive evaluation is warranted. However, perhaps the greatest advantage of psychological testing is that it can monitor change even when formal diagnostic criteria are not present. In particular, psychological instruments can permit the detection of improvement in aspects of epilepsy management resulting from positive psychotropic effects of certain medications (Ettinger *et al.*, 2007) that may be present even in the absence a diagnosable mood disorder.

REFERENCES

Beck, A. T., Epstein, N., Brown, G., Steer, R. A. (1988). An inventory for measuring clinical anxiety: psychometric properties. *J Consult Clin Psychol* **56**(6), 893–897.

Beck, A. T. (1996). *BDI-II, Beck depression inventory: manual.* San Antonio, TX: The Psychological Corporation.

Beyenburg, S., Mitchell, A. J., Schmidt, D., Elger, C. E., Reuber, M. (2005). Anxiety in patients with epilepsy: systematic review and suggestions for clinical management. *Epilepsy Behav* **7**(2), 161–171.

Butcher, J. N., Dahlstrom, W. G., Graham, J. R. (1989). *Manual for the Restandardized Minnesota Multiphasic Personality Inventory: MMPI-2.* Minneapolis, MN: University of Minnesota Press.

Caplan, R., Siddarth, P., Gurbani, S., Hanson, R., Sankar, R., Shields, W. D. (2005). Depression and anxiety disorders in pediatric epilepsy. *Epilepsia* **46**(5), 720–730.

Cramer, J. A., Blum, D., Reed, M., Fanning, K. (2003). The influence of comorbid depression on quality of life for people with epilepsy. *Epilepsy Behav* **4**(5), 515–521.

Cramer, J. A., Blum, D., Fanning, K., Reed, M. A-. (2004). The impact of comorbid depression on health resource utilization in a community sample of people with epilepsy. *Epilepsy Behav* **5**(3), 337–342.

Devinsky, O., Barr, W. B., Vickrey, B. G., Berg, A. T., Bazil, C. W., Pacia, S. V., *et al.* (2005). Changes in depression and anxiety after resective surgery for epilepsy. *Neurology* **65**(11), 1744–1749.

Ettinger, A. B., Reed, M. L., Goldberg, J. F., Hirschfeld, R. M. A. (2005). Prevalence of bipolar symptoms in epilepsy vs other chronic health disorders. *Neurology* **65**(4), 535–540.

Ettinger, A. B., Kustra, R. P., Hammer, A. E. (2007). Effect of lamotrigine on depressive symptoms in adult patients with epilepsy. *Epilepsy Behav* **10**(1), 148–154.

First, M. B., Spitzer, R. L., Gibbon, M., Williams, J. (2001). *Structured clinical interview for DSM-V-TR Axis I disorders – non-patient ed. (SCI-I/NP – 2/2001 Revision).* New York: Biometric Research Department.

Gaitatzis, A., Trimble, M. R., Sander, J. W. (2004). The psychiatric comorbidity of epilepsy. *Acta Neurol Scand* **110**(4), 207–220.

Gilliam, F., Kanner, A. M. (2002). Treatment of depressive disorders in epilepsy patients. *Epilepsy Behav* **3**(5 Suppl 1), 2–9.

Gilliam, F. G., Barry, J. J., Hermann, B. P., Meador, K. J., Vahle, V., Kanner, A. M. (2006). Rapid detection of major depression in epilepsy: a multicentre study. *Lancet Neurol* **5**(5), 399–405.

Hamilton, M. (1960). A rating scale for depression. *J Neurol Neurosurg Psychiatry* **23**, 56–62.

Harden, C. L. (2002). The co-morbidity of depression and epilepsy: epidemiology, etiology, and treatment. *Neurology* **59**(6 Suppl 4), S48–S55.

Hitiris, N., Mohanraj, R., Norrie, J., Brodie, M. J. (2007). Mortality in epilepsy. *Epilepsy Behav* **10**(3), 363–376.

Jones, J. E., Hermann, B. P., Barry, J. J., Gilliam, F., Kanner, A. M., Meador, K. J. (2005). Clinical assessment of Axis I psychiatric morbidity in chronic epilepsy: a multicenter investigation. *J Neuropsychiatry Clin Neurosci* **17**(2), 172–179.

Jones, J. E., Hermann, B. P., Woodard, J. L., Barry, J. J., Gilliam, F., Kanner, A. M., *et al.* (2005). Screening for major depression in epilepsy with common self-report depression inventories. *Epilepsia* **46**(5), 731–735.

Kanner, A. M. (2003). Depression in epilepsy: prevalence, clinical semiology, pathogenic mechanisms, and treatment. *Biol Psychiatry* **54**(3), 388–398.

Kanner, A. M. (2005). Should neurologists be trained to recognize and treat comorbid depression of neurologic disorders? Yes. *Epilepsy Behav* **6**(3), 303–311.

Kellner, C. H., Knapp, R. G., Petrides, G., Rummans, T. A., Husain, M. M., Rasmussen, K., *et al.* (2006). Continuation electroconvulsive therapy vs pharmacotherapy for relapse prevention in major depression: a multisite study from the Consortium for Research in Electroconvulsive Therapy (CORE). *Arch Gen Psychiatry* **63**(12), 1337–1344.

Loring, D. W., Meador, K. J., Lee, G. P. (2004). Determinants of quality of life in epilepsy. *Epilepsy Behav* **5**(6), 976–980.

Mason, B. J., Kocsis, J. H., Leon, A. C., Thompson, S., Frances, A. J., Morgan, R. O., *et al.* (1993). Measurement of severity and treatment response in dysthymia. *Psychiatric Ann* **23**, 625–631.

Miller, P. R., Dasher, R., Collins, R., Griffiths, P., Brown, F. (2001). Inpatient diagnostic assessments: 1. Accuracy of structured vs. unstructured interviews. *Psychiatry Res* **105**(3), 255–264.

Montgomery, S. A., Åsberg, M. (1979). A new depression scale designed to be sensitive to change. *Br J Psychiatry* **134**, 382–389.

Overall, J. E., Gorham, D. R. (1962). The brief psychiatric rating scale. *Psychol Rep* **10**, 799–812.

Pignone, M. P., Gaynes, B. N., Rushton, J. L., Burchell, C. M., Orleans, C. T., Mulrow, C. D., *et al.* (2002). Screening for depression in adults: a summary of the evidence for the US preventive services task force. *Ann Intern Med* **136**(10), 765–776.

Pintor, L., Bailles, E., Fernandez-Egea, E., Sanchez-Gistau, V., Torres, X., Carreno, M., *et al.* (2007). Psychiatric disorders in temporal lobe epilepsy patients over the first year after surgical treatment. *Seizure* **16**(3), 218–225.

Radloff, L. S. (1977). The CES-D scale: a self-report depression scale for research in the general population. *Appl Psychol Meas* **1**, 385–401.

Rogers, R. (2003). Standardizing DSM-IV diagnoses: the clinical applications of structured interviews. *J Pers Assess* **81**(3), 220–225.

Sheehan, D. V., Lecrubier, Y., Sheehan, K. H., Amorim, P., Janavs, J., Weiller, E., *et al.* (1998). The Mini-International Neuropsychiatric Interview (M.I.N.I.): the development and validation of a structured diagnostic psychiatric interview for DSM-IV and ICD-10. *J Clin Psychiatry* **20**, 22–33.

Tracy, J. I., Dechant, V., Sperling, M. R., Cho, R., Glosser, D. (2007). The association of mood with quality of life ratings in epilepsy. *Neurology* **68**(14), 1101–1107.

Zimmerman, M., Mattia, J. I. (2001). A self-report scale to help make psychiatric diagnoses: the Psychiatric Diagnostic Screening Questionnaire. *Arch Gen Psychiatry* **58**(8), 787–794.

Zung, W. W. K. (1965). A self-rating depression scale. *Arch Gen Psychiatry* **12**, 63–70.

CHAPTER 15

Should the Screening for Depression, Anxiety, Attention Deficit – Hyperactivity Disorder and Learning Disorders Be Part of Neurological Evaluations of All Patients with Epilepsy?

Marlis Frey

INTRODUCTION

As clinicians, we continuously strive to improve the quality of care we provide to our patients and thus improve their quality of life. At the same time, there are ever present pressures for this care to be cost-effective and efficient. When providing care for patients with epilepsy, is it therefore justified to simply review seizure diaries and medication side effects, adjust or switch anticonvulsant medications and send patients on their way until the next appointment? In fact, epilepsy is associated with a variety of comorbid disorders that occur with a relatively high prevalence, all of which impact in a negative way the quality of life of patients. They include depression and anxiety disorders, attention deficit–hyperactivity disorder (ADHD) and learning disabilities (LD). Thus, should we allocate resources to evaluate patients for these comorbid disorders at the initial visit and if necessary, at each visit? There are many questions regarding the relation between epilepsy and these comorbid disorders. In particular, are they related to a common central nervous system dysfunction expressing itself in both conditions, or are these independent conditions? This chapter will review the multifactorial aspects of depression, anxiety, ADHD and LD in patients with epilepsy and the subsequent possible consequences. Each section will review the phenomenology of the disorder, its

Psychiatric Controversies in Epilepsy

suspected pathophysiology and its symptomatology. In the Clinical Implications section, we will discuss our experience with evaluation and management of this patient population.

DEPRESSION

Depressive disorders are the most frequently diagnosed psychiatric disorders in patients with epilepsy (Robertson *et al.*, 1987; Hermann *et al.*, 2000). Many studies have investigated the relationship between epilepsy and depression in academic healthcare patient populations and results indicate comorbid prevalence rates between 15% and 55% (Blumer, 1991; Devinsky and Bear, 1991; Smith *et al.*, 1991; Kanner, 2003; Vazquez and Devinsky, 2003). A community-based 2004 HealthStyles Survey found that participants with active epilepsy self-reported suffering from depression the previous year twice as frequently as participants without a diagnosis of epilepsy (Kobau *et al.*, 2006). Suicide rates for patients with epilepsy and depression have been reported to be five times higher compared to the general public (Robertson, 1987). Studies focusing on depression in children with epilepsy are more recent. Depression rates of almost 23% were found in an 8–18 years age group with new-onset epilepsy (Jones *et al.*, 2007). Approximately, 20% of this cohort met criteria for depressive disorder prior to onset of first seizure.

The presence of depressive disorders prior to the onset of epilepsy has also been reported in adults over the age of 55 years with new-onset seizure. In this patient population, new-onset seizure patients were 3.7 times more likely to have a history of depression than controls (Hesdorffer *et al.*, 2000). The findings from these studies have led to the hypothesis that both depression and seizures may be an expression of the same underlying pathologic process.

Depression may develop at the time of the seizure onset or be part of the seizure itself. The diagnosis of epilepsy has been shown to contribute to endogenous symptoms of depression (Betts, 1981). These symptoms have been suggested to be associated with a perceived loss of external locus of control (Hermann and Wyler, 1989).

Symptoms of depression may not develop until after the onset of the seizure disorder. These symptoms may be related to the negative effects of anticonvulsant medications (antiepileptic drugs – AEDs) such as phenobarbital, primidone, vigabatrin, topiramate and levetiracetam. However, these symptoms of depression should be temporary and disappear with removal of the causative factor.

Pathophysiology

A biological model of depression is based on decreased serotonergic, noradrenergic, dopaminergic and GABAergic functions. The clinical efficacy of selective serotonin reuptake inhibitors (SSRIs) suggests a strong validity to this model. Dysfunction at multiple levels of these neurotransmitter pathways has been identified (Altschuler *et al.*, 1999).

Alterations in the structure and function of various neuroanatomic substrates have also been implicated in the development of depressive disorders. Limbic structures located within the temporal lobes are involved in regulating pleasure and mood. These structures are also often involved in the generation of seizures. Due to this dual function of the limbic system, numerous studies have addressed the relationship between temporal lobe epilepsy and affective disorders. Some studies have not found an increased risk for developing depressive symptoms in patients with a temporal lobe seizure focus vs. patients with extratemporal seizures (Mendez *et al.*, 1993), while other studies have demonstrated such an increased risk (Altschuler *et al.*, 1999).

With regard to heritability, depression has been classified as a genetically complex disorder as it does not follow a classic Mendelian model and therefore cannot be traced back to one single gene (Landers and Schork, 1994; Sullivan *et al.*, 2000). Rather, developmental alterations in neurotransmitter-specific pathways may predispose an individual to the risk of developing depression (Pezawas *et al.*, 2005). These susceptibility genes will leave individuals vulnerable to developing depression when exposed to environmental factors such as grief, drug effects or stress (Murphy *et al.*, 2004). The co-occurrence of genetic susceptibility and environmental triggers results in a heritability of about 70% over one's lifetime (Wong and Licinio, 2001).

Taken together, these findings have led to a theory of bidirectional causative events in mood disorders and epilepsy (Hesdorffer *et al.*, 2000). This bidirectional model of depression and epilepsy suggests that individuals with either disorder are at higher risk of developing the other disorder. It does not suggest a causal relationship between depression and epilepsy but rather an underlying cerebral dysfunction of which both seizures and depression may be an expression (Kanner, 2005).

Symptomatology

The most widely used classification of psychiatric disorders is set forth in the Diagnostic and Statistical Manual of Mental Disorders (DSM) (American Psychiatric Association, 2000). This classification system is based on clinical and descriptive features specific for neuropsychiatric disorders. In epilepsy a similar system has been proposed by the International League Against Epilepsy (Krishnamoorthy *et al.*, 2007). This group is concerned that available classification systems such as the DSM-IV, which solely cover psychiatric disorders, are not acceptable for classifying clinical symptoms associated with psychiatric disorders in epilepsy. This concern has been expressed because it is not obvious that patients with epilepsy and comorbid psychiatric disorders have the same clinical presentation as patients with psychiatric disorders alone (Gaitatzis *et al.*, 2004).

Comparisons of clinical characteristics between inpatients with epilepsy and depression and inpatients with depression alone have demonstrated similar levels of psychomotor retardation and other neurovegetative signs. Whereas the epilepsy group reported more psychotic traits, such as paranoia, irritability, delusional thinking and persecutory auditory hallucinations, the nonepilepsy group reported more neurotic traits.

Depressive disorders in epilepsy have also been classified according to their temporal relation with seizure occurrence, with preictal referring to the period immediate prior to the seizure, ictal referring to the period during the seizure, and postictal referring to the period immediately following the seizure. In evaluating preictal depressive events, investigators have described a decline in mood during the hours or days before the seizure with an improvement following the seizure. Sudden mood changes, anhedonia, guilt and suicidal ideation may be expressions of simple partial seizures. Symptoms of postictal depression have been reported in 43 of 100 consecutively assessed patients, with a median duration of 24 hours(h) for most symptoms (Kanner *et al.*, 2004).

Interictal, the time period between seizures, depression is the most commonly recognized form of depression in patients with epilepsy (Barry *et al.*, 2000). However, up to 50% of episodes may not meet DSM criteria for depression (Mendez *et al.*, 1986). This has led to the terms "interictal dysphoric disorder" (Blumer, 1991) and "dysthymic-like disorder of epilepsy" (Kanner, 2005) to describe atypical interictal mood disorders in patients with epilepsy. Characteristically, symptoms of depressive mood, irritability, anxiety, insomnia, anergia and pains are the hallmarks of these waxing and waning disorders.

On rare occasions psychiatric symptoms such as depression may be exacerbated by or present initially in a seizure free state (Savard *et al.*, 1998). Typically, a patient with a history of frequent seizures may become seizure free for the first time and start exhibiting worsening of an existing, or develop a *de novo*, mood disorder. The underlying mechanisms are poorly understood and more research into risk factors and prevalence of this forced normalization phenomenon is warranted.

Clinical implications

The screening of patients with epilepsy for mood disorder or mood changes has become standard care in our patient management. Patients new to our practice are carefully evaluated for any past or present psychiatric issues. The temporal relationship of such symptoms with concurrent seizure medication regimen, seizure events (i.e., are the symptoms related to the preictal, postictal or interictal period or forced normalization) and life stressors is reviewed. Has the patient received treatment in the past for these symptoms and if so for how long and what were the reasons for discontinuation? What treatment strategies have and have not worked in the past, and for what reasons? If the patient reports symptoms of depression at the initial visit, duration and treatment of present symptoms are reviewed. As a quantitative indicator patients are asked to rate their mood on a scale of 1–7 with seven being their euthymic state. At each follow-up visit, possible mood changes since their last visit are discussed with patients. This seems especially important when AEDs with mood altering properties were either added or discontinued.

As previously discussed, patients with epilepsy often will not meet DSM criteria for depression but rather report intermittent affective–somatoform symptoms. In our experience, these vague or waxing and waning symptoms may lead to patients being under-treated. Additionally, due to their vague or intermittent nature, these symptoms may be assumed by patients to be an acceptable state.

Concerns about antidepressant medications lowering seizure threshold or potential drug interactions with AEDs may contribute to a reluctance to treat on the part of the physician as well as the patient. Our first question when choosing a treatment is whether we can minimize medication exposure by taking advantage of the mood stabilizing properties of one of the AEDs. When adding an antidepressant we have found SSRIs and selective serotonin–noradrenaline reuptake inhibitors (SNRIs) to be safe, effective and well tolerated when started at the smallest possible dose with increments every 10–14 days. Patients need to be assured of the generally transient nature of side effects and are encouraged to call with any concerns after starting the medication. This allows us to further individualize the schedule for increments, such as a smaller dose increase or longer intervals between increases. These adjustments help ensure better medication compliance and ultimately better outcome. Short-term cognitive–behavioral therapy alone or in conjunction with antidepressant medication has been useful in some patients with specific issues, such as job, family or lifestyle changes leading to dysthymic symptoms. Our treatment goal is elimination of all signs and symptoms of depressive disorder. Therefore, with patients in whom a trial of 2–3 different SSRIs/SNRIs does not result in an euthymic state, an evaluation with a collaborating psychiatrist is recommended.

ANXIETY

Although anxiety disorders are a recognized diagnostic entity, in most psychiatric diagnostic schemes they are often studied in combination with depression rather than as independent comorbidities of epilepsy. Subsequently, less scientific literature is available on the relationship between epilepsy and anxiety. A strong comorbidity between anxiety and depression has been demonstrated in general psychiatric patients (Murphy et al., 2004). Anxiety, in the absence of depression, is significantly more prevalent than depression alone in patients without concomitant chronic illness. In one of the few studies of anxiety in epilepsy without depression, the prevalence of epilepsy with anxiety was reported to be approximately 25% (Jacoby et al., 1996).

Risk factors contributing to anxiety in patients with epilepsy have been identified as neurological, pharmacological and psychosocial (Vazquez and Devinsky, 2003). In a small study, increased right-sided amygdala volumes were associated with anxiety disorder (Satishchandra *et al.*, 2003). A study of 23 psychotropic naïve pediatric patients found a positive correlation between anxiety and increased amygdala-to-hippocampal volume ratios (Macmillan *et al.*, 2003). Pharmacological agents such as steroids, thyroid medications, stimulants and antidepressants should not be underestimated as causative agents in the development of anxiety disorders. Psychosocial contributions to anxiety may reflect adjustments a patient has to make to having seizures as well as to the unpredictability and fear of recurrent seizures.

Pathophysiology

Anxiety has been described as a product of an evolutionary process of detecting environmental dangers (Pine, 2003). Neurocircuitry involving the Papez circuit, including the prefrontal cortex, the septum and the amygdala, allows for the physiological response and avoidance of perceived external threats. The development of anxiety has also been reviewed in a historical framework, evolving from psychodynamic to learning theories and cognitive–behavioral to neurobiologic theories (Goldstein and Harden, 2000). In psychodynamic theory, anxiety is thought to signify a real or a perceived threat to the integrity of personal identity. Therefore, bolstering the integrity of the personal identity would decrease symptoms of anxiety. Learning and cognitive–behavior theory view anxiety as a conditioned but faulty response to the stimulus of a perceived external threat. Accordingly, a realistic assessment of the threat and development of appropriate coping skills will decrease symptoms of anxiety. In neurobiologic theory, functional neuroimaging has shown neurophysiological abnormalities involving serotonin, γ-aminobutyric acid (GABA), norepinephrine (NE) and corticotropin-releasing hormone activities (Charney, 2003). Therefore, regulation or modulation of these neurotransmitter systems will alter the phenomenon of anxiety.

Structurally, the amygdala and its afferent and efferent connections play a major role in mediating fear and anxiety and their physiological responses (Phillips and LeDoux, 1992). Work investigating two distinct types of anxiety demonstrates involvement of specific and different limbic regions during emotion processing (Engels *et al.*, 2007). Anxious apprehension, defined as persistent worrisome verbal rumination without resolution, is often associated with physical symptoms. Symptoms for anxious arousal, in contrast, defined as having a more immediate quality of threat, include symptoms of shortness of breath, sweating, dizziness and palpitations. When both groups were presented with the same emotional threat stimuli while undergoing functional magnetic resonance imaging (fMRI), the anxious apprehension group demonstrated activation of left inferior frontal regions including Brocca's area. The anxious arousal group, on the other hand, activated areas in the inferior temporal area in the right hemisphere. These distinct and localized activations suggest specific pathways involved in different phenotypes of anxiety.

Animal models demonstrate differential plasticity of the immature fear circuit in the development of anxiety. Early life influences constrain the range of possible fear behavior in the mature animal model (Pine, 2007). The structural regions of the amygdala mature early in primates, allowing for adult-like responses to fear before reaching adolescence (Pine, 2003). The posterior temporal association cortex and prefrontal cortex connections are slower to mature and subsequently will influence responses to anxiety according to the developmental level.

Symptomatology

The DSM-IV (American Psychiatric Association, 2000) sets forth extensive criteria for over ten anxiety disorders. It is the intensity, duration and irrational proportion of the feeling that warrants the diagnosis of anxiety disorder. Psychiatric or physiological symptoms may be expressions of anxiety. Psychiatric symptoms may include hypervigilance, irritability or difficulties with concentration. Physiological symptoms may be either motor signs, such as restlessness and trembling or autonomic signs, such as sweating and dizziness.

In the patient with epilepsy, anxiety has been variously described as a reaction to the perceived psychosocial impact of having seizures or as a physical part of the peri-ictal experience (Betts, 1981). As in depression, anxiety may have a temporal relationship to the ictal event. Fear and anxiety have been reported to increase for hours to days prior to a seizure (Blanchet and Frommer, 1986). Fear, in extreme cases mimicking panic attacks (Young et al., 1995), has been described as part of an aura in up to 15% of patients with seizures originating in the temporal lobe or cingulate gyrus (Torta and Keller, 1999). Such fear will be stereotypical in its presentation, brief in duration and may progress to a complex partial seizure (Hughes et al., 1993). In contrast, a panic attack is "... a discrete period of intense fear and or discomfort ... peak within 10 min"(American Psychiatric Association, 2000, p. 209) during which at least 4 symptoms out of 13 are present. In contrast to seizures, no loss of consciousness is associated with panic attacks and age of onset is typically between 20 and 30 years of age (Handal et al., 1995). Interictal anxiety has been reported in patients with seizures of limbic or generalized onset (Devinsky and Vazquez, 1993). This anxiety may be present in the form of generalized anxiety disorder, post-traumatic stress disorder, panic attacks, obsessive-compulsive disorder (Vazquez and Devinsky, 2003) or agoraphobia (Gaitatzis et al., 2004). It has been stipulated that anticipatory anxiety related to having seizures in public may lead to agoraphobia. However, only rare cases of seizure phobia have been reported (Newsom-Davis et al., 1998).

Clinical implications

Any assessment for a possible mood disorder in a patient with epilepsy must include an evaluation for symptoms of anxiety. Just as with depression, we address previous history and treatment, temporal relationship with seizure events, family history, current symptoms and treatment needed. Careful review of current medications, AEDs and concomitant medications, as well as other medical conditions needs to be done. Patients have reported anxiety when being tapered off a sedating AED. Assurance of the transient nature of such symptoms has been helpful to our patients. Indeed, improved seizure control may lead to decreased anxiety in patients with seizure associated anxiety.

A patient's history of past or current anxiety needs to be taken into consideration when choosing an anticonvulsant medication. The more psycho-activating qualities of felbamate and lamotrigine may not be appropriate for a patient who already has issues with anxiety. On the other hand, a medication with anxiolytic properties such as pregabalin may be better suited. If anxiety does not seem to have a causal relationship to seizures, or concomitant medical condition, the symptoms may need to be treated separately. Several SSRIs are indicated for the treatment of anxiety disorder in the United States. Starting with the smallest possible dose of medication and increasing slowly has generally worked well for our patients. Since it can take 3–6 weeks for SSRIs to achieve therapeutic levels, we often offer our patients a short course (about 3 weeks) of a small dose of alprazolam to cover the symptoms of anxiety. This should allow for an immediate elimination of the symptoms and decrease the impairment for the patient. After a 3-week period, we slowly taper off the alprazolam as the anxiolytic properties of the SSRI should have started to take effect.

As with depression, cognitive–behavioral therapy alone or in conjunction with a selective serotonin reuptake inhibitor (SSRI) can be very effective in learning to eliminate anxiety related to a noxious stimulus. If none of these treatments decrease symptoms of anxiety we generally recommend evaluation by a collaborating psychiatrist.

ATTENTION DEFICIT–HYPERACTIVITY DISORDER (ADHD)

ADHD is a clinically complex, heterogeneous psychiatric disorder (Biederman, 2005) diagnosed in childhood with approximately 50% of patients continuing to exhibit symptoms throughout adult life (Barkley et al., 2002). It is the most commonly diagnosed psychiatric disorder in childhood (Wolraich et al., 2005) with an estimated prevalence of 3–7% (Rappley, 2005). The diagnosis continues to be predominantly made in boys, however, a better understanding of symptomatology in girls has led to an increase in sensitivity to gender issues in the diagnosis. Severity of childhood ADHD symptoms is the most significant predictor of persistent adult impairment (Kessler, 2004), and adult prevalence rates have been reported around 4% (Faraone and Doyle, 2001). In children, the prevalence rates of inattention, hyperactivity or impulsivity may be affected by severity of epilepsy (Davies *et al.*, 2003), with increasing rates seen in children with severe epilepsy. These increased rates are found regardless of whether parent/teacher questionnaires (Hoare and Kerley, 1991) or standardized neuropsychological testing were used (Sanchez-Carpintero and Neville, 2003). Prevalence rates may also be affected by type of epilepsy. Generalized epilepsies have been associated with impairment of sustained attention (Dunn and Kronenberger, 2006) while nondominant hemispheric foci are thought to also have an affect on attention (Aldenkamp and Arends, 2004). Thirty-three percent of children with complex partial seizures are reported to have ADHD symptoms (Semrud-Clikeman and Wical, 1999). Additionally, 37.7% are reported to have predominantly inattentive type (Dunn and Kronenberger, 2006), and 29.1% have inattentive combined type (Thome-Souza *et al.*, 2004). The challenge with evaluating these studies is in combining the various measures that have been used to determine the criteria for presence and severity of ADHD.

Pathophysiology

Our understanding of the pathophysiology of ADHD has increased in recent years with various factors implicated in the disorder. Structural and functional neural substrates and genetic and environmental indices are thought to be involved to varying degrees. Theories and models of neuronal system involvement in ADHD range from deficit in executive function (Willcutt *et al.*, 2005) due to deficient inhibitory control (Kipp, 2005; Sonuga-Barke, 2005) to response inhibition. Various studies have supported the involvement of executive cognitive processing and ADHD along with involvement of the frontal-striatal network and other brain regions, including the parietal lobes. For example, altered brain activation and decreased gray matter volume have been reported in the right inferior frontal cortex (Aaron and Poldrack, 2005). Longitudinal neurological structural MRI studies of children with ADHD have revealed persistent thinning of the left medial prefrontal cortex in the most impaired patients (Shaw *et al.*, 2006) and significantly smaller global brain volumes, including cerebellum, that persisted through childhood and adolescence (Castellanos *et al.*, 2002). Dysregulation of neurotransmitter modulation, involving NE and dopamine (DA), has been identified as an important factor in the pathophysiology. Imaging studies have found dopamine transporter (DAT) to play an important part in psychomotor activity and reward-seeking behavior. Mice without the DAT gene exhibit hyperactivity and inhibitory behavior deficits similar to that seen in humans with ADHD (Giros *et al.*, 1996). Pharmacologically, DAT is the main target for stimulants in the treatment of ADHD, with blockade of reuptake of both NE and DA. Family (Faraone and Khan, 2006), twin (Coolidge *et al.*, 2002) and adoption (Sprich *et al.*, 2000) studies have supported a strong genetic influence in the development of ADHD. Gene variants related to DA receptor D4, D5 and DAT have been studied most extensively (Faraone *et al.*, 2005) with some studies showing a strong correlation between the genetic alterations with ADHD, only to be challenged by other studies (Biederman, 2005). Taken together, treatment and genetic studies suggest a strong role for DA levels or receptors in the pathophysiology of ADHD.

Symptomatology

Criteria for the diagnosis of ADHD in children have been established in the DSM with three subtypes: ADHD predominantly inattentive, hyperactive–impulsive and ADHD combined type (American Psychiatric Association, 2000). The 2000 published clinical practice guidelines by the American Academy of Pediatrics (AAP; American Academy of Pediatrics, Subcommittee on Attention-Deficit/ Hyperactivity Disorder and Committee on Quality Improvement, 2000) stress the importance of using DSM criteria, and provide a clinical algorithm for the assignment of diagnoses (Mendez *et al.*, 1986). Oppositional defiant disorder, conduct disorder, anxiety disorder, and depressive disorder are the most common comorbid psychiatric conditions in children with ADHD. The DSM-IV-R criteria for adult diagnosis of ADHD contains three critical elements: childhood onset, presence of significant symptoms and impairment from these symptoms in at least two domains: school/work, social interaction, or home life. In epilepsy, ADHD inattentive type has been reported to be more common regardless of gender (Charney, 2003; Biederman, 2005). A study of attention performance of newly diagnosed children with either partial or generalized seizures showed 21% of patients meeting criteria for attentional problems prior to onset of treatment (Barkley *et al.*, 2002) with that number doubling at 1-year follow-up despite complete seizure control. Prognosis of persistent ADHD symptoms was strongly related to history of active seizures for at least 6 months and persistent abnormal EEG activity with behavioral and emotional disorder at time of diagnosis. A history of ADHD has also been shown to increase the risk for a first unprovoked seizure by 2.5-fold (Hesdorffer *et al.*, 2004).

Clinical implications

For patients with ADHD, treatment of individualized target symptoms with measurable outcomes remains the goal. In patients with seizures this goal may be complicated by finding a balance between seizure control and increased attention when decreased attention may be secondary to anticonvulsant medication therapy. An evaluation of a patient's AEDs is essential to determine their effects on inattention or behavior. Methylphenidate and amphetamine compounds remain the most commonly used stimulants. Their safety and efficacy have been well established and they work in a dose-dependent manner. We have not encountered any cases of decrease in seizure threshold due to stimulant use. The effectiveness of stimulants in controlling symptoms has varied according to different trials from as high as 68–80% in children (Rappley, 2005), about 70% in adolescents (Wolraich et al., 2005) and 30–50% in children and adults (Biederman et al., 1996).

Because of recent deaths suggestive of cardiac origin associated with stimulant therapy, we obtain ECGs, with clearance by a cardiologist if necessary, in our patients prior to initiating stimulant pharmacotherapy. Whether to choose a short- or long-acting medication is often a matter of preference. Short-acting medications may be targeted to specific activities whereas long-acting medications may eliminate an often stigmatizing trip to the school nurse during the school day. We follow the general recommendations of starting on a low dose and titrate weekly to an effective dose without side effects when treating adults or children.

Atomoxetine, a NE-reuptake inhibitor, non-stimulating agent, has shown a 25–30% improvement in approximately 60% of patients with ADHD (Kelsey et al., 2004). In addition to pharmacological therapy we work on a behavioral approach with parents and teachers to meet target behavioral goals. The treatment of children and adults with seizures and ADHD requires a comprehensive medical and psychosocial approach in which skills need to be acquired to deal with the developmental challenges at each stage of life.

LEARNING DISORDERS

The prevalence of learning disorders (LD) in the general population ranges from 2% to 10% (American Psychiatric Association, 2000). In children attending public schools in the United States, the prevalence of LD is approximately 5% with reading disorders being the most frequent deficit (Lang, 1996). Reports on the prevalence of learning disorders in children of normal intelligence with epilepsy range from 25% (Lhatoo and Sander, 2001) to 50% (Dunn and Austin, 1999). The spectrum of these disabilities ranges from mild underachievement to major challenges in learning requiring specialized academic education (Beghi et al., 2006). Generally, academic challenges are not confined to one area but rather may include spelling, reading comprehension and math (Williams, 2003).

Various levels of LD and cognitive impairments are associated with different epilepsy syndromes. West syndrome, Dravet syndrome, Doose syndrome and Lennox-Gastaut syndrome are generally associated with mental retardation, LD, behavioral problems and poor prognosis (Filippini et al., 2006). Better outcomes with early and aggressive treatment have been shown for both Landau-Kleffner syndrome (LKS) and electrical status epilepticus of slow-wave sleep (ESES) (Robinson et al., 2001). Benign childhood epilepsy with centrotemporal spikes (BCECTS)/rolandic epilepsy and Janz syndrome/juvenile myoclonic epilepsy have traditionally been considered not to be associated with any cognitive or behavioral issues (Besag, 2006). More recently, however, controlled studies have shown these syndromes to have negative effects on both IQ and behavior. Poorer school achievements have been shown in adults with a history of seizure disorder in childhood with subsequent seizure abatement and medication freedom in adulthood (Sillanpaa et al., 1998). Cognitive and behavioral disorders in children prior to onset of a seizure disorder have been documented by several studies (Austin et al., 2001; Berg et al., 2005). These findings confirm the complexity of the relationship between LD and epilepsy.

Pathophysiology

LD are most often part of a medical condition such as fetal alcohol syndrome, lead poisoning or fragile X syndrome (American Psychiatric Association, 2000). LD associated with epilepsy may be an epiphenomena of an underlying neurological disorder (Brown, 2006). Abnormal electrical activity and/or structural abnormality in an eloquent brain region during maturation may lead to permanent functional deficit. Subsequently, severe cognitive deficits are more likely seen with early seizure onset (Dam, 1990). LKS has been suggested as a model for evaluation of the effect of an EEG abnormality on cognition (Besag, 2006). Here, EEG abnormalities of continuous spike–wave discharges over temporal or temporoparietal regions during slow-wave sleep may be associated with an acquired aphasia (Genton et al., 1996). Improved speech functions have been reported with abolishment of the abnormal electrical activity with either pharmacological or surgical interventions (Morrell et al., 1989; Robinson et al., 2001). In these cases, early intervention to abolish abnormal electrographic activity was associated with better outcome.

BCECTS with rolandic discharge (RD) is often put on the other end of the spectrum of EEG impact on cognition from LKS. Semiologically, the seizures associated with this syndrome are brief, simple partial hemifacial motor seizures, which during sleep may evolve to secondarily generalized tonic seizures. The electroencephalogram (EEG) in these patients will typically show high-voltage focal centrotemporal discharges with activation of slow waves by sleep. Historically, BCECTS has been viewed as a benign, time-limited disorder without any cognitive or neurological deficits (Gobbi et al., 2006). More recently, however, learning, language as well as behavioral deficits have been documented (Yun et al., 2000) with the severity of impairment related to increased EEG activation during sleep (Baglietto et al., 2001). A recent study of BCECTS could not demonstrate a correlation between improved EEG and improved cognition in a treatment group, however (Stephani and Carlsson, 2006).

Overall, the pathology of learning disorders in patients with epilepsy remains open to investigation. However, cerebral dysfunctions of specific areas show correlations to LD in these domains.

Symptomatology

According to the DSM (American Psychiatric Association, 2000), an individual may be diagnosed with LD when standardized testing results fall two standard deviations below the results expected for the individual's level of schooling and intelligence. Often there is a link between LD and behavioral problems such as ADHD, depression and dysthymic disorder. Generally, mathematics disorder and written expression disorder occur in combination with reading disorder. These same domains of impairment have been identified as common in a cohort of children with epilepsy and academic underachievement (Williams, 2003).

In patients with epilepsy, LD may present as a permanent or state-dependent phenomena (Besag, 1995). Perinatal insults, chromosomal abnormalities and epileptic encephalopathies are frequently associated with permanent LD (Beghi et al., 2006). In these cases, LD and seizures may be a consequence of the same distinct, permanent underlying neurologic or medical disorder. State-dependent LD, on the other hand, may be treatable and therefore reversible. Cognitive side effects of first generation AEDs have been well established. Of the newer medications, topiramate has been shown to negatively affect verbal fluency, processing speed and working memory in healthy volunteers (Meador et al., 2005) as well as in patients (Bootsma et al., 2006). Other medication side effects, such as sedation, sleepiness, or blurred vision, may interfere with learning and result in a sudden negative impact on school performance. Improved learning and school performance may also be associated with improved seizure control. For example, frequent absence seizures will cause interruption of focusing and attention and may result in poor school performance. Improved control of these seizures may eliminate this state-dependent LD.

Clinical implications

Parents often verbalize concerns about the effects of seizures and their treatment on their child's ability to learn. Each child new to our practice is evaluated for possible LD. Specifically, we look at intellectual and physical development chronologically and try to determine whether deviations correlate with variables such as seizure frequency and possible medication side effects. Whenever possible, we eliminate AEDs with sedating properties and simplify drug regimens. Ideally, our goal is to let children have their best possible seizure control on medication with no or minimal cognitive impairment. Over the years, we have learned to not underestimate the effects poor seizure control and medication side effects have on the individual's ability to learn. If academic difficulties persist regardless of improved seizure control and decreased side effects, we will recommend the child undergo neuropsychological testing to determine academic strengths and challenges. Unfortunately, many children continue to have seizures in spite of being treated with multiple AEDs. In these cases, neuropsychological testing will allow for an objective evaluation of the effects of disease and treatment on cognitive functioning as well as available compensatory mechanisms. In the best case scenario, testing outcomes will alleviate concerns parents may have had about their child's current level of function and assure the child he is functioning within expected norms. Test results revealing significant deficits in specific areas often allow patients to qualify for individualized educational plans and specialized services thru public schools. We work in conjunction with neuropsychologists and schools on the implementation and continuous evaluation and modification of these services.

At each follow-up visit to our clinic, a child's academic progress is reviewed, possible causes for set-backs are investigated and modifications are discussed with parents and child. Realistic goal setting based on objective testing results can eliminate tension between parents, child and school and allow the child to reach his potential.

CONCLUSIONS

Chronicity and severity of epilepsy have long been the established parameters for predicting psychiatric comorbidity. The longer the duration and the higher the severity of the seizure disorder, the more likely psychiatric problems are expected to occur. These comorbidities include depression, anxiety, ADHD and LD. Models of psychiatric comorbidities range from reactive, proposing that depression and anxiety develop in reaction to the diagnosis of epilepsy or as a consequence of living with the disorder, to endogenous, proposing a biologic cause to the psychiatric symptoms related to the seizure disorder. More recently, however, bidirectional models looking at psychiatric comorbidities and epilepsy suggest a complex, temporal relationship between pre-existing psychiatric status, genetic influences, neurologic substrates and focality of epilepsy in their development. New ways of looking at epilepsy as a system expressing itself as seizures and mood disorder, ADHD or LD have garnered scientific attention. How a neuropathology will express itself in a given individual, whether initially as a seizure or a psychiatric disorder, is poorly understood at this time. For the clinician, however, it has to be clear that patients with epilepsy may present with more than just seizures. Given our current scientific knowledge, each of these patients deserves to be evaluated and treated for possible comorbidities at each visit. Ultimately, these ongoing evaluations will be more cost-effective in the long term as they allow for early detection and interventions. As well-informed clinicians we are able to make significant contributions to improve the well-being of our patients.

REFERENCES

Aaron, A. R., Poldrack, R. A. (2005). The cognitive neuroscience of response inhibition: relevance for genetic research in attention-deficit/hyperactivity disorder. *Biol Psychiatry* **57**(11), 1285–1292.

Aldenkamp, A., Arends, J. (2004). The relative influence of epileptic EEG discharges, short nonconvulsive seizures, and type of epilepsy on cognitive function. *Epilepsia* **45**(1), 54–63.

Altschuler, L., Rausch, R., Delrahim, S., *et al.* (1999). Temporal lobe epilepsy, temporal lobectomy, and major depression. *J Neuropsychiatry Clin Neurosci* **11**(4), 436–443.

American Academy of Pediatrics, Subcommittee on Attention-Deficit/Hyperactivity Disorder and Committee on Quality Improvement (2000). Clinical practice guideline: diagnosis and evaluation of the child with attention-deficit/hyperactivity disorder. *Pediatrics* **105**(5), 1158–1170.

American Psychiatric Association (2000). *Quick reference to the diagnostic criteria from DSM-IV-TR.* Washington, DC: American Psychiatric Press.

Austin, J. K., Harezlak, J., Dunn, D. W., *et al.* (2001). Behavior problems in children before first recognized seizures. *Pediatrics* **107**, 115–122.

Baglietto, M. G., Battaglia, F. M., Nobili, L., *et al.* (2001). Neuropsychological disorders related to interictal epileptic discharges during sleep in benign epilepsy of childhood with centrotemporal spikes. *Dev Med Child Neurol* **43**, 407–412.

Barkley, R. A., Fischer, M., Smallish, L., *et al.* (2002). The persistence of attention-deficit/hyperactivity disorder into young adulthood as a function of reporting source and definition of disorder. *J Abnorm Psychol* , 111–289.

Barry, J. J., Huynh, H., Lembke, A. (2000). Depression in individuals with epilepsy. *Curr Treat Options Neurol* **2**, 571–585.

Beghi, M., Cornaggia, C. M., Frigeni, B., *et al.* (2006). Learning disorders in epilepsy. *Epilepsia* **47** (Suppl 2), 14–18.

Berg, A. T., Smith, S. N., Frobish, D., *et al.* (2005). Special education needs of children diagnosed with epilepsy. *Dev Med Child Neurol* **47**, 749–753.

Besag, F. C. M. (1995). Epilepsy, learning and behavior in childhood. *Epilepsia* **36**(Suppl 1), 58–63.

Besag, F. M. C. (2006). Cognitive and behavioral outcomes of epileptic syndromes: implications for education and clinical practice. *Epilepsia* **47**(Suppl 2), 119–125.

Betts, T. A. (1981). Depression, anxiety and epilepsy. In *Epilepsy and Psychiatry* (E. H. Reynolds, M. R. Trimble, eds), pp. 60–71. London: Churchill Livingstone.

Biederman, J., Faraone, S. V., Milberger, S., *et al.* (1996). Predictors of persistence and remission of ADHD: results from a four-year prospective follow-up study of ADHD children. *J Am Acad Child Adolesc Psychiatry* **35**, 343–351.

Biederman, J. (2005). Attention-deficit/hyperactivity disorder: a selective overview. *Biol Psychiatry* **57**(11), 1215–1220.

Blanchet, P., Frommer, G. P. (1986). Mood change preceding epileptic seizures. *J Nerv Men Dis* **174**, 471–476.

Blumer, D. (1991). Epilepsy and disorders of mood. In *Neurobehavioral Problems in Epilepsy* (D. Smith, D. Treiman, M. R. Trimble, eds), pp. 185–195. New York: Raven Press.

Bootsma, H.-P. R., Aldenkamp, A. P., Diepman, L., *et al.* (2006). The effect of antiepileptic drugs on cognition: patient perceived cognitive problems of topiramate versus levetiracetam in clinical practice. *Epilepsia* **47**(Suppl 2), 24–27.

Brown, S. (2006). Deterioration. *Epilepsia* **47**(Suppl 2), 19–23.

Castellanos, F. X., Lee, P. P., Sharp, W., *et al.* (2002). Developmental trajectories of brain volume abnormalities in children and adolescents with attention-deficit/hyperactivity disorder. *JAMA* **288**(4), 1740–1748.

Charney, D. S. (2003). Neuroanatomical circuits modulating fear and anxiety behaviors. *Acta Psychiatr Scand* **108**(s417), 38–50.

Coolidge, F. L., Thede, L. L., Young, S. E. (2002). Heritability and the comorbidity of attention deficit hyperactivity disorder with behavioral disorders and executive function deficits: a preliminary investigation. *Dev Neuropsychol* **17**, 273–287.

Dam, M. (1990). Children with epilepsy: the effects of seizures, syndromes and etiological factors on cognitive functioning. *Epilepsia* **31**(Suppl 4), S26–S29.

Davies, S., Heyman, I., Goodman, R. (2003). A population survey of mental health problems in children with epilepsy. *Dev Med Child Neurol* **45**, 292–295.

Devinsky, O., Bear, D. M. (1991). Varieties of depression in epilepsy. *Neuropsychiatry Neuropsychol Behav Neurol* **4**, 49–61.

Devinsky, O., Vazquez, B. (1993). Behavioral changes associated with epilepsy. *Neurol Clin* **11**, 127–149.

Dunn, D. W., Austin, J. K. (1999). Behavioral issues in pediatric epilepsy. *Neurology* **53**(Suppl 2), S96–S100.

Dunn, D. W., Kronenberger, W. G. (2006). Childhood epilepsy, attention problems, and ADHD: review and practical considerations. *Semin Pediatr Neurol* **12**, 222–228.

Engels, A. S., Heller, W., Mohanty, A., et al. (2007). Specificity of regional brain activity in anxiety during emotion processing. *Psychophysiology* **44**, 352–363.

Faraone, S. V., Doyle, A. E. (2001). The nature and heritability of attention-deficit/hyperactivity disorder. *Child Adolesc Psychiatr Clin N Am* **10**, 299–316.

Faraone, S. V., Perlis, R. H., Doyle, A. E., et al. (2005). Molecular genetics of attention deficit hyperactivity disorder. *Biol Psychiatry* **7**, 1313–1323.

Faraone, S. V., Khan, S. A. (2006). Candidate gene studies of attention-deficit/hyperactivity disorder. *J Clin Psychiatry* **67**(Suppl 8s), 13–20.

Filippini, M., Boni, A., Dazzani, G., et al. (2006). Neuropsychological findings: myoclonic astatic epilepsy (MAE) and Lennox-Gastaut syndrome (LGS). *Epilepsia* **47**(Suppl 2), 56–59.

Gaitatzis, A., Trimble, M. R., Saner, J. W. (2004). The psychiatric comorbidity of epilepsy. *Acta Neurol Scand* **110**, 207–220.

Genton, P., Bureau, M., Dravet, C., et al. (1996). Less common epileptic syndromes. In *The treatment of epilepsy: principles and practice* (E. Wyllie, ed.) 2nd edition. Baltimore, MD: Williams and Wilkins.

Giros, B., Jaber, M., Jones, S. R., et al. (1996). Hyperlocomotion and indifference to cocaine and amphetamine in mice lacking the dopamine transporter. *Nature* **379**, 606–612.

Gobbi, G., Boni, A., Filippini, M. (2006). The spectrum of idiopathic rolandic epilepsy syndromes and idiopathic occipital epilepsies: from the benign to the disabling. *Epilepsia* **47**(Suppl 2), 62–66.

Goldstein, M. A., Harden, C. L. (2000). Epilepsy and anxiety. *Epilepsy Behav* **1**, 228–234.

Handal, N. M., Masand, P., Weilburg, J. B. (1995). Panic disorder and complex partial seizures: a truly complex relationship. *Psychosomatics* **36**, 498–501.

Hermann, B. P., Seidenberg, M., Bell, B. (2000). Psychiatric comorbidity in chronic epilepsy: identification, consequences, and treatment of major depression. *Epilepsia* **41**(Suppl 2), S31–S41.

Hermann, B. P., Wyler, A. R. (1989). Depression, locus of control, and the effects of epilepsy surgery. *Epilepsia* **30**, 332–338.

Hesdorffer, D. C., Hauser, W. A., Annegers, J. F., et al. (2000). Major depression is a risk factor for seizures in older adults. *Ann Neurol* **47**, 246–249.

Hesdorffer, D. C., Ludvigsson, P., Olafsson, E., et al. (2004). ADHD as a risk factor for incident unprovoked seizures and epilepsy in children. *Arch Gen Psychiatry* **61**(7), 731–736.

Hoare, P., Kerley, S. (1991). Psychosocial adjustment of children with chronic epilepsy and their families. *Dev Med Child Neurol* **33**, 201–215.

Hughes, J., Devinsky, O., Feldmann, et al. (1993). Premonitory symptoms in epilepsy. *Seizure* **2**(3), 201–203.

Jacoby, A., Baker, G. A., Steen, N., et al. (1996). The clinical course of epilepsy and its psychosocial correlates: findings from a UK community study. *Epilepsia* **37**, 148–161.

Jones, J. A., Watson, R., Sheth, R., et al. (2007). Psychiatric comorbidity in children with new onset epilepsy. *Dev Med Child Neurol* **49**, 493–497.

Kanner, A. M. (2003). Depression in epilepsy: a frequently neglected multifaceted disorder. *Epilepsy Behav* **4**(Suppl 9), 11–19.

Kanner, A. M. (2005). *Depression in neurological disorders*. Cambridgeshire, United Kingdom: The Lundbeck Institute.

Kanner, A. M., Soto, A., Gross-Kanner, H. (2004). Prevalence and clinical characteristics of postictal psychiatric symptoms in partial epilepsy. *Neurology* **62**, 708–713.

Kelsey, D. K., Summer, C. R., Casat, C. D., *et al.* (2004). Once-daily atomoxetine treatment for children with attention-deficit/hyperactivity disorder, including an assessment of evening and morning behavior: a double-blind, placebo-controlled trial. *Pediatrics* **114**, e1–e8.

Kessler, R. C. (2004). *Prevalence of ADHD in the US: Results From the NCS-R.* Washington, DC: American Psychiatric Association.

Kipp, K. (2005). A developmental perspective on the measurement of cognitive deficits in attention-deficit/hyperactivity disorder. *Biol Psychiatry* **57**(11), 1256–1260.

Kobau, R., Gilliam, F., Thurman, D. J. (2006). Prevalence of self-reported epilepsy or seizure disorder and its association with self-reported depression and anxiety: results from the 2004 HealthStyles Survey. *Epilepsia* **47**(11), 1915–1921.

Krishnamoorthy, E. S., Trimble, M. R., Blumer, D. (2007). The classification of neuropsychiatric disorders in epilepsy: a proposal by the ILAE commission on psychobiology of epilepsy. *Epilepsy Behav* **10**(3), 349–353.

Landers, E. S., Schork, N. J. (1994). Genetic dissection of complex traits. *Science* **265**, 2047–2048.

Lang, G. R. (1996). Learning disabilities. *Future Child* **6**, 54–76.

Lhatoo, S. D., Sander, A. S. (2001). The epidemiology of epilepsy and learning disability. *Epilepsia* **42**(Suppl 1), 6–9.

MacMillan, S., Szeszko, P. R., Moore, G. J., *et al.* (2003). Increased amygdala:hippocampal volume ratios associated with severity of anxiety in pediatric major depression. *J Child Adolesc Psychopharmacol* **13**, 65–73.

Meador, K. J., Loring, D. W., Vahle, V. J., *et al.* (2005). Cognitive and behavioral effects of lamotrigine and topiramate in healthy volunteers. *Neurology* **64**, 2108–2114.

Mendez, M. F., Cummings, J., Benson, D., *et al.* (1986). Depression in epilepsy. Significance and phenomenology. *Arch Neurol* **43**, 766–770.

Mendez, M. F., Doss, R. C., Taylor, J. L., *et al.* (1993). Depression in epilepsy: relationship to seizures and anticonvulsant therapy. *J Nerv Ment Dis* **181**, 444–447.

Morrell, F., Whisler, W. W., Bleck, T. P. (1989). Multiple subpial transection: a new approach to the surgical treatment of focal epilepsy. *J Neurosurg* **70**, 231–239.

Murphy, J. M., Horton, N. J., Laird, N. M., *et al.* (2004). Anxiety and depression: a 40-year perspective on relationships regarding prevalence, distribution, and comorbidity. *Acta Psychiatr Scand* **109**, 355–375.

Newsom-Davis, I., Goldstein, L. H., Fitzpatrick, D. (1998). Fear of seizures: an investigation and treatment. *Seizure* **7**, 101–106.

Pezawas, L., Meyer-Lindenberg, A., Drabant, E. M., *et al.* (2005). 5-HTTLPR polymorphism impacts human cingulate-amygdala interactions: a genetic susceptibility mechanism for depression. *Nature Neurosci* **8**(6), 828–834.

Phillips, R. G., LeDoux, J. E. (1992). Differential contribution of amygdala and hippocampus to cued and contextual fear conditioning. *Behav Neurosci* **106**, 274–285.

Pine, D. S. (2003). Developmental psychobiology and response to threats: relevance to trauma in children and adolescents. *Biol Psychiatry* **53**, 796–808.

Pine, D. S. (2007). Research review: a neuroscience framework for pediatric anxiety disorders. *J Child Psychol Psychiatry* **48**(7), 631–648.

Rappley, M. D. (2005). Attention-deficit/hyperactivity disorder. *New Engl J Med* **352**, 165–173.

Robertson, M. (1987). Carbamazepine and depression. *Int Clin Psychopharmacol* **2**, 23–35.

Robertson, M. M., Trimble, M. R., Townsend, H. R. A. (1987). Phenomenology of depression in epilepsy. *Epilepsia* **8**, 364–372.

Robinson, R. O., Baird, G., Robinson, G., *et al.* (2001). Landau-Kleffner syndrome: course and correlates with outcome. *Dev Med Child Neurol* **43**, 243–247.

Sanchez-Carpintero, Neville, B. G. R. (2003). Attentional ability in children with epilepsy. *Epilepsia* **44**, 1340–1349.

Satishchandra, E. S., Krishnamoorthy, E. S., van Elst, L. T., *et al.* (2003). Mesial temporal structures and comorbid anxiety in refractory partial epilepsy. *J Neuropsychiatry Clin Neurosci* **15**, 450–452.

Savard, G., Andermann, L. F., Reutens, D., *et al.* (1998). Epilepsy, surgical treatment and postoperative psychiatric complications: a re-evaluation of the evidence. In *Forced normalization and alternative psychosis of epilepsy* (M. R. Trimble, B. Schmitz, eds), pp. 179–192. Petersfield: Wrightson Biomedical Publishing Ltd.

Semrud-Clikeman, M., Wical, B. (1999). Components of attention in children with complex partial seizures with and without ADHD. *Epilepsia* **40**, 211–215.

Shaw, P., Lerch, J., Greenstein, D., *et al.* (2006). Longitudinal mapping of cortical thickness and clinical outcome in children and adolescents with attention-deficit/hyperactivity disorder. *Arch Gen Psychiatry* **63**, 540–549.

Sillanpaa, M., Jalava, M., Kaleva, O., *et al.* (1998). Long-term prognosis of seizures with onset in childhood. *New Engl J Med* **338**, 1715–1722.

Smith, D. F., Baker, G. A., Dewey, M., *et al.* (1991). Seizure frequency, patient perceived seizure severity and the psychosocial consequences of intractable epilepsy. *Epilepsy Res* **9**, 231–241.

Sonuga-Barke, E. J. S. (2005). Causal models of attention-deficit/hyperactivity disorder: from common simple deficits to multiple developmental pathways. *Biol Psychiatry* **57**(11), 1231–1238.

Sprich, S., Biederman, J., Crawford, M. H., *et al.* (2000). Adoptive and biological families of children and adolescents with ADHD. *J Am Acad Child Adolesc Psychiatry* **39**, 1432–1437.

Stephani, U., Carlsson, G. (2006). The spectrum from BCECTS to LKS: the rolandic EEG trait – impact on cognition. *Epilepsia* **47**(Suppl 2), 67–70.

Sullivan, P. F., Neale, M. C., Kendler, K. S. (2000). Genetic epidemiology of major depression: review and meta-analysis. *Am J Psychiatry* **157**, 1552–1562.

Thome-Souza, S., Kucznyski, E., Assumcao, F., *et al.* (2004). Which factors may play a pivotal role on determining the type of psychiatric disorder in children and adolescents with epilepsy? *Epilepsy Behav* **5**, 988–994.

Torta, R., Keller, R. (1999). Behavioral, psychotic, and anxiety disorders in epilepsy: etiology, clinical features, and therapeutic implications. *Epilepsia* **40**(Suppl 10), S2–S20.

Vazquez, B., Devinsky, O. (2003). Epilepsy and anxiety. *Epilepsy Behav* **4**(Suppl 4), 20–25.

Willcutt, E. G., Doyle, A. E., Nigg, J. T., *et al.* (2005). Validity of the executive function theory of attention-deficit/hyperactivity disorder: a meta-analytic review. *Biological Psychiatry* **57**(11), 1336–1346.

Williams, J. (2003). Learning and behavior in children with epilepsy. *Epilepsy Behav* **4**, 107–111.

Wolraich, M. L., Wibbelsman, C. J., Brown, T. E., *et al.* (2005). Attention-deficit/hyperactivity disorder among adolescents: a review of the diagnosis, treatment, and clinical implications. *Pediatrics* **115**, 1734–1746.

Wong, M. L., Licinio, J. (2001). Research and treatment approaches to depression. *Nature Rev Neurosci* **2**, 343–351.

Young, G. B., Chandrara, P. C., Blumer, W. T., *et al.* (1995). Mesial temporal lobe seizures presenting as anxiety disorders. *J Neuropsychiatry Clin Neurosci* **7**, 352–357.

Yun, W. Y., Park, Y. D., Cohen, M. J., *et al.* (2000). Cognitive and behavioral problems in children with centrotemporal spikes. *Pediatr Neurol* **23**, 391–395.

CHAPTER 16

Should a Psychiatric Evaluation Be Included in Every Pre-surgical Work-up?

Andres M. Kanner

INTRODUCTION

Every patient with pharmaco-resistant epilepsy that is considered for epilepsy surgery must undergo a detailed pre-surgical evaluation. The goals of this evaluation are to identify the location of the epileptogenic area and determine whether it is accessible to a resection with minimal or no risks. Every surgical candidate undergoes a detailed neuropsychological evaluation to ascertain the risk of developing post-surgical cognitive deficits, particularly for those who are being considered for an antero-temporal lobectomy (ATL).

Post-surgical psychiatric complications of epilepsy surgery have been recognized for more than 50 years. For example, in 1957, Hill *et al.* (1957) described post-surgical depressive episodes occurring independently of seizure outcome that remitted within 18 months. Since the 1970s, a significant number of case series and studies on prevalence rates and risks of developing post-surgical psychiatric complications have been published in the literature (Taylor, 1972). Furthermore, the psychiatric issues that need to be considered in surgical candidates are complex, and include:

1. The risk of developing post-surgical psychiatric complications and its relation with a pre-surgical psychiatric history.
2. The relation of post-surgical psychiatric complications with post-surgical seizure outcome.
3. The impact of pre-surgical psychiatric comorbidities on post-surgical seizure outcome and psychosocial adjustment.
4. The impact of epilepsy surgery on the course of pre-surgical psychiatric comorbidities.

Given the above, it is logical to expect that pre-surgical psychiatric evaluations be carried out in every surgical candidate. Yet, that is not the case. The aim of this chapter is to (i) address the question of whether a pre-surgical psychiatric evaluation is warranted in all surgical candidates, (ii) analyze the reasons behind the failure to incorporate it in the pre-surgical protocols followed in most epilepsy centers and (iii) provide the evidence in support for its use in all patients.

USE OF PSYCHIATRIC EVALUATIONS IN MAJOR EPILEPSY CENTERS

As shown below, most if not all epilepsy centers have one or more neuropsychologists that are part of the epilepsy team. Such is not the case with psychiatrists, which are seldom part of the epilepsy team. In fact, in many centers, in addition to performing an in-depth evaluation of cognitive functions, neuropsychologists have been given the responsibility to "screen" for comorbid psychiatric disease. In a significant number of centers, neuropsychological self-rating questionnaires developed to screen for psychiatric symptoms (e.g., Beck Depression Inventory) and provide a personality profile (e.g., the Minnesota Multiphasic Personality Inventory) have "replaced" psychiatric evaluations.

To examine the way major epilepsy centers make use of psychiatric and neuropsychological evaluations in candidates for ATL, a survey was sent to the 88 epilepsy centers belonging to the National Association of Epilepsy Centers. Forty-seven centers (53%) completed the survey. The survey consisted of seven questions.

In the first question, the centers were asked the frequency with which they order a psychiatric evaluation in surgical candidates for ATL. The data collected revealed that only 10 centers (21%) routinely perform a psychiatric evaluation in *every* patient. Three centers (6%) perform a psychiatric evaluation only in case of a previous psychiatric history, 16% if recommended by the neuropsychologist and 45% follow either of the latter two criteria.

The second question of the survey inquired about the *availability* and *use* of psychiatrists in each center. Among the 47 centers, only 12 (26%) had a psychiatrist in their epilepsy team. Consultations were provided by the hospital's psychiatry liaison and consultation service (LCS) in the other 35 centers: in 10 (21%) the *same* psychiatrist was always available to do the consultation while in 24 (51%) consultations were carried out by *different* psychiatrists. In one center, consultations were done by psychiatry residents who then staffed the cases with their attending.

Centers concerned with the risk of post-surgical psychiatric complications may, in theory, attempt to get pre-surgical psychiatric evaluations. To test this hypothesis, the third question inquired whether psychiatric complications following ATL were considered to be frequent enough to warrant a pre-surgical psychiatric evaluation. The answers to this question revealed a total lack of consensus, as 21 centers (45%) considered it to be a problem, while 26 (55%) did not.

The availability of an epilepsy team psychiatrist appeared to correlate with the concerns about post-surgical psychiatric complications and on the use of psychiatric evaluations in every patient. Thus, 75% of centers with an epilepsy team psychiatrist voiced a concern of frequent post-surgical psychiatric complications, while this was true in only one-third of centers in which consultations were carried out by psychiatrists from the LCS. Of note there was no difference among centers where the same (30%) or different (33%) psychiatrists performed the evaluation. These data suggest that psychiatrists with special expertise in psychiatric aspects of epilepsy are more "attuned" to potential post-surgical psychiatric complications.

As suspected, the lack of an epilepsy team psychiatrist was associated with a reliance on neuropsychological evaluations to identify comorbid psychopathology. For example, among the 24 centers that relied on a different LCS psychiatrist for their evaluations, 23 (96%) based their decisions to order a psychiatric evaluation on the basis of the recommendations of a neuropsychologist or if the patient was known to have a psychiatric history, while this was the case in eight of the 10 centers (80%) who have the same LCS psychiatrist available. In contrast only four of the 12 centers (33%) with an epilepsy team psychiatrist relied on these criteria.

The survey's fourth question asked whether neuropsychological evaluations were adequate to identify patients at risk for post-surgical psychiatric complications. Seventy percent did not believe that they could rely on neuropsychological evaluations alone. Clearly, it appears that a significant number of epilepsy centers would order a pre-surgical psychiatric evaluation if the resources (e.g., psychiatrists) were available. Of note 75% of centers opined that it was "difficult" to find psychiatrists interested in providing thorough psychiatric evaluations in their epilepsy centers.

The question at hand is whether a neuropsychological evaluation can be a substitute for a psychiatric evaluation. The answer is no and here are some of the reasons: most neuropsychological evaluations rely on screening instruments aimed at identifying psychiatric symptoms occurring during the prior 1–4 weeks, depending on the instrument used; thus, these instruments are likely to fail to detect any past psychiatric history that may be in remission at the time of the pre-surgical evaluation and to capture the complexity of the comorbid psychiatric disorders that are so common in these patients. In short, screening instruments are not sufficient to identify the complexity of such psychiatric comorbidities.

In fact, only with a comprehensive psychiatric evaluation that investigates present and a *lifetime* history can clinicians have the necessary information to formulate a correct psychiatric diagnosis, recommend the appropriate treatment and make estimations on the risk for potential post-surgical psychiatric complications. Furthermore, a positive family psychiatric history of mood, anxiety and attention deficit–hyperactivity disorder (ADHD) is a pivotal risk factor of each one of these disorders. Unfortunately, these data are rarely investigated in a neuropsychological evaluation. In addition, epileptologists and neuropsychologists cannot rely on a patient's spontaneous self-report of past or concurrent psychiatric history, as more often than not they are unlikely to volunteer such information on their own. Consequently, the most frequent psychiatric comorbidities in epilepsy patients (depression, anxiety, ADHD) are very often unrecognized by the treating epileptologist until they become severe enough to warrant an inpatient psychiatric hospitalization. Indeed, patients do not volunteer information on a comorbid psychiatric disorder out of misinterpretation of such disorders being "a normal reaction" to a life with epilepsy, while some patients may hide such history out of fear that it would disqualify them from consideration for epilepsy surgery. For example, a failure to recognize a chronic depressive disorder is illustrated in a study of 97 patients with partial epilepsy and a depressive episode severe enough to warrant the consideration of pharmacotherapy. Among these patients, 60% had been symptomatic for more than 1 year before any treatment had been suggested (Kanner et al., 2000). Only one third of the 97 patients had been treated within 6 months of the onset of their symptoms.

SURGICAL CANDIDATES: A POPULATION AT RISK OF COMORBID PSYCHIATRIC DISORDERS

Patients with pharmaco-resistant epilepsy constitute a population at high-risk for psychiatric comorbidity, ranging between 30% and 70% (Kanner, 2003). In a review of the literature Koch-Stoecker (2002) found prevalence rates ranging from 43% to 80% among seven case series. In the case series from the Bethel Epilepsy Center, 43% of patients met criteria for a psychiatric syndrome according to the criteria included in the Diagnostic and Statistical Manual of Mental Disorder (DSM), Third Edition, Revised, while an additional 29% met criteria for a psychiatric syndrome and a personality disorder.

Mood and anxiety disorders are the most common psychiatric comorbidities in adults followed by psychotic disorders, while in children ADHD is the most frequently recognized psychiatric comorbidity, but mood disorders often go under-recognized or misdiagnosed as behavioral disorders (Toone et al., 1982; Kessler et al., 1994; McDermott et al., 1995; Koch-Stoecker, 2002; Kanner, 2003). On the other hand, the prevalence rate of ADHD in adults with epilepsy is yet to be established.

Primary mood disorders comprise a heterogeneous group of conditions with respect to their clinical manifestations, course and response to treatment. Mood disorders in epilepsy may be identical to primary mood disorders that meet diagnostic criteria suggested in the DSM and include major depressive disorders, dysthymia, minor depression, bipolar disorder and cyclothymia. Yet, a significant percentage of mood disorders fails to meet any of the DSM-IV-TR diagnostic criteria in patients with epilepsy. That is, they present with atypical manifestations, which further compounds the heterogeneity of clinical manifestation of mood disorders alluded to above in these patients. In fact, atypical mood disorders are relatively frequent in patients with epilepsy. Clearly, the atypical manifestations of mood disorders are unlikely to be detected with screening instruments developed for primary mood disorders of patients without epilepsy (Blumer and Altshuler, 1998). As shown below, recognition of mood disorders before an ATL is of the essence as they have been identified as principal risk factors for post-surgical depressive episodes.

Furthermore, depressive disorders usually do not occur in isolation, both in patients with and without epilepsy. Rather they tend to occur in association with anxiety disorders. For example, in a study of 174 consecutive outpatients with epilepsy from five epilepsy centers, 73% of patients who met DSM-IV diagnostic criteria for a depressive disorder also met criteria for an anxiety disorder (Jones et al., 2005). Generalized anxiety disorder, panic disorder, phobias, agoraphobia without panic disorder and obsessive compulsive disorder are the most frequently identified anxiety disorders in patients with epilepsy. The recognition of anxiety disorders pre-surgically is important as it may interfere in a significant manner with the patients' ability to collaborate with the overall pre-surgical evaluation, above all when it calls for the use of invasive recordings with intracranial electrodes. Furthermore, comorbid anxiety and mood disorders significantly increase the suicidal risk, which has been recognized as one of the post-surgical psychiatric complications.

Psychotic disorders have lower prevalence rates than mood and anxiety disorders, and ADHD in patients with epilepsy, but they are still significantly higher than those of the general population (Slater et al., 1963; Logsdail and Toone, 1988). Consideration of epilepsy surgery in patients with comorbid psychotic disorders has been the source of great controversy among epilepsy centers. In a large number of programs, the presence of a psychotic disorder constitutes a criterion for exclusion from pre-surgical evaluation; other centers, on the other hand, do not rule out a surgical option, as long as the patient is able to cooperate with the pre-surgical evaluation and understands the risks of the surgical procedure and the benefits and limitations of this type of treatment. The availability of (good) psychiatric consultants may be a potential factor that may increase the likelihood that epilepsy centers would offer surgical options to patients with comorbid psychotic disorders.

Interictal psychotic disorders can be indistinguishable from primary schizophreniform disorders and present with delusions, hallucinations, referential thinking and thought disorders. Yet, in a significant proportion of patients with epilepsy, they may differ in significant ways. For example, Slater demonstrated that psychotic disorders of epilepsy are remarkable for the absence of negative symptoms (e.g., flat affect, apathy), better premorbid history, and less common deterioration of the patient's personality (Slater et al., 1963). This form of psychosis is less severe and more responsive to therapy, a very relevant point for patients who are being considered for surgery, as it makes it more likely that these patients can cooperate during a pre-surgical evaluation with the proper support.

Postictal psychosis (PIP) can be present in the form of isolated symptoms or as a cluster of symptoms mimicking psychotic disorders and represents approximately 25% of psychotic disorders of epilepsy. The yearly incidence of postictal psychiatric disorders among patients with partial epilepsy that are undergoing video-EEG was estimated to be 6.4% (Kanner et al., 1996). The occurrence of PIP in potential surgical candidates is significant as several studies have found a greater risk of bilateral independent ictal foci (Lancman *et al.*, 1994; Devinsky *et al.*, 1995; Umbricht *et al.*, 1995; Kanemoto *et al.*, 1996; Kanner *et al.*, 1996; Logsdail and Toone, 1988). In a study completed at the Rush Epilepsy Center, the occurrence of PIP was found to predict the presence of bilateral independent ictal foci with an 89% probability (Kanner and Ostrovskaya, 2008a). By the same token, patients with recurrent PIP are at significant risk of developing interictal psychosis (Kanner and Ostrovskaya, 2008b). To minimize this risk, clinicians must carefully weigh the option of "palliative" surgery, particularly in patients with mesial temporal sclerosis (MTS), provided that most seizures originate from the side of the atrophic hippocampus and the neuropsychological data are concordant with the intended surgical target.

Clearly, patients with refractory epilepsy being considered for surgery are likely to suffer from comorbid psychiatric disorders, which can increase the risk of post-surgical complications, interfere with their ability to collaborate in the pre-surgical evaluations and distort their expectations of epilepsy surgery.

Post-surgical psychiatric complications

Post-surgical psychiatric complications can be the expression of (i) a *de novo* psychiatric disorder, (ii) a *recurrence* of psychiatric disorder that had been in remission for a period of time prior to surgery and (iii) an *exacerbation* in severity of a psychiatric disorder that was present in a sub-clinical form or that was mild enough in severity that it went unrecognized by patient, family and clinician or that was identified because of a more careful evaluation of the patient.

The most frequent post-surgical psychiatric complications include: (i) depressive and anxiety disorders, (ii) psychotic disorder and (iii) psychogenic non-epileptic events (PNEE). These are illustrated in a study completed at the Rush Epilepsy Center (Kanner *et al.*, 2006a). The study included 100 consecutive patients (60 men) who had undergone an ATL and had a minimal post-surgical follow-up period of 2 years (median follow-up duration: 8.3 ± 3.3 years; range, 3–14 years). Their mean age was 31 ± 10.7 years (range, 8–59 years) and the mean duration of their seizure disorder was 19.9 ± 8.8 years (range, 4–40 years). Among the 100 patients, 63 had temporal lobe epilepsy (TLE) secondary to MTS, 18 had lesional TLE (LTLE) and 19 idiopathic TLE (ITLE), documented with volumetric measurements of mesial structures. Fifty-six patients (56%) had a lifetime psychiatric history before surgery, 47 of whom had a mood disorder, which consisted of depression alone in 22 patients, and mixed depression and anxiety disorders in 25, while three patients also had ADHD. Among these 47 patients, 27 had experienced a major depressive disorder, 14 a dysthymic disorder and six bipolar illness. Nine patients had other psychiatric disorders that included ADHD or pure anxiety disorder.

Thirty-four patients met our criteria for post-surgical psychiatric complications: 15 patients with *de novo* depressive/anxiety disorders and four with *de novo* psychotic episodes. Sixteen patients experienced an exacerbation in severity of pre-surgical depressive/anxiety disorders; these complications occurred during the first 12 months after surgery in all patients. In addition, seven of these 31 patients with mood disorders developed *de novo* PNEE. At the last contact, the post-surgical psychiatric symptomatology of 16 patients had failed to remit despite multiple pharmacologic trials; three of these patients had developed a *de novo* post-surgical depressive disorder.

Univariate analyses identified pre-surgical history of any psychiatric disorder, and in particular of depression, and persistent seizures as predictors of post-surgical psychiatric complications. Interestingly enough, having failed to obtain gainful employment after surgery was not a predictor of post-surgical psychiatric complications (see below).

A review of the literature yielded several studies that also identified the risk and prevalence rates of post-surgical psychiatric complications that included mood, anxiety and psychotic disorders. Several examples follow:

Post-surgical mood/anxiety disorders

The prevalence rates of post-surgical depressive disorders vary widely among the different studies, ranging from 5% to 63%, with a mean of 26%. Wrench *et al.* (2004) published a study of 62 patients who underwent epilepsy surgery; 43 had an ATL and 19 an extratemporal resection (ETR). Both groups had comparable pre-surgical histories of depression and anxiety (33% and 23%, respectively for ATL and 53% and 18%, respectively for ETR). At 1 month after surgery, symptoms of anxiety and/or depression were reported by 66% of ATL patients and 19% of ETR patients. At 3 months, 54% of ATL and 33% of ETR patients were still symptomatic with 30% of ATL and 17% of ETR patients still experiencing a depressive episode. Furthermore, at the 3-month evaluation, 13% of ATL patients had developed a *de novo* depression and 15% a *de novo* anxiety disorder, while 18% had developed other *de novo* psychiatric disorders. In contrast, only 17% of ETR patients had developed *de novo* anxiety, but not depression or other psychopathology. At 3-month follow-up, there was no significant association between post-surgical psychopathology and seizure outcome.

In a study of 274 patients, Bruton (1988) found a 20-fold increase in the prevalence rates of depression after surgery, varying in severity from mild dysphoric to major depressive episodes associated with suicidal attempts. More often than not, these post-surgical complications were an expression of a recurrence or exacerbation of pre-surgical comorbid disorders. Furthermore, *de novo* psychiatric disorders were less frequent and were also likely to occur in the first 6 months after surgery.

Glosser *et al.* (2000) published a study of 44 patients who underwent an ATL; in the first month after surgery, 12 patients (31%) developed *de novo* depression and/or anxiety disorders or recurrence of a psychiatric disorder that had been in remission during the 6 months preceding the surgical procedure. By 6 months, they were still symptomatic but significantly improved and by 1 year, all but two patients had become free of symptoms. Blumer *et al.* (1998) reported much higher prevalence rates of *de novo* psychiatric complications in a study of 50 consecutive patients, 44 of whom underwent an ATL and six a frontal lobe resection; 14 patients (32%) developed *de novo* psychiatric disorders presenting as an interictal dysphoric disorder in six patients, depressive episodes in two and a psychotic disorder in six, while only three patients (7%) experienced an exacerbation of a pre-surgical interictal dysphoric disorder. In all but two patients the psychiatric complications occurred within 2 months after surgery. All psychiatric complications remitted with psychotropic treatment. Blumer *et al.* associated the development of post-surgical psychiatric complications with persistent seizures.

In a study of 49 patients who underwent an ATL and were followed for a period of almost 11 years, Altshuler *et al.* (1999) found that five (10%) developed *de novo* depressive episodes, four within the first post-surgical year. Similar findings were reported by Ring *et al.* (1998) in a study of 60 consecutive patients, who underwent an ATL and had a psychiatric evaluation prior to surgery, at six weeks and 3 months after surgery. At six weeks, 45% of all patients were experiencing emotional lability that reflected a *de novo* psychiatric complication in half of the patients. By 3 months, the emotional lability and symptoms of anxiety had remitted or improved significantly but not so the depressive states.

Pre-surgical *ictal* fear or panic has also been associated with post-surgical psychiatric complications. For example, Kohler *et al.* (2001) studied the association of ictal fear with mood and anxiety disorders before and 1 year after ATL. They compared 22 patients with ictal fear with two matched groups of patients with other types of auras and no auras at all. Pre-surgical and post-surgical evaluations at 1–2 months and 1 year after ATL were carried out to identify mood and anxiety disorders and the use of psychotropic medication. The majority of patients in the three groups experienced mood and anxiety disorders before surgery with comparable frequencies. Mood and anxiety disorders declined in the two control groups, but not in the ictal fear group after surgery. Postoperative mood and anxiety disorders were more common in patients with persistent seizures and in those in the ictal fear group who were seizure-free. Furthermore, a majority of patients with ictal fear required the use of psychotropic medication after surgery.

In addition, Kanemoto *et al.* (2001) identified an association between pre-surgical PIP and post-surgical mood disorders in a study of 52 patients who underwent an ATL. Post-surgical mood disorders presented as manic and depressive episodes during the first two post-surgical years.

Post-surgical psychosis

The prevalence rates of post-surgical psychotic complications have been estimated to range between 1% and 10% among patients undergoing ATL. Unfortunately, the data of post-surgical psychosis studies are significantly less robust than those of post-surgical mood and anxiety disorders and consist more often than not of small case series or anecdotal reports. Several case series have included a mixture of patients with pre-surgical and *de novo* post-surgical psychotic disorders.

De novo post-surgical psychotic episodes may present as schizophreniform-like disorders, manic episodes and postictal psychotic episodes. In the study carried out at the Rush Epilepsy Center cited above, Kanner *et al.* (2000) identified a *de novo* psychotic episode in four patients within the first 6 months after surgery consisting of a manic episode in two and a paranoid episode in the other two patients. Symptoms remitted in three patients with pharmacotherapy without the need of hospitalization while the other two had to be hospitalized in a psychiatric unit. In the third patient, symptoms remitted after two inpatient hospitalizations. Two of these patients had lesional epilepsy caused by dysembryoplastic neuroepithelioma (DNET) in one and a ganglioglioma in the second.

Stevens (1990) identified a *de novo* psychotic disorder in two patients within the first 12 months after surgery among a group of 14 patients who had undergone an ATL and who were followed for a period of 20–30 years. Both patients were seizure-free. Among 74 patients who underwent an ATL, Jensen and Larsen (1979) identified nine who developed a *de novo* psychotic disorder. Six of the latter nine patients began experiencing psychotic symptoms after they became seizure-free.

In a study of 57 consecutive patients who underwent an ATL, Leinonen *et al.* (1994) identified five (8.8%) who developed postoperative psychotic episodes. Two (3.5%) patients had experienced PIP before surgery which they continued to have post-surgically. Among the other three patients, two (3.5%) experienced a definite and one (1.8%) a probable *de novo* schizophreniform psychotic disorder.

Shaw *et al.* (2004) identified 11 patients who developed *de novo* post-surgical schizophreniform psychosis among 320 consecutive patients (3.2%) who underwent an ATL. Psychotic symptomatology became apparent within the first year in all patients. These 11 patients were compared to a control group of 33 patients. The psychotic patients were more likely to have bilateral epileptiform activity, a smaller amygdala in the non-operated side and pathologies other than MTS.

Some investigators have associated the risk of post-surgical psychotic episodes with a right temporal seizure focus. For example, Mace and Trimble (1991) reported seven consecutive patients who developed a *de novo* psychotic disorder following an ATL; six were right-sided. Nonetheless, the relation between side of seizure focus and the risk of developing post-surgical psychosis cannot be established on the bases of these small case series.

Other authors have associated the presence of gangliogliomas or DNET with the development of *de novo* post-surgical psychotic disorders. Andermann *et al.* (1999) reported six patients who experienced a *de novo* psychotic disorder from four centers. The psychotic disorders consisted of schizophreniform-like episodes with paranoid and depressive symptomatology. These investigators estimated a risk of 2.5% for the development of *de novo* psychosis (1 of 39) in patients with these types of lesions who undergo an ATL. Such associations remain to be established in larger studies, however.

As stated above, post-surgical manic episodes can be psychiatric complications of ATL. For example, Carran *et al.* (2003) reported 16 patients who developed a *de novo* manic episode following an ATL. These patients were identified from a case series of 415 consecutive patients (i.e., 3.8% of patients) who had undergone an ATL. They were compared to a control group of asymptomatic patients matched for age and gender and a second group of 30 patients who experienced a post-surgical depression. The manic episode occurred within the first year after the ATL and was short-lived in all but one patient. Compared to the two control groups, patients with post-surgical mania were more likely to display bilateral electrographic abnormalities, and to have a right temporal seizure focus, though this difference did not reach significance when compared to the depressed group. Post-surgical symptomatic patients were more likely to have experienced GTC seizures before surgery and to fail to achieve seizure freedom post-surgically.

Postictal psychotic episodes can also occur *de novo* after ATL. Christodoulou *et al.* (2002) reported three cases (1%) among 282 consecutive patients who had undergone an ATL. All three patients had seizures predominantly from the contralateral (non-surgical) side or had bilateral independent seizures, while none of the patients who continued to have seizures from the side of the surgery developed *de novo* PIP. Manchanda *et al.* (1993) identified four patients (1.3%) who developed a *de novo* PIP among a group of 298 consecutive patients who had undergone an ATL. All four patients had a right-sided resection and had no preoperative psychiatric history.

Post-surgical PNEE

The development of these events has been attributed to the "stress" associated with a "seizure-free" life in patients with chronic epilepsy that are not "emotionally, physically or economically ready" to face their own or their families increased expectations. While this "hypothesis" makes intuitive sense, the available data does not seem to support it. Here is some of the evidence. (1) The prevalence rates of post-surgical PNEE among the different studies are relatively low, ranging between 1.8% and 10% among the different case series. In the study carried out at the Rush Epilepsy Center cited above, 7% of the patients developed *de novo* PNEE [(Kanner and Frey, 2007) and Kanner *et al.*, submitted]. A pre-surgical (lifetime) psychiatric history was associated with the development of post-surgical PNEE. Interestingly enough, PNEE were not reported in seizure-free patients; in fact, persistent seizures were significantly associated with the development of *de novo* PNEE in every one of the seven patients. Furthermore, failure to obtain gainful employment was not associated with the development of PNEE.

Ney *et al.* (1998) reported the occurrence of post-surgical *de novo* PNEE in five of 96 patients (5.2%) who underwent epilepsy surgery. Low full-scale IQ, preoperative psychiatric comorbidity and major surgical complications were identified as risk factors. Reuber *et al.* (2002) identified 13 patients with both epileptic and PNEE and investigated their post-surgical outcome: 11 of the 13 patients had significant clinical improvement post-surgically. However, in 2 out of 13 patients the severity of the PNEE (including pseudo-status epilepticus) increased postoperatively despite a significant improvement of their epileptic seizures. Both patients had a pre-surgical psychiatric history.

Disclosure of post-surgical psychiatric complications

Clearly, epilepsy surgery is associated with the risk of post-surgical psychiatric complications, which should be openly discussed with patients and family members with as much detail as the other surgical risks. Based on the data reviewed in the previous section, a careful psychiatric evaluation can help identify those patients at greater risk of developing post-surgical psychiatric complications, particularly those at risk of post-surgical depressive and anxiety episodes among those with a previous history of mood disorder. Furthermore, a psychiatrist can provide significant help in counseling patients and their family on the potential post-surgical psychiatric risks.

IMPACT OF PRE- AND POST-SURGICAL PSYCHIATRIC ILLNESS ON POST-SURGICAL SEIZURE OUTCOME

To date, three studies have suggested that a *pre-surgical* psychiatric history is associated with a worse post-surgical seizure outcome. In one study of 121 patients who underwent an ATL, Anhouri *et al.* (2000) reported a worse post-surgical seizure outcome for patients with a lifetime psychiatric history compared with those without. Koch-Stoecker investigated the post-surgical seizure outcome among 100 consecutive patients who underwent an ATL; 78 had a pre-surgical lifetime psychiatric history. Among patients without comorbid psychiatric history, 89% were seizure-free after surgery while this occurred in only 43% of patients with pre-surgical psychiatric history (unpublished data). In the study cited above, Kanner *et al.* (2006a) used a logistic regression model to identify predictors of post-surgical seizure outcome in the 100 consecutive patients who had undergone an ATL. Covariates included: (i) cause of TLE (MTS, LTLE and ITLE), (ii) history of secondarily generalized tonic–clonic seizures (never, only at the onset of the seizure disorder, 1–2/year, >2/year), (iii) duration of seizure disorder, (iv) lifetime history of depression (may have included major depressive disorder, dysthymia), (v) extent of resection of mesial temporal structures and (vi) presence of neuropathologic abnormalities. A lifetime history of depression and extent of resection of mesial structures were the only predictors of persistent auras in the absence of disabling seizures, while the cause of the TLE, duration of the seizure disorder, having recurrent generalized tonic–clonic (GTC) seizures and a lifetime history of depression were significant predictors of failure to achieve freedom from disabling seizures.

Most of the studies that have investigated the relationship between post-surgical psychiatric disturbances and seizure outcome have also found a significant association between persistent post-surgical seizures and post-surgical depressive disorders. This is not a surprising finding since pre-surgical psychiatric history is the most frequent predictor of post-surgical psychiatric complications.

Vickrey et al. (1995a, b) also demonstrated an association between persistent seizures, including persistence of only auras and worse quality of life post-surgically compared to patients who are free of seizures and auras. The negative impact of persistent auras on the quality of life had been attributed to the need to keep patients on antiepileptic drugs (AEDs) or even the use of higher doses of AEDs to assuage the concern of recurrent seizures. Nonetheless, these data are counterintuitive since patients with auras but no disabling seizures can function normally in all areas. In fact, the study by Kanner et al. (2006a) cited above raises the question of whether the worse quality of life in patients with persistent auras is driven by a concurrent mood disorder and not (only) by the auras.

Psychosocial outcome

Family dynamics

Epilepsy surgery is expected to have a positive impact on the patient's life. With the achievement of seizure freedom, patients can become more independent, not only as it pertains to their ability to drive but in other areas of their life. Paradoxically, in some cases seizure freedom can have a negative impact on the family dynamics. Indeed, some family members become accustomed to the patients' limitations and have difficulties giving up their role of "caretaker." Unfortunately, these dysfunctional "family dynamics" are not rare in families of patients with a chronic illness like epilepsy and invariably are bound to lead to conflict when patients try to become more independent. In fact, divorce is not an uncommon "complication" of successful epilepsy surgery. Thus, all couples and families need to be evaluated for the eventual risk of these types of family problems.

Gainful employment

Obtaining gainful employment is one of the goals of a successful surgical treatment in patients whose persistent seizures had precluded them from working. Unfortunately, such is not always the case. A review of the literature reveals that the main factors associated with post-surgical employment are: reduction of seizures or seizure freedom, pre-surgical cognitive ability, psychiatric comorbidity, pre-surgical employment and improvement of neuropsychological function (Kanner and Balabanov, 2006). In a study recently completed at the Rush Epilepsy Center in 88 patients who underwent an ATL, working before surgery was the strongest predictor of post-surgical gainful employment followed by achieving a seizure-free state, a negative lifetime history of depression and being a woman (Kanner et al., 2006b). Lendth et al. (1997) found that a young age at the time of the surgery and improvement of general neuropsychological functioning and especially attention are associated with employment after surgery. In another study, Reeves et al. (1997) found that being a student or working full time within a year before the surgery, driving after the surgery and obtaining further education after surgery are associated with full time work postoperatively. Clearly, these data are indicative of the need to carry out vocational evaluations before surgery among unemployed patients.

IMPACT OF EPILEPSY SURGERY ON PRE-SURGICAL PSYCHIATRIC DISORDERS

While this chapter has focused on the risk of post-surgical psychiatric complications, it is important to note that epilepsy surgery improves pre-surgical psychiatric comorbidities. In the study carried out at the Rush Epilepsy Center, a lifetime psychiatric history prior to surgery had been identified in 56 patients of whom 51 were symptomatic at the time of the psychiatric evaluation (Kanner et al., 2006a). At the last contact, 16 continued to be symptomatic despite multiple treatment strategies and an additional 15 patients were symptom-free on psychotropic medication. Thus, epilepsy surgery resulted in total remission off psychotropic medication in 45% of patients. In the study reported by Altshuler et al. (1999), 17 of 49 patients (35%) had a lifetime history of at least one major depressive episode. Eight of these patients never experienced another major depressive episode post-surgically. In this study, like in our study, the only predictor for post-surgical depressive disorder was a pre-surgical history of depression. Devinsky et al. (2005) reported the results of a study of 360 patients from seven epilepsy centers in the United States who underwent epilepsy surgery; 89% had an ATL. Psychiatric syndromes were identified at baseline and 2 years after surgery with the structured interview Composite International Diagnostic Interview (CIDI). Pre-surgically, 75 patients (22%) met criteria for a diagnosis of depression, 59 (18%) of anxiety disorders and 12 (4%) of other psychiatric disorders including bipolar illness and schizophrenia. At the 2-year post-surgical evaluation, only 26 patients (9%) met diagnostic criteria for depression and 20 (10%) for anxiety, while three patients (1%) met criteria for other psychiatric diagnoses. Thus, epilepsy surgery had resulted in symptom remission in more than 50% of patients. In this study, the presence of an anxiety or depressive disorder post-surgically was not associated with seizure outcome.

In summary, the relatively high psychiatric comorbidity in surgical candidates, and its negative impact on post-surgical seizure outcome and an increased risk of post-surgical psychiatric complications requires a careful pre-surgical psychiatric evaluation in *every* surgical candidate. Such evaluation can help avert unnecessary post-surgical psychiatric problems and a better adjustment of the patient and family to a seizure-free life, and is important in the process of counseling patients on the post-surgical psychiatric risks.

REFERENCES

Altshuler, L., Rausch, R., DeIrahim, S., Kay, J., Crandall, P. (1999). Temporal lobe epilepsy, temporal lobectomy and major depression. *J Neuropsychiatry Clin Neurosci* **11**(4), 436–443.

Andermann, L. F., Savard, G., Meencke, H. J., McLachlan, R., Moshe, S., Andermann, F. (1999). Psychosis after resection of ganglioglioma or DNET: evidence for an association. *Epilepsia* **40**(1), 83–87.

Anhouri, S., Brown, R. J., Krishnamoorthy, E. S., Trimble, M. R. (2000). Psychiatric outcome following temporal lobectomy: a predictive study. *Epilepsia* **41**, 1608–1615.

Blumer, D., Altshuler, L. L. (1998). Affective disorders. In *Epilepsy: A Comprehensive Textbook*, **Vol. II** (J. Engel, T. A. Pedley, eds), pp. 2083–2099. Philadelphia, PA: Lippincott-Raven.

Blumer, D., Wakhlu, S., Davies, K., et al. (1998). Psychiatric outcome of temporal lobectomy for epilepsy: incidence and treatment of psychiatric complications. *Epilepsia* **39**, 478–486.

Bruton, C. J. (1988). *The Neuropathology of Temporal Lobe Epilepsy*, Oxford: Oxford University Press. Maudsley Monographs 31.

Carran, M. A., Kohler, C. G., O'Connor, M. J., Bilker, W. B., Sperling, M. R. (2003). Mania following temporal lobectomy. *Neurology* **61**, 770–774.

Christodoulou, C., Koutroumanidid, M., Hennessy, M. J., et al. (2002). Postictal psychosis after temporal lobectomy. *Neurology* **59**(9), 1432–1435.

Devinsky, O., Abrahmson, H., Alper, K., et al. (1995). Postictal psychosis: a case control study of 20 patients and 150 controls. *Epilepsy Res* **20**, 247–253.

Devinsky, O., Barr, W. B., Vickrey, B. G., et al. (2005). Changes in depression and anxiety after resective surgery for epilepsy. *Neurology* **65**(11), 1744–1942.

Glosser, G., Zwill, A. S., Glosser, D. S., et al. (2000). Psychiatric aspects of temporal lobe epilepsy before and after anterior temporal lobectomy. *J Neurol Neurosurg Psychiatry* **68**, 53–58.

Hill, D., Pond, D. A., Mitchell, W., Falconer, M. A. (1957). Personality changes following temporal lobectomy for epilepsy. *J Ment Sci* **103**, 18–27.

Jensen, I., Larsen, J. K. (1979). Mental aspects of temporal lobe epilepsy. Follow-up of 74 patients after resection of a temporal lobe. *J Neurol Neurosurg Psychiatry* **42**, 256–265.

Jones, J. E., Hermann, B. P., Barry, J. J., Gilliam, F., Kanner, A. M., Meador, K. J. (2005). Clinical assessment of Axis I psychiatric morbidity in chronic epilepsy: a multicenter investigation. *J Neuropsychiatry Clin Neurosci* **17**(2), 172–179.

Kanemoto, K., Kawasaki, J., Kawai, J. (1996). Postictal psychosis: a comparison with acute interictal and chronic psychoses. *Epilespia* **37**, 551–556.

Kanemoto, K., Kim, Y., Miyamoto, T., Kawasaki, J. (2001). Presurgical postictal and acute interictal psychoses are differentially associated with postoperative mood and psychotic disorders. *J Neuropsychiatry Clin Neurosci* **13**(2), 243–247.

Kanner, A. M. (2003). Depression in epilepsy: prevalence, clinical semiology, pathogenic mechanisms and treatment. *Biol Psychiatry* **54**, 388–398.

Kanner, A. M., Balabanov, A. J. (2006). Neurorehabilitation in epilepsy. In *Textbook of Neural Repair and Rehabilitation*, **Vol. 2** (M. E. Selzer, L. Cohen, F. H. Gage, eds), pp. 542–559. Cambridge: Cambridge University Press.

Kanner, A. M., Frey, M. (2007). Predictors of post-surgical *de novo* psychogenic non-epileptic events following an antero-temporal lobectomy: an unexpected finding. *Neurology* **68**(Suppl 1), A358.

Kanner, A. M., Ostrovskaya, A. (2008). Long-term predictors of postictal psychotic episodes I: Are they predictive of bilateral ictal foci? *Epilepsy Behav* **12**(1), 150–153.

Kanner, A. M., Ostrovskaya, A. (2008). Long-term predictors of postictal psychotic episodes II: are they predictive of interictal psychotic disorders? *Epilepsy Behav* **12**(1), 154–156.

Kanner, A. M., Stagno, S., Kotagal, P., Morris, H. H. (1996). Postictal psychiatric events during prolonged video-electroencephalographic monitoring studies. *Arch Neurol* **53**, 258–263.

Kanner, A. M., Kozak, A. M., Frey, M. (2000). The use of sertraline in patients with epilepsy: Is it safe? *Epilepsy Behav* **1**(2), 100–105.

Kanner, A. M., Byrne, R., Smith, M. C., Balabanov, A. J., Frey, M. (2006). Does a life-time history of depression predict a worse postsurgical seizure outcome following a temporal lobectomy? *Ann Neurol* **60**(Suppl 10), 19.

Kanner A. M., Frey, M., Byrne, R. (2006b). Predictors of postsurgical gainful employment following an anterotemporal lobectomy. *Presented at the Annual Meeting of the American Academy of Neurology*, San Diego, CA, May, 2006. *Neurology*, 2006, **66**(Suppl 1).

Kessler, R. C., McGonagle, K. A., Zhao, S., *et al.* (1994). Lifetime and 12-month prevalence of DSM-III-R psychiatric disorders in the United States: results from the national comorbidity study. *Arch Gen Psychiatry* **51**, 8–19.

Koch-Stoecker, S. (2002). Psychiatric effects of surgery for temporal lobe epilepsy. In *The Neuropsychiatry of Epilepsy* (M. Trimble, B. Schmitz, eds), pp. 266–282. Cambridge: Cambridge University Press.

Kohler, C. G., Carran, M. A., Bilker, W., *et al.* (2001). Association of fear auras with mood and anxiety disorders after temporal lobectomy. *Epilepsia* **42**(5), 674–681.

Lancman, M. E., Craven, W. J., Asconape, J. J., Penry, J. K. (1994). Clinical management of recurrent postictal psychosis. *J Epilepsy* **7**, 47–51.

Leinonen, E., Tuunainen, A., Lepola, U. (1994). Postoperative psychoses in epileptic patients after temporal lobectomy. *Acta Neurol Scand* **90**(6), 394–399.

Lendth, M., Helmstaedter, C., Elger, C. E. (1997). Pre- and postoperative socio-economic development of 151 patients with focal epilepsies. *Epilepsia* **38**, 1330–1337.

Logsdail, S. J., Toone, B. K. (1988). Postictal psychosis. A clinical and phenomenological description. *Br J Psychiatry* **152**, 246–252.

Mace, C. J., Trimble, M. R. (1991). Psychosis following temporal lobe surgery: a report of six cases. *J Neurol Neurosurg Psychiatry* **54**(7), 639–644.

Manchanda, R., Miller, H., McLachlan, R. S. (1993). Post-ictal psychosis after right temporal lobectomy. *J Neurol Neurosurg Psychiatry* **56**(3), 277–279.

McDermott, S., Mani, S., Krishnaswami, S. (1995). A population-based analysis of specific behavior problems associated with childhood seizures. *J Epilepsy* **8**, 110–118.

Ney, G. C., Barr, W. B., Napolitano, C., *et al.* (1998). New-onset psychogenic seizures after surgery for epilepsy. *Arch Neurol* **55**(5), 726–730.

Reeves, A. L., So, E. L., Evans, R. W., Cascino, G. D., Sharbrough, F. W., O'Brien, P. C., Trenerry, M. R. (1997). Factors associated with work outcome after anterior temporal lobectomy for intractable epilepsy. *Epilepsia* **38**(6), 689–695.

Reuber, M., Kurthen, M., Fernandez, G., *et al.* (2002). Epilepsy surgery in patients with additional psychogenic seizures. *Arch Neurol* **59**(1), 82–86.

Ring, H. A., Moriarty, J., Trimble, M. R. (1998). A prospective study of the early postsurgical psychiatric associations of epilepsy surgery. *J Neurol Neurosurg Psychiatry* **64**(5), 601–604.

Shaw, P., Mellers, J., Henderson, M., Polkey, C., David, A. S., Toone, B. K. (2004). Schizophrenia-like psychosis arising de novo following a temporal lobectomy: timing and risk factors. *J Neurol Neurosurg Psychiatry* **75**, 1003–1008.

Slater, E., Beard, A. W., Glithero, E. (1963). The schizophrenia-like psychoses of epilepsy. *Br J Psychiatry* **109**, 95–150.

Stevens, J. R. (1990). Psychiatric consequences of temporal lobectomy for intractable seizures: a 20–30-year follow-up of 14 cases. *Psychol Med* **20**(3), 529–545.

Taylor, D. C. (1972). Mental state and temporal lobe epilepsy. A correlative account of 100 patients treated surgically. *Epilepsia* **13**(6), 727–765.

Toone, B. K., Garralda, M. E., Ron, M. A. (1982). The psychoses of epilepsy and the functional psychoses: a clinical and phenomenological comparison. *Br J Psychiatry* **141**, 256–261.

Umbricht, D., Degreef, G., Barr, W. B., Lieberman, J. A., Pollack, S., Schaul, N. (1995). Postictal and chronic psychoses in patients with temporal lobe epilepsy. *Am J Psychiatry* **152**(2), 224–231.

Vickrey, B. G., Hays, R. D., Engel, J., Jr., Spritzer, K., Rogers, W. H., Rausch, R., Graber, J., *et al.* (1995). Outcome assessment for epilepsy surgery: the impact of measuring health-related quality of life. *Ann Neurol* **37**(2), 158–166.

Vickrey, B. G., Hays, R. D., Rausch, R., Engel, J., Jr., Visscher, B. R., Ary, C. M., Rogers, W. H., *et al.* (1995). Outcomes in 248 patients who had diagnostic evaluations for epilepsy surgery. *Lancet* **346**(8988), 1445–1449.

Wrench, J., Wilson, S. J., Bladin, P. F. (2004). Mood disturbance before and after seizure surgery: a comparison of temporal and extratemporal resections. *Epilepsia* **45**, 534–543.

CHAPTER 17

Do Antidepressants Improve or Worsen Seizures in Patients with Epilepsy?

Kenneth Alper

INTRODUCTION

The relevance of the question of antidepressants and seizure threshold in epilepsy

Alteration of seizure threshold is an important domain of risk in the use of psychopharmacological treatment in epilepsy. Depression is reportedly an even stronger determinant of quality of life in epilepsy than seizure control (Gilliam et al., 2003; Boylan et al., 2004), but is often undertreated in patients with epilepsy despite markedly elevated rates of depression and suicide (Jones et al., 2003; Kanner, 2003; Swinkels et al., 2005). Concern regarding seizure threshold can contribute to the problem of undertreatment of psychiatric disorders, particularly depression, in patients at risk for epilepsy and seizures. From a neurobiological perspective, the interaction of epilepsy, depression, and antidepressant treatment is of interest in the context of a pathophysiological commonality in epilepsy and psychiatric disorders.

Assessment of risk and clinical decision making

The prescribing information for clomipramine and bupropion provides warnings and statements regarding seizure risks that are specific to the respective compounds. For the remaining antidepressants, the prescribing information includes a generic statement regarding caution in patients with epilepsy or otherwise at risk for seizures. To a great extent, such concern is based on case reports involving tricyclic antidepressants (TCAs) at supratherapeutic levels (Preskorn and Fast, 1992; Dailey and Naritoku, 1996; Jobe and Browning, 2005). The generic statement regarding seizure risk that appears on the prescribing information for antidepressants is to a greater extent medicolegal boilerplate than the product of an objective evaluation of evidence of risks and benefits of antidepressant medication, particularly with regard to patients with epilepsy.

WHAT IS THE EVIDENCE THAT ANTIDEPRESSANTS LOWER SEIZURE THRESHOLD?

Tricyclic antidepressants

The association of TCAs with seizures has driven much of the concern regarding the effect of psychotropic drugs on seizure threshold. Seizures associated with TCAs are strongly related to plasma levels. Clomipramine has been regarded as being particularly likely to provoke seizures among the TCAs, with a much higher incidence of seizures at dosages greater than 300 mg/day (Robinson, 1978; Skowron and Stimmel, 1992; Rosenstein et al., 1993; Stimmel and Dopheide, 1996; Brodkin et al., 1997; Novartis, 2006).

More than three decades after the introduction of imipramine in 1960 as the first TCA approved in the US, and at the end of the era of TCAs as first line antidepressant monotherapy, Preskorn and Fast (1992) stated "…there are no case reports in the literature of TCA-induced seizures at therapeutic concentrations of TCAs". That paper systematically reviewed the literature and presented a series of eight cases with seizures associated with TCA treatment. The total tricyclics levels in these cases ranged from 433 to 1200 ng/ml, substantially in excess of the optimal therapeutic levels for most TCAs, which generally are estimated to fall within a range of 50–200 ng/ml (Janicak et al., 2001).

Preclinical evidence indicates that high levels of TCAs are epileptogenic, although at lower levels, as with other antidepressants, they are antiepileptic. It does not appear that the serotonergic and noradrenergic actions of TCAs decrease seizure threshold, and the antihistaminergic action of TCAs is viewed as mediating seizure risk (Ago et al., 2006).

Bupropion

The manufacturer's prescribing information states that bupropion is contraindicated in epilepsy. In clinical trials of antidepressants bupropion in the immediate release (IR) form, but not the sustained release (SR) form, was associated with higher seizure incidence (Alper et al., 2007). The seizure rate among patients treated with bupropion IR was approximately 1.6 times than that on placebo, which is a statistically significant but relatively small elevation. Bupropion is associated with a particularly high incidence of seizures in patients with eating disorders (Horne et al., 1988; Pope et al., 1989).

The incidence of seizures with bupropion is strongly related to dosage, and is increased approximately tenfold in patients receiving 600mg/day or more relative to patients on 450mg/day or less (Davidson, 1989). The lower rate of seizures in the clinical trials of SR vs. IR forms may be due to lower peak plasma concentrations with the SR form, suggesting the importance of pharmacokinetic factors in bupropion's proconvulsive effect (Dunner et al., 1998; Jefferson et al., 2005). As with other antidepressants in the animal model (Pisani et al., 1999), bupropion has been associated with anticonvulsant effects at lower serum levels and proconvulsant effects at higher levels (Tutka et al., 2004).

Serotonin and serotonin/norepinephrine reuptake inhibitors

The serotonin and serotonin/norepinephrine reuptake inhibitors (SRI/SNRIs) constitute most of the "second generation" antidepressants. The term second generation refers to replacement by SRI/SNRIs of TCAs and monoamine oxidase inhibitors as first line antidepressant therapy beginning in the US with the approval of fluoxetine in 1987. Case reports are difficult to evaluate as an indicator of risk because SRI/SNRI antidepressant treatment and unprovoked seizures may associate by chance; however, seizures associated with SRIs/SNRIs do appear to occur in a small subgroup of patients with epilepsy.

In a sample of 100 consecutive patients with epilepsy treated with sertraline for depression, Kanner et al. (2000) applied a set of criteria to assign probability of causality of antidepressant treatment to change in seizure activity. The categories of "definite" and "probable" degrees of causality for the apparent effect of treatment with sertraline on seizure worsening were defined on the basis of the occurrence of generalized tonic–clonic seizures or increased seizure frequency relative to the baseline seizure control prior to initiation of treatment with sertraline. One patient met definite criteria and five others met probable criteria for attributing a causal relationship of seizure worsening to sertraline. In all five of the probable cases antiepileptic drug (AED) adjustments allowed the continuation of sertraline while maintaining baseline seizure control. In four of these cases sertraline was continued at the same dose, and the fifth chose not to continue it.

Another study applied the same seizure causality criteria in a study of 36 children and adolescents treated with fluoxetine or sertraline (Thome-Souza et al., 2007). Worsening of seizure control occurred in a total of two cases, both of which were classified according to criteria as indicating a probable causal relationship. In one of these cases AED dosage was increased and sertraline was continued with subsequently improved seizure control relative to the pretreatment baseline. In the second case an increase in AED was refused and sertraline was discontinued, with a subsequent increase in seizure frequency.

Seizures in overdoses of antidepressants

A proconvulsant effect is observed with overdoses of both TCAs and SRI/SNRI antidepressants (Stimmel and Dopheide, 1996; Pisani *et al.*, 2002; Lee *et al.*, 2003; Cuenca *et al.*, 2004; Isbister *et al.*, 2004; Kelly *et al.*, 2004). Bupropion appears to be relatively frequently involved in drug-related seizures presenting to emergency services (Pesola and Avasarala, 2002; Balit *et al.*, 2003; Coughlin and Birkinshaw, 2003; Shepherd *et al.*, 2004). Seizures in patients on psychopharmacological medications generally tend to occur early in treatment, and in association with rapid upward titration of the dosage (Mendez *et al.*, 1986; Skowron and Stimmel, 1992; Pacia and Devinsky, 1994; Sajatovic and Meltzer, 1996; Stimmel and Dopheide, 1996; Spigset *et al.*, 1997), indicating that in addition to absolute levels, the rate of increase in plasma levels is also determinant of seizures associated with antidepressants.

Serotonin syndrome

Seizures may occur in patients treated with antidepressants due to the serotonin syndrome (Boyer and Shannon, 2005), a toxic state resulting from excessive agonist effects at serotonergic receptors in the central and peripheral nervous system. The major features of the serotonin syndrome are often described as a clinical triad of mental status changes, autonomic hyperactivity, and neuromuscular abnormalities, which are variably present across patients. Signs include tremor and diarrhea in mild cases, and delirium, hyperreflexia and clonus, neuromuscular rigidity, hyperthermia and seizures in more severe cases which can be life threatening. The incidence of serotonin syndrome with nefazodone has been reported to be 0.4 cases per 1,000 patient-months of regular treatment with similar rates observed with other SRIs (Mackay *et al.*, 1999). The serotonin syndrome is relatively frequent in overdoses of SRIs, in which it has been reported in approximately 15% of cases (Isbister *et al.*, 2004). Symptoms of the serotonin syndrome may often be easily overlooked when they are mild, and may worsen markedly with increased dosage of the original causal agent, or with the addition of other drugs that enhance serotonergic transmission.

Syndrome of inappropriate secretion of antidiuretic hormone

The syndrome of inappropriate secretion of antidiuretic hormone (SIADH) is a complication of treatment with antidepressants associated with risk for seizures that is mediated by hyponatremia. In a large ongoing European drug safety monitoring project, the incidence of hyponatremia defined as the rate of cases per antidepressant drug exposure is approximately 0.05% (Degner *et al.*, 2004). Hyponatremia was more strongly associated with treatment with SRIs than with TCAs, and of the 11 cases observed in the study, 4 presented as seizures with serum sodium concentrations ranging from 112 to 122 mmol/L. Risk factors for SIADH with SRIs include older age, female gender, concomitant use of diuretics, low body weight, and lower baseline serum sodium concentration (Jacob and Spinler, 2006). Hyponatremia associated with antidepressants tends to develop within the first few weeks of treatment (Jacob and Spinler, 2006).

PRECLINICAL EVIDENCE FOR AN ANTICONVULSANT EFFECT OF ANTIDEPRESSANTS

Extensive research utilizing animal models indicates an anticonvulsant effect of TCA and SRI/NRI antidepressants, most commonly fluoxetine or citalopram, which are reported to suppress or protect against seizures produced by proconvulsant drugs, lesions or electroshock in genetically epilepsy prone or nonepileptic rats or mice (Lange *et al.*, 1976; Sparks and Buckholtz, 1985; Dailey *et al.*, 1992; Leander, 1992; Pasini *et al.*, 1992; Prendiville and Gale, 1993; Kabuto *et al.*, 1994; Pisani *et al.*, 1999; Wada *et al.*, 1999; Ugale *et al.*, 2004; Kecskemeti *et al.*, 2005; Pericic *et al.*, 2006; Borowicz *et al.*, 2007; Richman and Heinrichs, 2007). A wide variety of antidepressants have been associated with antiepileptic effects in preclinical models, including bupropion (Tutka *et al.*, 2005) and clomipramine (Fischer and Muller, 1988).

Jobe and Browning (2005) and Dailey and Naritoku (1996) have reviewed preclinical studies related to antidepressants and seizure threshold. Major points of emphasis of these reviews are the attribution of the evident antiepileptic effect of antidepressants to serotonergic and noradrenergic actions, and that noradrenergic and serotonergic deficits contribute to seizure predisposition in preclinical models. They point out that investigation of possible serotonergic and noradrenergic antiepileptic mechanisms of action has been discouraged by the widely held assumption that such actions could only be proconvulsant, despite the substantial experimental pharmacological and clinical evidence that indicates that antidepressants commonly do not lower seizure threshold in epilepsy.

An apparent anticonvulsant effect of antidepressants at therapeutic dosages may be reconcilable with the proconvulsant effect observed with overdoses. A proconvulsant effect has been observed with very large increases in extracellular monoamine levels in brain tissue (Clinckers *et al.*, 2004b; Clinckers *et al.*, 2005). The apparent anticonvulsant effect of some antidepressants at conventional dosages may be associated with a relatively moderate increase in extracellular serotonin, with a tendency toward a proconvulsant effect at the much higher concentrations of extracellular serotonin that may be associated with supratherapeutic dosage or rapid increases in blood level of antidepressant (Pisani *et al.*, 1999; Clinckers *et al.*, 2004a, b; Clinckers *et al.*, 2005).

CLINICAL EVIDENCE FOR AN ANTICONVULSANT EFFECT OF ANTIDEPRESSANTS

Studies on seizure control in patients with epilepsy treated with antidepressants

Studies of seizure control in depressed patients with epilepsy treated with anti-depressants have reported no worsening (Harmant *et al.*, 1990; Gross *et al.*, 2000; Hovorka *et al.*, 2000; Kanner *et al.*, 2000; Kuhn *et al.*, 2003), or improvement (Fromm *et al.*, 1978; Ojemann *et al.*, 1983; Hurst, 1986; Ojemann *et al.*, 1987; Sakakihara *et al.*, 1995; Specchio *et al.*, 2004) in seizure control.

Antidepressants as adjunctive antiepileptic drugs

SRI antidepressants have been used as adjunctive antiepileptic therapy. In an open label study on fluoxetine given at a dose of 20 mg/day in patients with refractory epilepsy, 6 of 17 (35%) patients became seizure free with the remainder having 30% reductions in seizure frequency (Favale *et al.*, 1995). Sixteen of these patients had seizures at frequencies in the range daily to weekly prior to the study, and none were regarded as ever having experienced significant improvement under the usual antiepileptic treatment. No patient was reported to have an increase in seizure frequency. A possible pharmacokinetic interaction of fluoxetine and AEDs did not account for the observed improvement in seizure control because AED levels obtained at monthly intervals were unchanged throughout the study over a mean period of observation of 14 months.

The above study did not indicate whether fluoxetine was used for the indication of depression or solely for the treatment of epilepsy. However, in a subsequent study by the same group involving only patients with refractory epilepsy and no diagnosis of depression, citalopram given at a dose of 20 mg/day was effective as an adjunctive antiepileptic agent (Favale *et al.*, 2003). Nine of the 11 patients treated with citalopram experienced a reduction of seizure frequency of at least 50%, and no patient was reported to have had an increase in seizure frequency. A 37% overall reduction in seizure frequency was observed in a trial of citalopram for the treatment of depression in 39 patients with epilepsy (Specchio *et al.*, 2004).

Imipramine has also been used as an adjunctive antiepileptic agent. Based on preclinical observations in cats, Fromm *et al.* (1972) conducted an open label trial of imipramine in 16 patients with absence and "minor motor" seizures and noted initial reductions in seizure frequency in 10 patients that persisted in six patients at 1 year. In a subsequent double blind crossover study by the same group, imipramine produced initial statistically significant reductions in seizure frequency at 4 weeks in 5 of 10 patients with absence and myoclonic–astatic seizures that persisted at 1 year in two patients (Fromm *et al.*, 1978), and no patients responded to placebo. In this double blind crossover study, medications for the control of absence and myoclonic seizures had been discontinued, but medications for the control of partial or generalized tonic–clonic seizures were continued. In an additional 16 patients with absence and myoclonic–astatic seizures, all antiepileptic medications were continued and imipramine was added on an open label basis. Decreases in seizure frequency of at least 50% were observed initially in 10 of these 16 patients that persisted in four patients at 1 year. Serum imipramine levels in the patients who showed a long-term response ranged between 40 and 120 ng/ml, which is on the order of approximately one half of the concentrations associated with therapeutic effectiveness of imipramine for the indication of depression (Janicak *et al.*, 2001).

Another study of imipramine in 15 patients with refractory diverse seizure types found initial reductions in seizure frequency of at least 80% in eight patients which persisted at 1 year in five patients, four of whom remained completely seizure free (Hurst, 1986). In this study, drop attacks were most responsive to imipramine, while there was no clinical benefit for partial seizures. Given the frequently refractory nature and clinical significance of drop attacks, this work may merit replication.

Evidence for an anticonvulsant effect of antidepressants in clinical trials

Clinical trial data provide an approach to the investigation of the effects of psychopharmacological agents, and psychiatric disorders themselves, on seizure threshold. The analysis of clinical trial data may provide some methodological advantages including prospective evaluation of large numbers of depressed patients, and the systematic use of standardized methods of psychiatric diagnosis and assessment. A meta-analysis of FDA data from phases II and III clinical trials of antidepressants approved in the US between 1985 and 2004 indicated a reduction of seizure incidence of 52% (Alper *et al.*, 2007). This appears to be a similar magnitude of seizure reduction to that which has been reported in open label trials of SRIs in patients with epilepsy (see section "Antidepressants as adjunctive antiepileptic drugs").

DEPRESSION ITSELF MAY LOWER SEIZURE THRESHOLD

Epidemiological studies of depression and unprovoked seizures

Epidemiological studies suggest that depression or attempted suicide are themselves risk factors for seizures (Forsgren and Nystrom, 1990; Hesdorffer *et al.*, 2000; Hesdorffer *et al.*, 2006). Three community-based studies have examined depression as a risk factor for seizures by assessing depression retrospectively in incident cases of unprovoked seizures, and reported the following odds ratios for the occurrence of unprovoked seizures in depression: 1.7 (5.1 for suicide attempts) (Hesdorffer *et al.*, 2006); 3.7 (Hesdorffer *et al.*, 2000); and 7.0, which was increased to 17.2 in the subset of patients with "localized onset" (i.e. partial) seizures (Forsgren and Nystrom, 1990). Suicide attempts and migraine are associated with additive risk for seizures in mood disorders (Hesdorffer *et al.*, 2007). Attention deficit disorder, which is frequently comorbid with depression, is also associated with unprovoked seizures (Hesdorffer *et al.*, 2004).

Seizure control and mood

Greater severity of depression has been correlated with poorer seizure control (Cramer *et al.*, 2003), including a recent study which found evidence for a bidirectional causal association utilizing a modification of path analysis (Thapar *et al.*, 2005). Depression is also associated with pharmacoresistant epilepsy (Hitiris *et al.*, 2007).

Increased incidence of seizures in placebo groups in clinical trials of antidepressants

In depressed patients assigned to placebo in phases II and III trials of antidepressants, seizure incidence is greater than the published incidence of unprovoked seizures in community nonpatient samples. The incidence of unprovoked seizures in the general population is approximately 60 per 100,000 person exposure years (PEY) (Hauser *et al.*, 1993; Kotsopoulos *et al.*, 2002; Olafsson *et al.*, 2005). The reported incidence of seizures of 1,167 per 100,000 PEY in depressed patients treated with placebo in antidepressant clinical trials is 19 times the rate seen in the general population (Alper *et al.*, 2007).

Theoretical considerations

Depression and epilepsy are viewed as sharing a common attribute of increased neuronal excitability (Post, 2004; Kanner, 2006). A substantial literature reports evidence of alteration of neurotransmission involving the inhibitory neurotransmitter γ-aminobutyric acid (GABA) in both depression (Brambilla *et al.*, 2003; Sanacora *et al.*, 2004) and epilepsy (Cossart *et al.*, 2005). Imaging studies report common deficiencies of serotonergic transmission in depression and epilepsy (Theodore, 2004; Kanner, 2006).

Postmortem neuropathological examination of depressed patients has indicated hippocampal atrophy (Stockmeier *et al.*, 2004), which is a classic finding in partial epilepsy, although the patterns of neuron sprouting and loss may differ between epilepsy and mood disorders (Bausch, 2005; de Lanerolle and Lee, 2005). Antidepressants reportedly increase hippocampal cell proliferation and neurogenesis (Malberg and Schechter, 2005; Banasr *et al.*, 2006), which might offset the neuronal loss associated with epilepsy.

CONCLUSIONS

The available clinical and experimental evidence supports the hypotheses that psychiatric disorders themselves are associated with lowered seizure threshold, and suggests a possible antiepileptic effect of antidepressant medication. Clinical studies indicate that a small subgroup of patients with epilepsy has seizures that appear causally related to antidepressants, although seizure control is unchanged or improved in the great majority of patients. These data suggest a common attribute of increased neuronal excitability shared by epilepsy and depression (Post, 2004; Kanner, 2006). Prospective studies of antidepressant treatment in people with epilepsy and future clinical trials of psychotropic medications provide a significant opportunity to independently confirm and extend these findings. Investigation aimed at the development of new agents to treat epilepsy and seizures should include mechanisms of action involving serotonergic and noradrenergic neurotransmission.

REFERENCES

Ago, J., Ishikawa, T., Matsumoto, N., Ashequr Rahman, M., Kamei, C. (2006). Mechanism of imipramine-induced seizures in amygdala-kindled rats. *Epilepsy Res* **72**, 1–9.

Alper, K., Schwartz, K. A., Kolts, R. L., Khan, A. (2007). Seizure incidence in psychopharmacological clinical trials: an analysis of Food and Drug Administration (FDA) summary basis of approval reports. *Biol Psychiatry* **62**(4), 345–354.

Balit, C. R., Lynch, C. N., Isbister, G. K. (2003). Bupropion poisoning: a case series. *Med J Aust* **178**, 61–63.

Banasr, M., Soumier, A., Hery, M., Mocaer, E., Daszuta, A. (2006). Agomelatine, a new antidepressant, induces regional changes in hippocampal neurogenesis. *Biol Psychiatry* **59**, 1087–1096.

Bausch, S. B. (2005). Axonal sprouting of GABAergic interneurons in temporal lobe epilepsy. *Epilepsy Behav* **7**, 390–400.

Borowicz, K. K., Furmanek-Karwowska, K., Sawicka, K., Luszczki, J. J., Czuczwar, S. J. (2007). Chronically administered fluoxetine enhances the anticonvulsant activity of conventional antiepileptic drugs in the mouse maximal electroshock model. *Eur J Pharmacol* **567**, 77–82.

Boyer, E. W., Shannon, M. (2005). The serotonin syndrome. *New Engl J Med* **352**, 1112–1120.

Boylan, L. S., Flint, L. A., Labovitz, D. L., Jackson, S. C., Starner, K., Devinsky, O. (2004). Depression but not seizure frequency predicts quality of life in treatment-resistant epilepsy. *Neurology* **62**, 258–261.

Brambilla, P., Perez, J., Barale, F., Schettini, G., Soares, J. C. (2003). GABAergic dysfunction in mood disorders. *Mol Psychiatry* **8**, 721–737, 715.

Brodkin, E. S., McDougle, C. J., Naylor, S. T., Cohen, D. J., Price, L. H. (1997). Clomipramine in adults with pervasive developmental disorders: a prospective open-label investigation. *J Child Adolesc Psychopharmacol* **7**, 109–121.

Clinckers, R., Smolders, I., Meurs, A., Ebinger, G., Michotte, Y. (2004a). Anticonvulsant action of GBR-12909 and citalopram against acute experimentally induced limbic seizures. *Neuropharmacology* **47**, 1053–1061.

Clinckers, R., Smolders, I., Meurs, A., Ebinger, G., Michotte, Y. (2004b). Anticonvulsant action of hippocampal dopamine and serotonin is independently mediated by D and 5-HT receptors. *J Neurochem* **89**, 834–843.

Clinckers, R., Gheuens, S., Smolders, I., Meurs, A., Ebinger, G., Michotte, Y. (2005). *In vivo* modulatory action of extracellular glutamate on the anticonvulsant effects of hippocampal dopamine and serotonin. *Epilepsia* **46**, 828–836.

Cossart, R., Bernard, C., Ben-Ari, Y. (2005). Multiple facets of GABAergic neurons and synapses: multiple fates of GABA signalling in epilepsies. *Trends Neurosci* **28**, 108–115.

Coughlin, P. A., Birkinshaw, R. I. (2003). Zyban: increasing the workload in an accident and emergency department? *Eur J Emerg Med* **10**, 62–63.

Cramer, J. A., Blum, D., Reed, M., Fanning, K. (2003). The influence of comorbid depression on quality of life for people with epilepsy. *Epilepsy Behav* **4**, 515–521.

Cuenca, P. J., Holt, K. R., Hoefle, J. D. (2004). Seizure secondary to citalopram overdose. *J Emerg Med* **26**, 177–181.

Dailey, J. W., Naritoku, D. K. (1996). Antidepressants and seizures: clinical anecdotes overshadow neuroscience. *Biochem Pharmacol* **52**, 1323–1329.

Dailey, J. W., Yan, Q. S., Mishra, P. K., Burger, R. L., Jobe, P. C. (1992). Effects of fluoxetine on convulsions and on brain serotonin as detected by microdialysis in genetically epilepsy-prone rats. *J Pharmacol Exp Ther* **260**, 533–540.

Davidson, J. (1989). Seizures and bupropion: a review. *J Clin Psychiatry* **50**, 256–261.

de Lanerolle, N. C., Lee, T. S. (2005). New facets of the neuropathology and molecular profile of human temporal lobe epilepsy. *Epilepsy Behav* **7**, 190–203.

Degner, D., Grohmann, R., Kropp, S., Ruther, E., Bender, S., Engel, R. R., Schmidt, L. G. (2004). Severe adverse drug reactions of antidepressants: results of the German multicenter drug surveillance program AMSP. *Pharmacopsychiatry* **37**(Suppl 1), S39–S45.

Dunner, D. L., Zisook, S., Billow, A. A., Batey, S. R., Johnston, J. A., Ascher, J. A. (1998). A prospective safety surveillance study for bupropion sustained-release in the treatment of depression. *J Clin Psychiatry* **59**, 366–373.

Favale, E., Rubino, V., Mainardi, P., Lunardi, G., Albano, C. (1995). Anticonvulsant effect of fluoxetine in humans. *Neurology* **45**, 1926–1927.

Favale, E., Audenino, D., Cocito, L., Albano, C. (2003). The anticonvulsant effect of citalopram as an indirect evidence of serotonergic impairment in human epileptogenesis. *Seizure* **12**, 316–318.

Fischer, W., Muller, M. (1988). Pharmacological modulation of central monoaminergic systems and influence on the anticonvulsant effectiveness of standard antiepileptics in maximal electroshock seizure. *Biomed biochim acta* **47**, 631–645.

Forsgren, L., Nystrom, L. (1990). An incident case-referent study of epileptic seizures in adults. *Epilepsy Res* **6**, 66–81.

Fromm, G. H., Amores, C. Y., Thies, W. (1972). Imipramine in epilepsy. *Arch Neurol* **27**, 198–204.

Fromm, G. H., Wessel, H. B., Glass, J. D., Alvin, J. D., Van Horn, G. (1978). Imipramine in absence and myoclonic–astatic seizures. *Neurology* **28**, 953–957.

Gilliam, F., Hecimovic, H., Sheline, Y. (2003). Psychiatric comorbidity, health, and function in epilepsy. *Epilepsy Behav* **4**(Suppl 4), S26–S30.

Gross, A., Devinsky, O., Westbrook, L. E., Wharton, A. H., Alper, K. (2000). Psychotropic medication use in patients with epilepsy: effect on seizure frequency. *J Neuropsychiatry Clin Neurosci* **12**, 458–464.

Harmant, J., van Rijckevorsel-Harmant, K., de Barsy, T., Hendrickx, B. (1990). Fluvoxamine: an antidepressant with low (or no) epileptogenic effect. *Lancet* **336**, 386.

Hauser, W. A., Annegers, J. F., Kurland, L. T. (1993). Incidence of epilepsy and unprovoked seizures in Rochester, Minnesota: 1935–1984. *Epilepsia* **34**, 453–468.

Hesdorffer, D. C., Hauser, W. A., Annegers, J. F., Cascino, G. (2000). Major depression is a risk factor for seizures in older adults. *Ann Neurol* **47**, 246–249.

Hesdorffer, D. C., Ludvigsson, P., Olafsson, E., Gudmundsson, G., Kjartansson, O., Hauser, W. A. (2004). ADHD as a risk factor for incident unprovoked seizures and epilepsy in children. *Arch Gen Psychiatry* **61**, 731–736.

Hesdorffer, D. C., Hauser, W. A., Olafsson, E., Ludvigsson, P., Kjartansson, O. (2006). Depression and suicide attempt as risk factors for incident unprovoked seizures. *Ann Neurol* **59**, 35–41.

Hesdorffer, D. C., Luethvigsson, P., Hauser, W. A., Olafsson, E., Kjartansson, O. (2007). Co-occurrence of major depression or suicide attempt with migraine with aura and risk for unprovoked seizure. *Epilepsy Res* **75**, 220–223.

Hitiris, N., Mohanraj, R., Norrie, J., Sills, G. J., Brodie, M. J. (2007). Predictors of pharmacoresistant epilepsy. *Epilepsy Res* **75**, 192–196.

Horne, R. L., Ferguson, J. M., Pope, H. G., Jr., Hudson, J. I., Lineberry, C. G., Ascher, J., Cato, A. (1988). Treatment of bulimia with bupropion: a multicenter controlled trial. *J Clin Psychiatry* **49**, 262–266.

Hovorka, J., Herman, E., Nemcova, I. I. (2000). Treatment of interictal depression with citalopram in patients with epilepsy. *Epilepsy Behav* **1**, 444–447.

Hurst, D. L. (1986). The use of imipramine in minor motor seizures. *Pediatr Neurol* **2**, 13–17.

Isbister, G. K., Bowe, S. J., Dawson, A., Whyte, I. M. (2004). Relative toxicity of selective serotonin reuptake inhibitors (SSRIs) in overdose. *J Toxicol Clin Toxicol* **42**, 277–285.

Jacob, S., Spinler, S. A. (2006). Hyponatremia associated with selective serotonin-reuptake inhibitors in older adults. *Ann Pharmacother* **40**, 1618–1622.

Janicak, P. G., Davis, J. M., Preskorn, S. H., Ayd, F. J. (2001). *Principles and Practice of Psychopharmacotherapy* 3rd edition. Philadelphia, PA: Lippincott Williams and Wilkins.

Jefferson, J. W., Pradko, J. F., Muir, K. T. (2005). Bupropion for major depressive disorder: pharmacokinetic and formulation considerations. *Clin Ther* **27**, 1685–1695.

Jobe, P. C., Browning, R. A. (2005). The serotonergic and noradrenergic effects of antidepressant drugs are anticonvulsant, not proconvulsant. *Epilepsy Behav* **7**, 602–619.

Jones, J. E., Hermann, B. P., Barry, J. J., Gilliam, F. G., Kanner, A. M., Meador, K. J. (2003). Rates and risk factors for suicide, suicidal ideation, and suicide attempts in chronic epilepsy. *Epilepsy Behav* **4**(Suppl 3), S31–S38.

Kabuto, H., Yokoi, I., Takei, M., Kurimoto, T., Mori, A. (1994). The anticonvulsant effect of citalopram on El mice, and the levels of tryptophan and tyrosine and their metabolites in the brain. *Neurochem Res* **19**, 463–467.

Kanner, A. M. (2003). Depression in epilepsy: prevalence, clinical semiology, pathogenic mechanisms, and treatment. *Biol Psychiatry* **54**, 388–398.

Kanner, A. M. (2006). Epilepsy, suicidal behaviour, and depression: Do they share common pathogenic mechanisms? *Lancet Neurol* **5**, 107–108.

Kanner, A. M., Kozak, A. M., Frey, M. (2000). The use of sertraline in patients with epilepsy: is it safe? *Epilepsy Behav* **1**, 100–105.

Kecskemeti, V., Rusznak, Z., Riba, P., Pal, B., Wagner, R., Harasztosi, C., Nanasi, P. P., Szucs, G. (2005). Norfluoxetine and fluoxetine have similar anticonvulsant and Ca^{2+} channel blocking potencies. *Brain Res Bull* **67**, 126–132.

Kelly, C. A., Dhaun, N., Laing, W. J., Strachan, F. E., Good, A. M., Bateman, D. N. (2004). Comparative toxicity of citalopram and the newer antidepressants after overdose. *J Toxicol Clin Toxicol* **42**, 67–71.

Kotsopoulos, I. A., van Merode, T., Kessels, F. G., de Krom, M. C., Knottnerus, J. A. (2002). Systematic review and meta-analysis of incidence studies of epilepsy and unprovoked seizures. *Epilepsia* **43**, 1402–1409.

Kuhn, K. U., Quednow, B. B., Thiel, M., Falkai, P., Maier, W., Elger, C. E. (2003). Antidepressive treatment in patients with temporal lobe epilepsy and major depression: a prospective study with three different antidepressants. *Epilepsy Behav* **4**, 674–679.

Lange, S. C., Julien, R. M., Fowler, G. W. (1976). Biphasic effects of imipramine in experimental models of epilepsy. *Epilepsia* **17**, 183–195.

Leander, J. D. (1992). Fluoxetine, a selective serotonin-uptake inhibitor, enhances the anticonvulsant effects of phenytoin, carbamazepine, and ameltolide (LY201116). *Epilepsia* **33**, 573–576.

Lee, K. C., Finley, P. R., Alldredge, B. K. (2003). Risk of seizures associated with psychotropic medications: emphasis on new drugs and new findings. *Expert Opin Drug Saf* **2**, 233–247.

Mackay, F. J., Dunn, N. R., Mann, R. D. (1999). Antidepressants and the serotonin syndrome in general practice. *Br J Gen Pract* **49**, 871–874.

Malberg, J. E., Schechter, L. E. (2005). Increasing hippocampal neurogenesis: a novel mechanism for antidepressant drugs. *Curr Pharm Des* **11**, 145–155.

Mendez, M. F., Cummings, J. L., Benson, D. F. (1986). Psychotropic drugs and epilepsy. *Stress Med* **2**, 325–332.

Novartis (2006). Anafranil® Clomipramine Hydrochloride Capsules Prescribing Information Rev. 020605.

Ojemann, L. M., Friel, P. N., Trejo, W. J., Dudley, D. L. (1983). Effect of doxepin on seizure frequency in depressed epileptic patients. *Neurology* **33**, 646–648.

Ojemann, L. M., Baugh-Bookman, C., Dudley, D. L. (1987). Effect of psychotropic medications on seizure control in patients with epilepsy. *Neurology* **37**, 1525–1527.

Olafsson, E., Ludvigsson, P., Gudmundsson, G., Hesdorffer, D., Kjartansson, O., Hauser, W. A. (2005). Incidence of unprovoked seizures and epilepsy in Iceland and assessment of the epilepsy syndrome classification: a prospective study. *Lancet Neurol* **4**, 627–634.

Pacia, S. V., Devinsky, O. (1994). Clozapine-related seizures: experience with 5,629 patients. *Neurology* **44**, 2247–2249.

Pasini, A., Tortorella, A., Gale, K. (1992). Anticonvulsant effect of intranigral fluoxetine. *Brain Res* **593**, 287–290.

Pericic, D., Strac, D. S., Vlainic, J. (2006). Zimelidine decreases seizure susceptibility in stressed mice. *J Neural Transm* **113**, 1863–1871.

Pesola, G. R., Avasarala, J. (2002). Bupropion seizure proportion among new-onset generalized seizures and drug related seizures presenting to an emergency department. *J Emerg Med* **22**, 235–239.

Pisani, F., Spina, E., Oteri, G. (1999). Antidepressant drugs and seizure susceptibility: from *in vitro* data to clinical practice. *Epilepsia* **40**(Suppl 10), S48–S56.

Pisani, F., Oteri, G., Costa, C., Di Raimondo, G., Di Perri, R. (2002). Effects of psychotropic drugs on seizure threshold. *Drug Saf* **25**, 91–110.

Pope, H. G., Jr., McElroy, S. L., Keck, P. E., Hudson, J. I., Ferguson, J. M., Horne, R. L. (1989). Electrophysiological abnormalities in bulimia and their implications for pharmacotherapy: a reassessment. *Int J Eat Disorder* **8**, 191–201.

Post, R. M. (2004). Neurobiology of seizures and behavioral abnormalities. *Epilepsia* **45**(Suppl 2), 5–14.

Prendiville, S., Gale, K. (1993). Anticonvulsant effect of fluoxetine on focally evoked limbic motor seizures in rats. *Epilepsia* **34**, 381–384.

Preskorn, S. H., Fast, G. A. (1992). Tricyclic antidepressant-induced seizures and plasma drug concentration. *J Clin Psychiatry* **53**, 160–162.

Richman, A., Heinrichs, S. C. (2007). Seizure prophylaxis in an animal model of epilepsy by dietary fluoxetine supplementation. *Epilepsy Res* **74**, 19–27.

Robinson, M. L. (1978). Epileptic fit after clomipramine. *Br J Psychiatry* **132**, 525–526.

Rosenstein, D. L., Nelson, J. C., Jacobs, S. C. (1993). Seizures associated with antidepressants: a review. *J Clin Psychiatry* **54**, 289–299.

Sajatovic, M., Meltzer, H. Y. (1996). Clozapine-induced myoclonus and generalized seizures. *Biol Psychiatry* **39**, 367–370.

Sakakihara, Y., Oka, A., Kubota, M., Ohashi, Y. (1995). Reduction of seizure frequency with clomipramine in patients with complex partial seizures. *Brain Dev* **17**, 291–293.

Sanacora, G., Gueorguieva, R., Epperson, C. N., Wu, Y. T., Appel, M., Rothman, D. L., Krystal, J. H., Mason, G. F. (2004). Subtype-specific alterations of gamma-aminobutyric acid and glutamate in patients with major depression. *Arch Gen Psychiatry* **61**, 705–713.

Shepherd, G., Velez, L. I., Keyes, D. C. (2004). Intentional bupropion overdoses. *J Emerg Med* **27**, 147–151.

Skowron, D. M., Stimmel, G. L. (1992). Antidepressants and the risk of seizures. *Pharmacotherapy* **12**, 18–22.

Sparks, D. L., Buckholtz, N. S. (1985). Combined inhibition of serotonin uptake and oxidative deamination attenuates audiogenic seizures in DBA/2J mice. *Pharmacol Biochem Behav* **23**, 753–757.

Specchio, L. M., Iudice, A., Specchio, N., La Neve, A., Spinelli, A., Galli, R., Rocchi, R., Ulivelli, M., de Tommaso, M., Pizzanelli, C., Murri, L. (2004). Citalopram as treatment of depression in patients with epilepsy. *Clin Neuropharmacol* **27**, 133–136.

Spigset, O., Hedenmalm, K., Dahl, M. L., Wiholm, B. E., Dahlqvist, R. (1997). Seizures and myoclonus associated with antidepressant treatment: assessment of potential risk factors, including CYP2D6 and CYP2C19 polymorphisms, and treatment with CYP2D6 inhibitors. *Acta Psychiatry Scand* **96**, 379–384.

Stimmel, G. L. D., Dopheide, J. A. (1996). Psychotropic drug-induced reductions in seizure threshold: incidence and consequences. *CNS Drugs* **5**, 37–50.

Stockmeier, C. A., Mahajan, G. J., Konick, L. C., Overholser, J. C., Jurjus, G. J., Meltzer, H. Y., Uylings, H. B., Friedman, L., Rajkowska, G. (2004). Cellular changes in the postmortem hippocampus in major depression. *Biol Psychiatry* **56**, 640–650.

Swinkels, W. A., Kuyk, J., van Dyck, R., Spinhoven, P. (2005). Psychiatric comorbidity in epilepsy. *Epilepsy Behav* **7**, 37–50.

Thapar, A., Roland, M., Harold, G. (2005). Do depression symptoms predict seizure frequency – or vice versa? *J Psychosom Res* **59**, 269–274.

Theodore, W. H. (2004). Epilepsy and depression: imaging potential common factors. *Clin EEG Neurosci* **35**, 38–45.

Thome-Souza, M. S., Kuczynski, E., Valente, K. D. (2007). Sertraline and fluoxetine: safe treatments for children and adolescents with epilepsy and depression. *Epilepsy Behav* **10**, 417–425.

Tutka, P., Barczynski, B., Wielosz, M. (2004). Convulsant and anticonvulsant effects of bupropion in mice. *Eur J Pharmacol* **499**, 117–120.

Tutka, P., Mroz, T., Klucha, K., Piekarczyk, M., Wielosz, M. (2005). Bupropion-induced convulsions: preclinical evaluation of antiepileptic drugs. *Epilepsy Res* **64**, 13–22.

Ugale, R. R., Mittal, N., Hirani, K., Chopde, C. T. (2004). Essentiality of central GABAergic neuroactive steroid allopregnanolone for anticonvulsant action of fluoxetine against pentylenetetrazole-induced seizures in mice. *Brain Res* **1023**, 102–111.

Wada, Y., Hirao, N., Shiraishi, J., Nakamura, M., Koshino, Y. (1999). Pindolol potentiates the effect of fluoxetine on hippocampal seizures in rats. *Neurosci Lett* **267**, 61–64.

CHAPTER 18

Do Central Nervous System Stimulants Lower Seizure Threshold?

Raj D. Sheth and Edgar A. Samaniego

INTRODUCTION

Central nervous system (CNS) stimulants comprise a broad class of drugs that can be used to increase motivation, alertness, mood, energy and wakefulness. Over the last decade their use has increased substantially, raising issues of adverse effects. One of the major concerns is whether stimulants decrease seizure threshold. There exists considerable controversy surrounding this issue. Many patients in need of stimulants have an underlying neurological disorder that is associated with a lower seizure threshold. Alternatively, stimulants may reduce seizure threshold in patients not otherwise predisposed to seizures. This review will examine the literature in an attempt to address this important question.

Classification of stimulants

CNS stimulants can be classified into the following broad pharmacological categories:

1. Sympathomimetics – including amphetamines (dextroamphetamine, methamphetamine), modafinil, and methylphenidate, all of which have multiple clinical indications.
2. Cocaine, cannabinoids, lysergic acid diethylamide (LSD), methylenedioxymethamphetamine (MDMA or "ecstasy"), and phencyclidine (PCP), have CNS effects that make them popular for illicit drug use.
3. Methylxanthines like caffeine and theophylline that are traditionally part of our daily consumption.

Psychiatric Controversies in Epilepsy

CNS stimulants are often described as "amphetamine like," since amphetamine is the prototypical stimulant agent. It is noteworthy that many of these drugs have a long history of efficacy and safety, whereas others are highly addictive substances associated with considerable morbidity and mortality (Table 18.1).

Mechanism of action

Stimulation may occur at cortical, brainstem or spinal levels and through different mechanisms. Most CNS stimulants are known to interact with monoamine neurons in the CNS. Neurons that synthesize, store, and release monoamine transmitters (norepinephrine, dopamine (DA), and serotonin) are widely distributed in the mammalian CNS (Rothman and Baumann, 2003). These neurons possess specialized plasma membrane proteins that transport previously released transmitter molecules from the extracellular space back into the cytoplasm (Masson et al., 1999). Psychostimulants target these monoamine transporters and increase extracellular DA (Wise, 1996; Volkow et al., 2001), serotonin (5-HT), and norepinephrine (Bymaster et al., 2002) by basically two mechanisms: reuptake inhibition and substrate-type release (Masson et al., 1999). Other CNS stimulants like caffeine block adenosine A1 and A2A receptors (Daly and Fredholm, 1998).

DA-receptor blockage has been correlated with seizure occurrence. The discovery of multiple DA-receptor families (mainly D1 and D2, but also D3, D4, and D5), sometimes mediating opposing influences on neuronal excitability, heralded a new era of DA epilepsy research (Sultan et al., 1990). In this context, there is a growing awareness that seizures might be precipitated as a consequence of treating other neurological disorders with D2 antagonists (schizophrenia) or D1 agonists (parkinsonism) (Starr, 1996). Animal studies showed that the pretreatment of rats with SCH 23390, a D1 antagonist, did not alter the incidence of seizures induced by high doses of cocaine, d-amphetamine and methamphetamine, suggesting that the mechanism of seizure induction by these drugs is complex and dose-dependent (Derlet et al., 1990).

CNS stimulants may induce seizures by the blockade of several receptors, by a hormonal mechanism, or by a kindling activity.

ATTENTION DEFICIT ASSOCIATED WITH NEUROLOGICAL DISORDERS

Epilepsy is primarily considered a disorder of paroxysmal events, but it is a broader-spectrum disease (Neville, 1999) that includes a range of disabilities such as autistic disorders (Taylor et al., 1999), attention deficit–hyperactivity disorder (ADHD) (Dunn et al., 2003), learning difficulties, and motor impairments (Neville and Boyd, 1995). In some patients these "comorbid" conditions may become the main clinical problem.

TABLE 18.1 Central nervous system stimulants

	Composition	Names	Mechanism of action	Indication	Epileptogenic evidence
Amphetamines	Methylphenidate hydrochloride	Ritalin Concerta Metadate	Unknown.	ADHD, Narcolepsy	Uncertain. PDR recommends discontinuation in the presence of seizures. May interact with AEDs.
	D-amphetamine saccharate, D,L-amphetamine aspartate, D-amphetamine sulfate, and D,L-amphetamine sulfate	Adderal	Dopamine reuptake inhibition.	ADHD, Narcolepsy	Uncertain. Case reports.
	Dextroamphetamine sulfate	Dexedrin	Dopamin reuptake inhibition.	ADHD, Narcolepsy	Uncertain. It was used in the past to treat epilepsy.
Modafinil		Provigil	Unknown.	Improve wakefulness in patients with excessive sleepiness associated with narcolepsy, obstructive sleep apnea/hypopnea syndrome, and shift work sleep disorder.	No evidence.
Cocaine			Monoamine transporters inhibition.	None.	Strong association with seizures (1–10%).
Cannabinoids	Delta-9-tetrahydrocannibinol (THC) and cannabidiol (CDB) –among others–.		Probable interaction with cannabinoid and NMDA receptors.	Anorexia, Antiemetic	Uncertain. Case reports of cannabinoid – induced seizures. Experimental evidence of antiseizure effect.
MDMA	3,4-Methylenedioxymethamphetamine	Ecstasy	Increased DA and 5-HT release.	None.	Seizures are a common CNS complication.
PCP	Phencyclidine		Noncompetitive antagonist of the NMDA receptor. Interacts with DA and 5-HT receptors.	None.	Evidence shows epileptogenic and antiepileptic properties.
Methylxanthines	Caffeine		Adenosine receptor blockage.		High doses decrease seizure threshold.
	Theophylline		Unknown.	Asthma	Case reports of seizures with asthma treatment.

Attention is part of the neurobehavioral spectrum of cognitive functions impaired in epilepsy patients. Attention deficits are easily recognized in children because of continuous parent supervision and its impact in learning and development. Attention, mental speed, memory, and language are affected in adults with epilepsy as well (Devinsky, 2004). Attention is also impaired in patients with Parkinson disease and Lewy body dementia (Ballard *et al.*, 2002); studies further suggest that reduction of noradrenaline impairs attention and DA depletion slows responses in Parkinson's disease (Riekkinen *et al.*, 1998). Furthermore, methylphenidate increases the motor effects of L-Dopa in Parkinson's disease (Camicioli *et al.*, 2001), probably by blocking DA's transporter. Studies using *in vivo* brain imaging techniques, such as positron emission tomography, have shown that DA transporter densities are affected in Parkinson's disease (Kim *et al.*, 1997).

Treatment of other conditions like narcolepsy-related sleepiness traditionally employs amphetamine and amphetamine-like stimulant drugs, such as methylphenidate (Mitler *et al.*, 1994). Stimulants are used to treat depressive symptoms, especially apathy, in patients with depression and other disorders involving the frontal lobes (Marin *et al.*, 1995).

ADHD is a behavioral phenotype frequently seen in children with epilepsy (Lindsay *et al.*, 1979). Some children may have both epilepsy and ADHD. Others may have an underlying CNS dysfunction that causes both epilepsy and difficulty with attention. Still others may have problems with attention secondary to their epilepsy (Dunn and Kronenberger, 2005). Noeker and Haverkamp (2003) have suggested that children with the combined type of ADHD and epilepsy may have a concurrent comorbidity in which inattention and epilepsy are both related to a common CNS disturbance.

About 65% of children will continue to have impairing ADHD symptoms into adulthood (Faraone *et al.*, 2000). CNS stimulants are an effective treatment frequently prescribed for ADHD, specially after the multimodal treatment study demonstrated long-term efficacy and superiority over other treatment modalities (The MTA Cooperative Group, 1999). Though the safety of stimulants in patients with well-controlled epilepsy and concurrent ADHD has been documented (Feldman *et al.*, 1989; Gross-Tsur *et al.*, 1997a, b), CNS stimulants have been related with decreasing seizure threshold, although there are no controlled studies that support this statement. One study addressed first time seizure onset on 234 children with ADHD (Hemmer *et al.*, 2001). Electroencephalograms (EEGs) obtained up to 8 weeks after being on CNS stimulants showed a 15.4% incidence of epileptiform activity, considerably higher than the estimated incidence (1.1%) of EEG abnormalities in normal children (Cavazzuti *et al.*, 1980).

Borgatti *et al.* (2004) evaluated children with epilepsy before starting an antiepileptic drug (AED) and after 1 year of treatment. They found problems with attention in 21% of the children at baseline and 42% one year later, suggesting that factors related to the seizure disorder or therapy had a negative effect on performance. There are numerous anecdotal reports of cognitive decline or behavioral dysfunction in association with many of the antiepileptic drugs, although it has been difficult to demonstrate with controlled studies (Williams et al., 1998). Most antiepileptic drugs do not adversely affect attention and behavior in therapeutic doses, with the exception of phenobarbital, gabapentin, and topiramate. Some antiepileptics, such as lamotrigine and carbamazepine, may even have beneficial effects (Schubert, 2005).

SEIZURE INCIDENCE AND CNS STIMULANTS

Few studies have correlated the use of CNS stimulant drugs and seizures. A retrospective survey of seizures associated with drug intoxication in the San Francisco Bay area over a 2-year period showed that the leading causes of seizures were tricyclic antidepressants in 29%, CNS stimulant drugs in 29%, antihistaminics in 14%, theophylline in 5%, and isoniazid in 5% of cases (Olson et al., 1994). Another retrospective study of recreational drug-induced seizures at the San Francisco General Hospital identified 49 cases in 47 patients between 1975 and 1987. The recreational drugs implicated were cocaine, amphetamine, heroin, and PCP (Alldredge et al., 1989). Epidemiological data has shown that heroin is a risk factor for first time seizures, marijuana had a protective effect and cocaine did not have any correlation (Ng et al., 1990).

Amphetamines

Amphetamines are non-catecholamine sympathomimetic amines with CNS stimulant activity. This group comprises amphetamine, dextroamphetamine, methamphetamine, and methylphenidate. They have Food and Drug Administration (FDA) approval for the treatment of ADHD and narcolepsy. Amphetamines act as substrate for monoamine transporters thereby stimulating non-exocytotic transmitter release and elevating synaptic levels of DA, norepinephrine, and 5-HT throughout the neuraxis (Kuczenski et al., 1995).

Epileptic seizures are relatively rare at therapeutic doses but may occur after the first dose. Amphetamine-related seizures appear to be less common than cocaine (Hanson et al., 1999). Seizures as a toxic effect of amphetamine-like drugs are often accompanied by other signs of overdose like fever, hypertension, cardiac arrhythmias, delirium, or coma (Alldredge et al., 1989). Methamphetamine-related seizures appear to be refractory to phenytoin pretreatment; the only AEDs that influenced seizure occurrence were diazepam and valproate (Hanson et al., 1999).

Methylphenidate

Methylphenidate is used as a therapeutic agent for ADHD and narcolepsy. Methylphenidate may inhibit the metabolism of anticonvulsants (phenobarbital, diphenylhydantoin, and primidone), and a decreased dose of these drugs may be required when given concomitantly with methylphenidate (Gross-Tsur *et al.*, 1997a, b). However, other authors have been unable to demonstrate an effect of methylphenidate on AED levels (Mirkin and Wright, 1971; Kupferberg *et al.*, 1972).

Methylphenidate is commonly believed to lower seizure threshold. However, no controlled studies have proved this hypothesis, and most evidence is anecdotal. Nevertheless, the *Physician's Desk Reference* (Montvale, 1998) and many other reference books discourage the use of stimulants in children with ADHD or even epileptiform discharges in the EEG alone (Aldenkamp *et al.*, 2006). However, studies have shown that methylphenidate does not increase the risk for development of seizures in children without epilepsy or in children with epileptiform EEG abnormalities (Gross-Tsur *et al.*, 1997a, b); some studies even report beneficial effects on the EEG (Gucuyener *et al.*, 2003). A retrospective study of adults and children with traumatic brain injury and active seizure disorders found a trend toward a lesser incidence of seizures during methylphenidate therapy (Wroblewski *et al.*, 1992).

Modafinil

Modafinil is a wakefulness-promoting agent. The precise mechanism(s) through which modafinil promotes wakefulness is unknown. Modafinil has wake-promoting actions like sympathomimetic agents including amphetamine and methylphenidate, although the pharmacologic profile is not identical to that of sympathomimetic amines. The mechanism of action has not been fully characterized, but there is evidence that a single injection of modafinil increases DA levels in the nucleus accumbens (Murillo-Rodriguez *et al.*, 2007).

Modafinil is indicated to improve wakefulness in patients with excessive sleepiness associated with narcolepsy, obstructive sleep apnea/hypopnea syndrome, and shift work sleep disorder. Its use in healthy people is being explored by the military to improve performance and alertness of aviators (Caldwell *et al.*, 2000). There is no evidence that modafinil induces seizures in healthy individuals or people with epilepsy.

Illegal substances

Cocaine

In contrast to amphetamine-like drugs that are substrate for monoamine transporters (Seiden *et al.*, 1993), cocaine and other chemically related drugs are nonselective, competitive inhibitors of monoamine transporters (Ritz *et al.*, 1987).

Cocaine, specially in high doses, has been associated with seizures (Pascual-Leone *et al.*, 1990), though it is not clear if cocaine use can reduce seizure threshold in patients with underlying epilepsy as a direct effect or indirectly by contributing to poor compliance with antiepileptic drugs, poor diet, or poor sleep habits

(Koppel *et al.*, 1996). The reported frequency of cocaine-associated seizures varies from 1% to 8% in retrospective clinical studies. A study of 500 cocaine addicts showed that 10% of subjects had a single seizure, and 3% had status epilepticus (Choy-Kwong and Lipton, 1989). Seizures occur frequently in first-time cocaine users; one study reported 40% seizures among 44 first-time users (Lowenstein *et al.*, 1987). Cocaine and its metabolite, benzoylecgonine, produced seizures when injected in the ventricular system of rats (Konkol *et al.*, 1992).

Cocaine-induced seizures often occur in the absence of other signs of toxicity. They can appear immediately or several hours after use, owing perhaps to pharmacologically active metabolites (Myers and Earnest, 1984; Lowenstein *et al.*, 1987). A focal signature to seizures should suggest a structural lesion such as cocaine-related intracerebral hemorrhage. Seizures are more likely to occur after smoking crack than after intranasal cocaine HCl, probably as a consequence of the much higher doses with the former mode of administration (Brust, 2006).

Cannabinoids

Marijuana is the most commonly used illegal drug in the United States. Approximately 60 cannabinoids and 260 non-cannabinoid constituents have been identified (Turner *et al.*, 1980). The main constituents of marijuana are delta-9-tetrahydrocannibinol (THC), the primary psychoactive constituent, and cannabidiol (CBD), the primary non-psychoactive constituent. Dronabinol, the synthetic form of THC, has FDA approval for the treatment of anorexia and as an antiemetic (Montvale, 1998).

Cannabinoid receptors are found in the brainstem, limbic, and neocortical areas that modulate seizure activity (Abood and Martin, 1996). The mechanism by which marijuana and the cannabinoids alter seizure threshold is not well defined. There may be a functional connection between cannabinoid and *N*-methyl-d-aspartate (NMDA) receptors because the two receptors are co-localized in many brain areas (Feigenbaum *et al.,* 1989). Cannabinoids have been shown to interact with glutamatergic transmission in the CNS through interaction with NMDA and non-NMDA receptors. The synthetic and non-psychotropic cannabinoid, dexanabinol (HU-211), blocks NMDA receptors in a stereospecific manner, blocking NMDA-induced tremor, seizures, and death in mice (Ameri, 1999). There is also electrophysiological evidence that cannabinoids are capable of inhibiting presynaptic release of glutamate in rat hippocampal cultures. By presynaptic inhibition of glutamate release, cannabinoids may exert anticonvulsant activity (Shen *et al.*, 1996).

In the 19th century, marijuana was used to treat epilepsy (Gordon and Devinsky, 2001). However, little medical attention was subsequently given to its possible anti-epileptic effects. Little is known about the extended effects of marijuana or its constituents on the brain. Short-term use of marijuana can decrease alpha amplitude and frequency, sleep duration, and rapid eye movement (REM) sleep. However, within an average of 10 days of continued administration, these functions returned to normal (Jones et al., 1981; Hughes, 1996). Further, there is a dose-dependent effect of cannabinoids on CNS excitability, with low doses producing activation and high doses reducing electrical activity (Pertwee, 1988).

The main problem of cannabinoids used as antiepileptic drugs concerns the separation of their psychoactive from anticonvulsant activity. However, the cannabinoid metabolite, CBD, devoid of psychotropic actions, has been reported to possess anticonvulsants effects comparable with those of phenytoin (based on similar spectra of anticonvulsant activity) in mice with electroshock convulsions (Karler and Turkanis, 1981). Moreover, clinical trials with CBD were undertaken in patients with complex partial seizures with secondary generalization. The data revealed that CBD in oral doses of 200–300 mg/day is, in fact, effective against this form of epilepsy (Cunha et al., 1980; Carlini and Cunha, 1981). One epidemiologic study of illicit drug use and new-onset seizures found that marijuana use appeared to be a protective factor against first seizures in men (Ng et al., 1990).

3,4-Methylenedioxymethamphetamine

MDMA is a ring-substituted amphetamine derivative that is also structurally related to the hallucinogenic compound mescaline. MDMA has both CNS stimulant and hallucinogenic properties. The mode of action for MDMA is based on its ability to bind 5-HT, DA, and norepinephrine transporters (Slikker et al., 1989), resulting in the release of monoamine neurotransmitters via monoamine transporter reversal (Rudnick and Wall, 1992). However, while enhanced DA neurotransmission is thought to predominantly mediate the behavioral effects of amphetamine, a unique contribution of 5-HT has been proposed to underlie the neuropsychopharmacology of MDMA (Callaway et al., 1991).

Overdose with MDMA can cause seizures, delirium, coma, and death (Kalant, 2001). A Danish study reported that seizures are among the most common CNS complications after the ingestion of ecstasy. The pathophysiology of seizures due to MDMA use seems related to severe hyponatremia (Hedetoft and Christensen, 1999).

Phencyclidine

Originally developed as an anesthetic in the 1950s, PCP was later abandoned because of a high frequency of postoperative delirium with hallucinations. PCP is a non-competitive antagonist at the NMDA receptor complex (Lodge and Anis, 1982), but it also acts on the dopaminergic (Chaudieu et al., 1989) and serotoninergic systems (Hori and Kanda, 1996). PCP intoxication is characterized by stupor or coma, muscular rigidity, rhabdomyolysis, and hyperthermia. Intoxicated patients may progress from aggressive behavior to coma, with elevated blood pressure and enlarged non-reactive pupils. PCP withdrawal syndrome has been observed in monkeys after interruption of daily access to the drug. It is characterized by somnolence, tremor, seizures, diarrhea, piloerection, bruxism, and vocalizations

(Zagnoni and Albano, 2002). A review of 1,000 cases of PCP toxicity reported 26 tonic-clonic seizures and five cases of status epilepticus (McCarron et al., 1981).

The PCP analog, metaphit, induces tonic–clonic seizures in mice exposed to audio stimulation. Metaphit blocks NMDA receptors, and seizures can be prevented by pretreating mice with PCP; thus in these experiments PCP showed a seizure protective role (Debler et al., 1993).

Methylxanthines: caffeine and theophylline

Caffeine and theophylline are consumed worldwide to enhance wakefulness, but the cellular mechanisms are poorly understood. Caffeine blocks adenosine receptors suggesting that adenosine decreases cortical arousal (Nehlig et al., 1992). Adenosine decreases the firing rate of neurons and exerts an inhibitory effect on synaptic transmission and on the release of most neurotransmitters, while caffeine increases the turnover of many neurotransmitters, including monoamines and acetylcholine (Nehlig, 1999). Theophylline's antiinflamatory mechanism has been extensively characterized, while its CNS properties are poorly understood.

Both caffeine and theophylline lower the convulsive threshold and, when administered in high doses, produce seizures (Chu, 1981; Czuczwar et al., 1987). High doses of caffeine in humans may cause nausea, trembling, nervousness, and seizures (Frucht et al., 2000). Intravenous caffeine given before electroconvulsive therapy can prolong seizure duration (Lurie and Coffey, 1990). Status epilepticus has been reported during asthma treatment with theophylline and aminophylline (Schwartz and Scott, 1974; Yarnell and Chu, 1975). It also has been shown that non-convulsive doses of theophylline markedly attenuated the anticonvulsant potential of topiramate (Luszczki et al., 2007).

Seizures and hallucinations have been reported in patients who received both quinolones and theophylline (Raoof et al., 1987). It has been postulated that quinolones have a γ-aminobutyric acid (GABA) receptor binding inhibition activity that enhances the excitatory effect of theophylline (Segev et al., 1988).

REFERENCES

Abood, M. E., Martin, B. R. (1996). Molecular neurobiology of the cannabinoid receptor. *Int Rev Neurobiol* **39**, 197–221.

Aldenkamp, A. P., Arzimanoglou, A., et al. (2006). Optimizing therapy of seizures in children and adolescents with ADHD. *Neurology* **67**(12 Suppl 4), S49–S51.

Alldredge, B. K., Lowenstein, D. H., et al. (1989). Seizures associated with recreational drug abuse. *Neurology* **39**(8), 1037–1039.

Ameri, A. (1999). The effects of cannabinoids on the brain. *Prog Neurobiol* **58**(4), 315–348.

Ballard, C. G., Aarsland, D., et al. (2002). Fluctuations in attention: PD dementia vs DLB with parkinsonism. *Neurology* **59**(11), 1714–1720.

Borgatti, R., Piccinelli, P., et al. (2004). Study of attentional processes in children with idiopathic epilepsy by Conners' Continuous Performance Test. *J Child Neurol* **19**(7), 509–515.

Brust, J. C. (2006). Seizures and substance abuse: treatment considerations. *Neurology* **67**(12 Suppl 4), S45–S48.

Bymaster, F. P., Katner, J. S., *et al.* (2002). Atomoxetine increases extracellular levels of norepinephrine and dopamine in prefrontal cortex of rat: a potential mechanism for efficacy in attention deficit/hyperactivity disorder. *Neuropsychopharmacology* **27**(5), 699–711.

Caldwell, J. A., Jr., Caldwell, J. L., *et al.* (2000). A double-blind, placebo-controlled investigation of the efficacy of modafinil for sustaining the alertness and performance of aviators: a helicopter simulator study. *Psychopharmacology (Berl)* **150**(3), 272–282.

Callaway, C. W., Johnson, M. P., *et al.* (1991). Amphetamine derivatives induce locomotor hyperactivity by acting as indirect serotonin agonists. *Psychopharmacology (Berl)* **104**(3), 293–301.

Camicioli, R., Lea, E., *et al.* (2001). Methylphenidate increases the motor effects of L-Dopa in Parkinson's disease: a pilot study. *Clin Neuropharmacol* **24**(4), 208–213.

Carlini, E. A., Cunha, J. M. (1981). Hypnotic and antiepileptic effects of cannabidiol. *J Clin Pharmacol* **21**(Suppl 8–9), 417S–427S.

Cavazzuti, G. B., Cappella, L., *et al.* (1980). Longitudinal study of epileptiform EEG patterns in normal children. *Epilepsia* **21**(1), 43–55.

Chaudieu, I., Vignon, J., *et al.* (1989). Role of the aromatic group in the inhibition of phencyclidine binding and dopamine uptake by PCP analogs. *Pharmacol Biochem Behav* **32**(3), 699–705.

Choy-Kwong, M., Lipton, R. B. (1989). Seizures in hospitalized cocaine users. *Neurology* **39**(3), 425–427.

Chu, N. S. (1981). Caffeine- and aminophylline-induced seizures. *Epilepsia* **22**(1), 85–94.

Cunha, J. M., Carlini, E. A., *et al.* (1980). Chronic administration of cannabidiol to healthy volunteers and epileptic patients. *Pharmacology* **21**(3), 175–185.

Czuczwar, S. J., Janusz, W., *et al.* (1987). Inhibition of aminophylline-induced convulsions in mice by antiepileptic drugs and other agents. *Eur J Pharmacol* **144**(3), 309–315.

Daly, J. W., Fredholm, B. B. (1998). Caffeine – an atypical drug of dependence. *Drug Alcohol Depend* **51**(1–2), 199–206.

Debler, E. A., Lipovac, M. N., *et al.* (1993). Metaphit-induced audiogenic seizures in mice: I. Pharmacologic characterization. *Epilepsia* **34**(2), 201–210.

Derlet, R. W., Albertson, T. E., *et al.* (1990). The effect of SCH 23390 against toxic doses of cocaine, d-amphetamine and methamphetamine. *Life Sci* **47**(9), 821–827.

Devinsky, O. (2004). Therapy for neurobehavioral disorders in epilepsy. *Epilepsia* **45**(Suppl 2), 34–40.

Dunn, D. W., Kronenberger, W. G. (2005). Childhood epilepsy, attention problems, and ADHD: review and practical considerations. *Semin Pediatr Neurol* **12**(4), 222–228.

Dunn, D. W., Austin, J. K., *et al.* (2003). ADHD and epilepsy in childhood. *Dev Med Child Neurol* **45**(1), 50–54.

Faraone, S. V., Biederman, J., *et al.* (2000). Toward guidelines for pedigree selection in genetic studies of attention deficit hyperactivity disorder. *Genet Epidemiol* **18**(1), 1–16.

Feigenbaum, J. J., Bergmann, F., *et al.* (1989). Nonpsychotropic cannabinoid acts as a functional N-methyl-D-aspartate receptor blocker. *Proc Natl Acad Sci USA* **86**(23), 9584–9587.

Feldman, H., Crumrine, P., *et al.* (1989). Methylphenidate in children with seizures and attention-deficit disorder. *Am J Dis Child* **143**(9), 1081–1086.

Frucht, M. M., Quigg, M., *et al.* (2000). Distribution of seizure precipitants among epilepsy syndromes. *Epilepsia* **41**(12), 1534–1539.

Gordon, E., Devinsky, O. (2001). Alcohol and marijuana: effects on epilepsy and use by patients with epilepsy. *Epilepsia* **42**(10), 1266–1272.

Gross-Tsur, V., Manor, O., *et al.* (1997a). Epilepsy and attention deficit hyperactivity disorder: is methylphenidate safe and effective? *J Pediatr* **130**(1), 40–44.

Gross-Tsur, V., Manor, O., *et al.* (1997b). Epilepsy and attention deficit hyperactivity disorder: is methylphenidate safe and effective? *J Pediatr* **130**(4), 670–674.

Gucuyener, K., Erdemoglu, A. K., *et al.* (2003). Use of methylphenidate for attention-deficit hyperactivity disorder in patients with epilepsy or electroencephalographic abnormalities. *J Child Neurol* **18**(2), 109–112.

Hanson, G. R., Jensen, M., *et al.* (1999). Distinct features of seizures induced by cocaine and amphetamine analogs. *Eur J Pharmacol* **377**(2–3), 167–173.

Hedetoft, C., Christensen, H. R. (1999). Amphetamine, ecstasy and cocaine. Clinical aspects of acute poisoning. *Ugeskr Laeger* **161**(50), 6907–6911.

Hemmer, S. A., Pasternak, J. F., *et al.* (2001). Stimulant therapy and seizure risk in children with ADHD. *Pediatr Neurol* **24**(2), 99–102.

Hori, Y., Kanda, K. (1996). Developmental alterations in NMDA receptor-mediated currents in neonatal rat spinal motoneurons. *Neurosci Lett* **205**(2), 99–102.

Hughes, J. R. (1996). A review of the usefulness of the standard EEG in psychiatry. *Clin Electroencephalogr* **27**(1), 35–39.

Jones, R. T., Benowitz, N. L., *et al.* (1981). Clinical relevance of cannabis tolerance and dependence. *J Clin Pharmacol* **21**(Suppl 8–9), 143S–152S.

Kalant, H. (2001). The pharmacology and toxicology of "ecstasy" (MDMA) and related drugs. *CMAJ* **165**(7), 917–928.

Karler, R., Turkanis, S. A. (1981). The cannabinoids as potential antiepileptics. *J Clin Pharmacol* **21**(Suppl 8–9), 437S–448S.

Kim, H. J., Im, J. H., *et al.* (1997). Imaging and quantitation of dopamine transporters with iodine-123-IPT in normal and Parkinson's disease subjects. *J Nucl Med* **38**(11), 1703–1711.

Konkol, R. J., Erickson, B. A., *et al.* (1992). Seizures induced by the cocaine metabolite benzoylecgonine in rats. *Epilepsia* **33**(3), 420–427.

Koppel, B. S., Samkoff, L., *et al.* (1996). Relation of cocaine use to seizures and epilepsy. *Epilepsia* **37**(9), 875–878.

Kuczenski, R., Segal, D. S., *et al.* (1995). Hippocampus norepinephrine, caudate dopamine and serotonin, and behavioral responses to the stereoisomers of amphetamine and methamphetamine. *J Neurosci* **15**(2), 1308–1317.

Kupferberg, H. J., Jeffery, W., *et al.* (1972). Effect of methylphenidate on plasma anticonvulsant levels. *Clin Pharmacol Ther* **13**(2), 201–204.

Lindsay, J., Ounsted, C., *et al.* (1979). Long-term outcome in children with temporal lobe seizures. III: Psychiatric aspects in childhood and adult life. *Dev Med Child Neurol* **21**(5), 630–636.

Lodge, D., Anis, N. A. (1982). Effects of phencyclidine on excitatory amino acid activation of spinal interneurones in the cat. *Eur J Pharmacol* **77**(2–3), 203–204.

Lowenstein, D. H., Massa, S. M., *et al.* (1987). Acute neurologic and psychiatric complications associated with cocaine abuse. *Am J Med* **83**(5), 841–846.

Lurie, S. N., Coffey, C. E. (1990). Caffeine-modified electroconvulsive therapy in depressed patients with medical illness. *J Clin Psychiatry* **51**(4), 154–157.

Luszczki, J. J., Jankiewicz, K., *et al.* (2007). Pharmacokinetic and pharmacodynamic interactions of aminophylline and topiramate in the mouse maximal electroshock-induced seizure model. *Eur J Pharmacol* **561**(1–2), 53–59.

Marin, R. S., Fogel, B. S., *et al.* (1995). Apathy: a treatable syndrome. *J Neuropsychiatry Clin Neurosci* **7**(1), 23–30.

Masson, J., Sagne, C., *et al.* (1999). Neurotransmitter transporters in the central nervous system. *Pharmacol Rev* **51**(3), 439–464.

McCarron, M. M., Schulze, B. W., *et al.* (1981). Acute phencyclidine intoxication: incidence of clinical findings in 1,000 cases. *Ann Emerg Med* **10**(5), 237–242.

Mirkin, B. L., Wright, F. (1971). Drug interactions: effect of methylphenidate on the disposition of diphenylhydantoin in man. *Neurology* **21**(11), 1123–1128.

Mitler, M. M., Aldrich, M. S., *et al.* (1994). Narcolepsy and its treatment with stimulants. ASDA standards of practice. *Sleep* **17**(4), 352–371.

Montrale, N. (1998). Physicians Desk reference. Medical Economics Company, Woodcliff Lake, NJ.

Murillo-Rodriguez, E., Haro, R., et al. (2007). Modafinil enhances extracellular levels of dopamine in the nucleus accumbens and increases wakefulness in rats. Behav Brain Res 176(2), 353–357.

Myers, J. A., Earnest, M. P. (1984). Generalized seizures and cocaine abuse. Neurology 34(5), 675–676.

Nehlig, A. (1999). Are we dependent upon coffee and caffeine? A review on human and animal data. Neurosci Biobehav Rev 23(4), 563–576.

Nehlig, A., Daval, J. L., et al. (1992). Caffeine and the central nervous system: mechanisms of action, biochemical, metabolic and psychostimulant effects. Brain Res Rev 17(2), 139–170.

Neville, B. G. (1999). Reversible disability associated with epilepsy. Brain Dev 21(2), 82–85.

Neville, B. G., Boyd, S. G. (1995). Selective epileptic gait disorder. J Neurol Neurosurg Psychiatry 58(3), 371–373.

Ng, S. K., Brust, J. C., et al. (1990). Illicit drug use and the risk of new-onset seizures. Am J Epidemiol 132(1), 47–57.

Noeker, M., Haverkamp, F. (2003). Neuropsychological deficiencies as a mediator between CNS dysfunction and inattentive behaviour in childhood epilepsy. Dev Med Child Neurol 45(10), 717–718.

Olson, K. R., Kearney, T. E., et al. (1994). Seizures associated with poisoning and drug overdose. Am J Emerg Med 12(3), 392–395.

Pascual-Leone, A., Dhuna, A., et al. (1990). Cocaine-induced seizures. Neurology 40(3 Pt 1), 404–407.

Pertwee, R. G. (1988). The central neuropharmacology of psychotropic cannabinoids. Pharmacol Ther 36(2–3), 189–261.

Raoof, S., Wollschlager, C., et al. (1987). Ciprofloxacin increases serum levels of theophylline. Am J Med 82(4A), 115–118.

Riekkinen, M., Kejonen, K., et al. (1998). Reduction of noradrenaline impairs attention and dopamine depletion slows responses in Parkinson's disease. Eur J Neurosci 10(4), 1429–1435.

Ritz, M. C., Lamb, R. J., et al. (1987). Cocaine receptors on dopamine transporters are related to self-administration of cocaine. Science 237(4819), 1219–1223.

Rothman, R. B., Baumann, M. H. (2003). Monoamine transporters and psychostimulant drugs. Eur J Pharmacol 479(1–3), 23–40.

Rudnick, G., Wall, S. C. (1992). The molecular mechanism of "ecstasy" [3,4-methylenedioxy-methamphetamine (MDMA)]: serotonin transporters are targets for MDMA-induced serotonin release. Proc Natl Acad Sci USA 89(5), 1817–1821.

Schubert, R. (2005). Attention deficit disorder and epilepsy. Pediatr Neurol 32(1), 1–10.

Schwartz, M. S., Scott, D. F. (1974). Aminophylline-induced seizures. Epilepsia 15(4), 501–505.

Segev, S., Rehavi, M., et al. (1988). Quinolones, theophylline, and diclofenac interactions with the gamma-aminobutyric acid receptor. Antimicrob Agents Chemother 32(11), 1624–1626.

Seiden, L. S., Sabol, K. E., et al. (1993). Amphetamine: effects on catecholamine systems and behavior. Annu Rev Pharmacol Toxicol 33, 639–677.

Shen, M., Piser, T. M., et al. (1996). Cannabinoid receptor agonists inhibit glutamatergic synaptic transmission in rat hippocampal cultures. J Neurosci 16(14), 4322–4334.

Slikker, W., Jr., Holson, R. R., et al. (1989). Behavioral and neurochemical effects of orally administered MDMA in the rodent and nonhuman primate. Neurotoxicology 10(3), 529–542.

Starr, M. S. (1996). The role of dopamine in epilepsy. Synapse 22(2), 159–194.

Sultan, S., Chouinard, G., et al. (1990). Antiepileptic drugs in the treatment of neuroleptic-induced supersensitivity psychosis. Prog Neuropsychopharmacol Biol Psychiatry 14(3), 431–438.

The MTA Cooperative Group (1999). A 14-month randomized clinical trial of treatment strategies for attention-deficit/hyperactivity disorder. The MTA Cooperative Group. Multimodal Treatment Study of Children with ADHD. Arch Gen Psychiatry 56(12), 1073–1086.

Taylor, D. C., Neville, B. G., et al. (1999). Autistic spectrum disorders in childhood epilepsy surgery candidates. Eur Child Adolesc Psychiatry 8(3), 189–192.

Turner, C. E., Elsohly, M. A., et al. (1980). Constituents of Cannabis sativa L. XVII. A review of the natural constituents. J Nat Prod 43(2), 169–234.

Volkow, N. D., Wang, G., *et al.* (2001). Therapeutic doses of oral methylphenidate significantly increase extracellular dopamine in the human brain. *J Neurosci* **21**(2), RC121.

Williams, J., Bates, S., *et al.* (1998). Does short-term antiepileptic drug treatment in children result in cognitive or behavioral changes? *Epilepsia* **39**(10), 1064–1069.

Wise, R. A. (1996). Addictive drugs and brain stimulation reward. *Annu Rev Neurosci* **19**, 319–340.

Wroblewski, B. A., Leary, J. M., *et al.* (1992). Methylphenidate and seizure frequency in brain injured patients with seizure disorders. *J Clin Psychiatry* **53**(3), 86–89.

Yarnell, P. R., Chu, N. S. (1975). Focal seizures and aminophylline. *Neurology* **25**(9), 819–822.

Zagnoni, P. G., Albano, C. (2002). Psychostimulants and epilepsy. *Epilepsia* **43**(Suppl 2), 28–31.

CHAPTER 19

The Psychotropic Effects of Vagus Nerve Stimulation in Epilepsy

Eric J. Foltz and David M. Labiner

INTRODUCTION

Better understanding of co-morbid conditions associated with epilepsy, particularly psychiatric disorders, has led to new treatment strategies. Based on similar sites and mechanisms of action, as well as effects on mood noted in the treatment of epilepsy, several antiepileptic medications have subsequently been studied for use in the treatment of psychiatric conditions. Carbamazapine, valproic acid, and lamotrigine, among others, have been used in the treatment of bipolar disorder. These agents have also been used as adjunctive treatment for depression. Temporal lobectomy in patients with epilepsy has been shown to improve depression (Altshuler *et al.*, 1999). Given these findings, it is worth considering the psychotropic effects of another treatment for epilepsy, namely vagus nerve stimulation (VNS).

THE VAGUS NERVE

The vagus nerve, cranial nerve X, originates in the medulla. For years the function of this nerve was not completely understood. The vagus nerve contains approximately 80% afferent fibers and 20% efferent fibers (Groves and Brown, 2005). Sensory information from the head, neck, thorax and abdomen is carried to the nucleus of the solitary tract (NTS). From the NTS, there are bilateral projections to several areas in the brain, including the cerebellum, the periaqueductal gray (PAG), the raphe nucleus, the locus coeruleus, the limbic system, and the cerebral cortex. Thus, the vagus nerve has a potential role in regulating seizure activity, emotion, mood, and pain (Nemeroff *et al.*, 2006).

The efferent fibers of the vagus nerve arise from the nucleus ambiguus. This nucleus receives input from the motor areas of the cortex via the reticular formation. Output from the nucleus ambiguus supplies the palatoglossus muscle of the

Psychiatric Controversies in Epilepsy

tongue, and striated muscles of the soft palate, larynx, and pharynx (Hermanowicz, 2003). The autonomic functions of the vagus nerve are carried out by parasympathetic fibers that originate in the dorsal vagal motor nucleus. These fibers travel to plexi near target organs in the neck, thorax, and abdomen. The autonomic functions of the vagus nerve include regulating secretions, bronchoconstriction, peristalsis in the gut, and slowing the heart rate. The right vagus acts on the heart's pacemaker, the sinoatrial node. The left vagus nerve, however, affects cardiac contraction force by innervating the atrioventricular node (Groves and Brown, 2005).

THE HISTORY OF VAGUS NERVE STIMULATION

In 1883, James L. Corning created an external device for vagus nerve stimulation (VNS) in order to suppress seizures. His belief was that seizures were caused by a condition which he termed venous hyperemia, and that suppression was achieved by decreasing heart rate through the compression of the carotid artery (Lanska, 2002). Over half a century later experiments with cats revealed that VNS produced synchronized electrical activity in the orbital cortex (Bailey and Bremer, 1938). Electroencephalogram (EEG) desynchronization, the opposite of synchronization seen with seizures, was produced by high intensity stimulation (Boon et al., 2001). It was believed that the VNS activated unmyelinated vagal C-fibers, which then activated the reticular activating system and produced desynchronization. It is now known that the VNS parameters that are effective in treating epilepsy are below the threshold required for activating the C-fibers. Zabara (1985a, b) showed that strychnine-induced seizures in dogs were reduced by VNS. In 1992 he also reported that seizure suppression outlasted acute vagal nerve stimulation (Zabara, 1992). Bilateral ablation of the locus coeruleus in rats significantly impaired VNS-induced seizure reduction (Krahl et al., 1998). The locus coeruleus is a major production site of norepinephrine, which is primarily an inhibitory neurotransmitter. In 2001, Krahl et al. found that stimulation of vagal myelinated A- and B-fibers suppressed seizures in rats, and that destruction of C-fibers did not cause loss of VNS effect (Krahl et al., 2001).

The vagus nerve stimulator is a programmable pulse generator and electrode lead for use in humans that was developed by Cyberonics (Cyberonics, Inc., Houston, TX, USA). The goal was to determine if adjunctive use of VNS to deliver electrical stimulation to the left vagus nerve in humans would be useful in reducing seizure frequency in patients with medically refractory epilepsy. The generator is surgically implanted subcutaneously in the left infraclavicular region. The spiral electrode is wrapped around the left vagus nerve via an incision above the sternocleidomastoid muscle. The portion of the lead that connects to the generator is then tunneled subcutaneously to the infraclavicular site (Reid, 1990). The VNS can be programmed using a handheld computer with manufacturer-supplied software and a programing wand. Adjustable parameters include output and magnet currents, frequency, pulse width, stimulation and magnet on times, stimulation off time, and magnet pulse width (Labiner and Ahern, 2007). The patient can use a handheld magnet to activate the generator in an attempt to abort or shorten a seizure (Ben-Menachem *et al.*, 1994).

The first reported implantation of VNS in a human was performed in 1988 (Penry and Dean, 1990). Two pilot studies were performed, known as the EO1 and EO2 studies. In these studies, 15 subjects had a VNS device implanted, of which 14 received VNS. A mean reduction in seizures of 46.6% was found, with the response ranging from no improvement to complete cessation of seizures (Boon *et al.*, 2001).

Full-scale experimentation in humans could not be performed in the typical placebo-control manner that is commonly used with medications. In order to pre-serve blinding, subjects in the early human studies of VNS would be randomized to either high- or low-frequency settings. The hypothesis to be tested was that high parameter settings would be more efficacious in the treatment of partial sei-zures when compared to low parameter settings (Table 19.1). In the EO3 study, a 14-week acute phase study of 67 patients, Ben-Menachem *et al.*(1994) showed that subjects in the high VNS group had a statistically significant greater mean reduc-tion in seizures compared to the low VNS group (30.9% vs. 11.3%, respectively, $p = 0.029$). Furthermore, there was a trend in the high VNS group to be clinical responders (i.e. to have a greater percentage of patients reporting at least a 50% reduction in seizures). In raw numbers the high group averaged 79 fewer seizures after VNS, compared to only four fewer average seizures in the low group. These results were replicated by Handforth *et al.* (1998) in a 3-month VNS study in an even larger population of 196 patients (EO5 trial). The side effects and tolerability in the earlier trial patients were reported by Ramsay *et al.* (1994). They found that

TABLE 19.1   Stimulation parameters

	Range	
	High	Low
Parameter		
Output current (mA)	0.25–3.0	0.25–3.0
Frequency (Hz)	25–50	1–2
Pulse width (μs)	500	130
On time (s)	30–90	30
Off time (min)	5–10	60–180
Magnet output current (mA)	0.5–3.0	0
Magnet on time (s)	30–90	n/a
Magnet pulse width	500	n/a

Source: Adapted from Ben-Menachem *et al.* (1994).

subjects in the high-stimulation group most commonly experienced hoarseness, coughing, and throat pain. Those in the low-stimulation group had hoarseness and throat pain. The only serious adverse event reported was left vocal cord paralysis due to a malfunction of the generator resulting in direct current delivery to the vagus nerve. No other subjects discontinued therapy.

A long-term extension study in the same 67 subjects from the EO3 study was carried out. All participants received high VNS for 16–18 months. For those who remained on high VNS settings, there was a mean 52% reduction in seizure frequency at the end of this period. Those subjects who were switched from low to high settings at the beginning of the extension experienced a mean seizure frequency reduction of 38.1% at the end of the 16–18 month time period. These results were statistically significant and showed that the efficacy of VNS achieved in the acute phase was maintained over the long term (George *et al.*, 1994); 195 subjects from the EO5 study entered a long-term follow-up study that looked at percentage change in total seizure frequency at 3 and 12 months of high-stimulation parameters. Compared to preimplantation baseline, there was a 34% median reduction in seizure frequency at 3 months and a 45% median reduction at 12 months ($p < 0.0001$). The initial high-stimulation group went from a median 23% reduction of seizures at the end of the double-blind trial to 37% at 3 months, and 46% at 12 months. Over the same period, the initial low-stimulation group progressed from 21% to 29%, and ultimately 40% ($p < 0.0001$ for both groups) (DeGiorgio *et al.*, 2000).

VNS caused no significant alterations of vital signs, cardiac rhythms, gastrin levels or serum concentrations of antiepileptic medications in any of these human trials.

DEPRESSION IN EPILEPSY

Depression is one of the most significant co-morbidities associated with epilepsy (Harden, 2002). Clinically significant depression and/or anxiety effects up to 50% of patients with epilepsy. People with epilepsy are more likely to have depression, and are more likely to have a severe form of depression, than patients with other medical conditions. Community-based studies in epilepsy patients have found rates of depression ranging from 9% to 22% (Edeh *et al.*, 1990; Jacoby *et al.*, 1996). In hospital studies of patients with epilepsy, including those with refractory epilepsy, these rates can approach 60% (Robertson *et al.*, 1994; Victoroff *et al.*, 1994).

As with primary depression, the causes of depression in epilepsy can be numerous. Age of onset of seizures, seizure frequency, quality of life, seizure type, marriage, genetics, employment, and duration of disease can all factor into the development of depression in epilepsy. A study performed by Robertson *et al.* (1987) found that over 50% of the patients with concomitant epilepsy and depression had a family history of psychiatric disease. Studies have shown that men with epilepsy may be at higher risk to develop depression than women (Mendez *et al.*, 1986; Strauss *et al.*, 1992). One study even reported that there was a correlation between male gender, left temporal lobe epilepsy (TLE), and an increased risk of developing depression (Altshuler *et al.*, 1990).

One theory regarding depression in epilepsy postulates that seizures act as an electrophysiologic form of electroconvulsive therapy, and depression may worsen as seizure frequency improves (Robertson *et al.*, 1987). Landolt coined the term "forced normalization", describing exacerbation of a psychiatric illness. The often debated theory suggests that the loss of electrophysiologic inhibition after a seizure with reduction in seizures was responsible for depression (Landolt, 1953, 1958).

Patients with TLE are more likely to experience depression when compared to those with extratemporal foci (Perini and Mendius, 1984; Mendez *et al.*, 1986; Piazzini *et al.*, 2001). The same is true for patients with complex partial seizures when compared to those with generalized seizures (Mendez *et al.*, 1986; Indaco *et al.*, 1992; Gureje, 1991; Piazzini *et al.*, 2001). A major component in the development of depression in patients with TLE is believed to be frontal lobe dysfunction. Positron emission tomography (PET) studies have shown decreased glucose metabolism in both inferior frontal lobes in patients with TLE and depression (Bromfield *et al.*, 1992). Beck Depression Inventory (BDI) scores were higher in patients with left TLE who exhibited reduced single-photon emission–computed tomography (SPECT) activity in the bilateral frontal lobes and the right temporal region (Schmitz *et al.*, 1997). The degree of frontal lobe dysfunction was significantly correlated to self reported dysphoria in patients with left TLE, as well (Hermann *et al.*, 1991; Seidenberg *et al.*, 1995). Mendez *et al.* (1994) postulated that dysfunction in the temporal lobe secondary to a seizure focus may cause frontal hypometabolism due to the temporal region's rich afferent input into the frontal lobes.

THE EFFECTS OF VNS THERAPY ON MOOD IN EPILEPSY: RESULTS FROM THE PILOT STUDIES

In their report, Ben-Menachem *et al.* (1994) stated that "the psychological impact of being able to stop a seizure is very beneficial and should not be underestimated". They were referring to the use of the handheld magnet to abort a seizure. The use of VNS in these studies surprisingly showed that it could have a psychological impact on a greater scale than this, however.

Ramsay *et al.* (1994) utilized a visual-analog "Global Scale" of patient well being in their study of safety, tolerability, and side effects with VNS use. This rating scale was administered to the investigator, the patient, and a companion at each visit. The goal was to determine the effects of VNS therapy on the patient's quality of life and overall functioning. The linear analog scale was a 100 mm line with

Place an "X" on the line at the point you believe indicates the status of your overall
well-being as compared to when you were admitted to the study (Visit 1).
If there is no change, place a mark through the circle in the middle of the line.

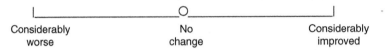

Considerably	No	Considerably
worse	change	improved

FIGURE 19.1   Visual analog "well-being scale". "Considerably worse" lies at a point 50 mm to the
left of the midpoint and "considerably improved" lies at a point 50 mm to the right of midpoint.
Source: Modified from Handforth *et al.* (1998).

the midpoint designated as 0. The furthest point to the left of 0 was designated
as −50, and the furthest point to the right of center designated +50. Worsened
overall condition resulted in a point to the left, and better condition resulted in
a point to the right. Distances from the midpoint were measured in millimeters
and recorded. Mean changes were calculated. Overall improvement was noted in
the low VNS group, but greater mean improvement in condition was noted by
all three raters in the high VNS group. The only statistically significant difference
between the high and low groups was found in the investigators' ratings (Ramsay
et al. 1994). Using the same scale (Figure 19.1), Handforth *et al.* (1998) found that
comparisons within each group (high and low VNS) showed significant improvement from the baseline visit to visits 7 through 9, after VNS implantation and
activation. This was true for the interviewer, the patient, and the patient's companion ($p < 0.001$). In the between-group comparisons, both blinded interview-
ers ($p = 0.02$) and patients ($p = 0.004$) rated the improvement in well being as
greater in the high-stimulation group vs. the low-stimulation group. There was
no statistical significance in the difference between the companions' ratings. This
clinical observation raised the interesting possibility that VNS has a positive psychotropic effect independent of seizure control.

VNS EFFECTS ON MOOD AND DEPRESSION IN EPILEPSY

Based on these findings in epilepsy patients treated with VNS, researchers sought
to examine if this therapy could have an effect on the co-morbid condition of
depression. A study carried out in addition to the randomized control trial and
EO3 study examined 11 patients who had VNS implanted for medically refractory epilepsy. They received low ($n = 5$) or high ($n = 6$) stimulation for 3 months,
followed by the low VNS group being converted to high stimulation. Mood was
assessed 4 weeks prior to implantation, and at 3 and 6 months after VNS therapy
was started (Elger *et al.*, 2000). Patients were interviewed by a psychiatrist and
completed two questionnaires, the Washington Psychosocial Seizure Inventory
(WPSI) (Dodrill *et al.*, 1980) and the Befindlichkeits-Skala (Bf-S) (Zersen *et al.*,

1970). The psychiatrist completed other rating scales, including the Montgomery-Åsberg Depression Rating Scale (MADRS; Montgomery and Åsberg, 1979) and the Scale for the Assessment of Negative Symptoms (SANS; Andreasen, 1981).

Seven of 11 patients had a baseline MADRS score between 10 and 20, indicating subdepressive mood states. Eight of 11 had baseline SANS scores ranging between 30 and 60, indicating mild negative symptoms. No patients met the criteria for clinical depression at baseline (Elger et al., 2000). Nine patients had a reduction in the MADRS score. Of the seven that had scores in the subdepressive range, this number decreased to four of eleven at 3 months and two of ten (one lost to follow-up) at 6 months. At 6 months, all 10 remaining patients showed improvement in their SANS scores. At 3 months only three patients continued to have scores in the 30–60 range, and this number decreased to one at 6 months. Although the high-stimulation group showed an accelerated mood improvement when compared to the low-stimulation group, the statistical power of this study was too small to confirm if this was a significant effect. When seizure control was reviewed with improvement in depressive symptoms, only two of eleven patients reached a level of 50% reduction in seizure frequency. At 6 months, mood elevation was noted in six of the eight seizure non-responder patients, as well as the two seizure responders (Elger et al., 2000).

A second study of individuals with epilepsy examined whether addition of VNS to a stable antiepileptic medication regimen had a greater effect on mood than a stable antiepileptic regimen alone. Twenty non-randomized subjects received VNS and 20 patients who declined VNS and medication changes were used as controls. Using the Cornell Dysthymia Rating Scale (CDRS; Mason et al., 1993a, b), the Beck Depression Inventory (BDI; Beck, 1967), and both the Hamilton Rating Scales for Depression and Anxiety (HAM-D and HAM-A, respectively; Hamilton, 1959, 1960) subjects underwent baseline testing the day following VNS implantation, prior to device activation. Testing with the aforementioned scales was repeated approximately 3 months later. Seizure frequency was also monitored. Seizure control was found to be significantly improved in the VNS group ($p = 0.01$). The authors found no correlation between seizure frequency and baseline CDRS, BDI, and HAM-D scores. Subjects with VNS showed significant improvement in the CDRS ($p = 0.001$), the BDI ($p = 0.045$), and the HAM-D ($p = 0.017$) when repeat testing was compared to baseline. There was no significant difference in scores between the VNS group and the control groups over time, but there was a trend toward significance in the VNS group with the BDI scores ($p = 0.07$). When the VNS group was broken into seizure responders (at least a 50% reduction in seizures) and non-responders, no significant difference for any scale was found over time (Harden et al., 2000).

Hoppe et al. (2001) studied 28 out of 40 patients who received VNS therapy using self-reporting questionnaires. The questionnaires were applied at least 4 weeks prior to VNS implantation and then again at least 6 months after implantation. The scales used were the BDI, the Bf-S, the Self-Rating on Anxiety Scale (SAS; Zung,

1971), the QOLIE-10, which is a modification of the Health-Related Quality of Life Scale (Cramer et al., 1996), and the Level of Activity scale from the Behavioral Psychosocial Scales on Epilepsy (BPSE; Helmstaedter and Elger, 1994). Seizure diaries were also kept by the patients. Out of 28 patients, 26 had complex partial seizures with secondary generalization. Sixteen of these patients had no signs of depression at baseline, five had mild depression (BDI score 12–17), and five had clinical depression (BDI score > 17). The results revealed statistically significant improvement in the Bf-S and the SAS ($p = 0.001$ and $p = 0.004$ respectively) over time. In contrast to the study of Harden et al. (2000), there was not a significant change in BDI scores from baseline to follow-up. The authors explained this using the floor effect. The baseline BDI scores were lower in this study when compared to Harden's study, so there was less room for improvement. They did find that those patients with a baseline BDI score > 11 had a significant improvement ($p < 0.05$), however. Additionally, this study showed that absolute seizure frequency reduction was correlated with mood outcome (Hoppe et al., 2001).

An investigation of 15 children, ages 4–17, who had VNS for refractory epilepsy targeted the effects on mood, quality of life (QoL), behavior, and cognition at 3 and 9 months after implantation. A correlation between improvement of quality of life, mood, and seizure severity was found. Eleven of 12 children with improvement in QoL had improvement in mood and seizure severity, five of whom showed improvement in depressive parameters (Hallböök et al., 2005).

VNS FOR TREATMENT OF DEPRESSION

Rush et al. used the results of the VNS effects on mood and depression in epilepsy as the impetus to examine if VNS therapy would be effective in treatment-resistant depression (2000). They used an open-label, non-randomized study of 30 patients with major depressive disorder (MDD) or bipolar I or II disorder (in depressed phase). Patients had to score 20 or more on the Hamilton Depression Rating Scale ($HDRS_{28}$; Hamilton, 1960 and 1967) at each of two baseline visits to be eligible for VNS implantation. Medications had to be stable for at least 4 weeks prior to initial baseline visit. The device was inactive for 2 weeks postimplantation, during which time patients had to score 18 or greater on the $HDRS_{28}$ at two more visits. This was followed by 2 weeks of adjustment following device activation. After this, device parameters were unchanged for 8 weeks. In total, there was a 40% response rate defined as a 50% or more reduction in the baseline $HDRS_{28}$ score. Seventeen percent had an $HDRS_{28}$ exit score of 10 or less (a remission). Furthermore, they found a 50% response rate on the MADRS (Rush et al., 2000).

These patients were followed for an additional 9 months. The original 40% response rate (on the $HDRS_{28}$) was maintained at the end of this period and remission rates increased significantly from 17% at 3 months to 29% at 1 year ($p = 0.045$) (Marangell et al., 2002). In a follow-up study of 59 patients with treatment-resistant depression who received VNS for 3 months, including the original 30 patients from Rush's study, Nahas and associates showed continued improvement at 1 and 2 years. Using the $HDRS_{28}$ they showed that a 31% response rate at 3 months increased to 44% at 1 year, and remained stable at 42% after 2 years of VNS therapy. Remission rates were 15%, 27%, and 22% over the same time periods. None of these changes were statistically significant, however (Nahas *et al.*, 2005); 205 patients who participated in a 3-month sham-control VNS study (Rush *et al.*, 2005a), had a doubling of response and remission rates at 1 year ($p < 0.005$) (Rush *et al.*, 2005b).

PROPOSED MECHANISMS OF ACTION OF VNS IN EFFECTS ON MOOD AND DEPRESSION

There are many proposed mechanisms by which VNS can affect depression and mood, and probably many more which have yet to be elucidated. The first possible avenue of VNS action in depression is its neurochemical effects. VNS is known to stimulate the locus coeruleus, which releases norepinephrine. Norepinephrine is known to play a role in depression. Enhancement of the noradrenergic system by antidepressant medication improves depression, and the same mechanism may be operative with VNS (Harden *et al.*, 2000). VNS produces long-lasting norepinephrine release in the basolateral amygdala (Van Bockstaele *et al.*, 1999). It is a similar situation with serotonin. The nucleus of the solitary tract projects to the raphe nucleus, which is nearly the exclusive site of serotonin production in the brain. VNS stimulation of the NTS, and hence the raphe nuclei, can lead to increased serotonin release. This was demonstrated in rats by Debonnel and Dorr (2004). Evaluation of cerebrospinal fluid (CSF) in patients treated with VNS for epilepsy revealed a 33% increase of 5-hydroxyindoleacetic acid (5-HIAA), the metabolite of serotonin. This result trended toward significance (Ben-Menachem *et al.*, 1995). A second study of sham vs. active VNS in depression showed a significant 21% increase in CSF homovanillic acid (HVA) (Carpenter *et al.*, 2004).

VNS may also act through neural substrates. *C-fos* is a marker of neuronal activity. VNS treated rats had increased *c-fos* expression in the brainstem (NTS, locus coeruleus, PAG) and the forebrain (paraventricular nuclei, neocortex, lateral hypothalamus, and hippocampal CA3 region) compared to sham controls (Kling *et al.*, 2003). This supports the hypothesis that VNS can act on the brainstem and affect limbic and cortical activity that determines mood.

Studies on regional cerebral blood flow (rCBF) can also shed some light on VNS and depression. Devous *et al.* learned that treatment with VNS improves rCBF abnormalities seen in the limbic region and cortex in depressed patients. They found a correlation between HAM-D scores and increased rCBF in the medial temporal cortex after VNS therapy (Devous *et al.*,2002). A study of two depressed subjects receiving VNS therapy revealed acute increases in rCBF in the orbitofrontal cortex, the medial and dorsolateral prefrontal cortex, cerebellum, and insula by functional magnetic resonance imaging (fMRI) (George *et al.*, 2002). Conway *et al.* (cited by George *et al.*, 2002) found increased glucose metabolism on PET scan in the orbitofrontal cortex, insula, amygdala, cingulate gyrus, and parahippocampal gyrus after 3 months of VNS therapy. Blood oxygenation level (BOLD) fMRI performed in depressed patients with chronic VNS and antidepressant medication displayed bilateral activation of the orbitofrontal and parieto-occipital cortex and further activation in the left amygdala, left temporal cortex, and the hypothalamus (Bohning *et al.*, 2001).

SUMMARY

Based on what is known about the mechanisms of actions of VNS and the extensive areas of the brain that are affected by this process, it is only logical to conclude that this therapy can have a positive psychotropic effect when used in the treatment of epilepsy. The findings of several researchers have proven this in studies of VNS effects in epilepsy, VNS effects on depression in epilepsy, and VNS therapy for primary depression. This treatment should be considered in any patient who suffers from both epilepsy and depression, especially if one or both of the conditions is not responsive to medication alone.

REFERENCES

Altshuler, L. L., Devinsky, O., Post, R. M., Theodore, W. (1990). Depression, anxiety, and temporal lobe epilepsy: laterality of focus and symptoms. *Arch Neurol* **47**, 284–288.

Altshuler, L., Rausch, R., Delrahim, S., Kay, J., Crandall, P. (1999). Temporal lobe epilepsy, temporal lobectomy, and major depression. *J Neuropsychiatry Clin Neurosci* **11**, 436–443.

Andreasen, N. C. (1981). *Scale for the assessment of negative symptoms (SANS)*. Iowa City, IA: University of Iowa.

Bailey, P., Bremer, F. A. (1938). Sensory cortical representation of the vagus nerve. *J Neurophysiol* **1**, 405–412.

Beck, A. T. (1967). *Depression: Clinical, Experimental, and Theoretical Aspects*. New York: Hoeber.

Ben-Menachem, E., Mañon-Espaillat, R., Ristanovic, R., Wilder, B. J., Stefan, H., Mirza, W., Tarver, W. B., Wernicke, J. F. (1994). First International Vagus Nerve Stimulation Study Group. Vagus nerve stimulation for treatment of partial seizures: 1. A controlled study of effect on seizures. *Epilepsia* **35**(3), 616–626.

Ben-Menachem, E., Hamberger, A., Hedner, T., Hammond, E. J., Uthman, B. M., Slater, J., Treig, T., Stefan, H., Ramsay, R. E., Wernicke, J. F., Wilder, B. J. (1995). Effects of vagus nerve stimulation on amino acids and other metabolites in the CSF of patients with partial seizures. *Epilepsy Res* **20**, 221–227.

Bohning, D. E., Lomarev, M. P., Denslow, S., Nahas, Z., Shastri, A., George, M. S. (2001). Feasability of vagus nerve stimulation-synchronized blood oxygenation level-dependent functional MRI. *Invest Radiol* **36**, 470–479.

Boon, P., Vonck, K., De Reuck, J., Caemaert, J. (2001). Vagus nerve stimulation for refractory epilepsy. *Seizure* **10**, 448–455.

Bromfield, E. B., Altshuler, L., Leiderman, D. B., Balish, M., Ketter, T. A., Devinsky, O., Post, R. M., Theodore, W. H. (1992). Cerebral metabolism and depression in patients with complex partial seizures. *Arch Neurol* **49**(6), 617–623.

Carpenter, L. L., Moreno, F. A., Kling, M. A., Anderson, G. M., Regenold, W. T., Labiner, D. M., Price, L. H. (2004). Effect of vagus nerve stimulation on cerebrospinal fluid monoamine metabolites, norepinephrine, and gamma-aminobutyric acid concentrations in depressed patients. *Biol Psychiatry* **56**, 418–426.

Cramer, J. A., Perrine, K., Devinsky, O., Meador, K. (1996). A brief questionnaire to screen for quality of life in epilepsy: the QOLIE-10. *Epilepsia* **37**, 577–582.

Debonnel, G., Dorr, A.E. (2004). Effect of Vagus Nerve Stimulation (VNS) on dorsal raphe serotonergic neurons: an electrophysiological study in the rat. *Poster presented at the Society for Neuroscience Annual Meeting*, San Diego, CA, October 23–27, 2004.

DeGiorgio, C. M., Schachter, S. C., Handforth, A., Salinsky, M., Thompson, J., Uthman, B., Reed, R., Collins, S., Tecoma, E., Morris, G. L., Vaughn, B., Naritoku, D. K., Henry, T., Labar, D., Gilmartin, R., Labiner, D., Osorio, I., Ristanovic, R., Jones, J., Murphy, J., Ney, G., Wheless, J., Lewis, P., Heck, C. (2000). Prospective long-term study of vagus nerve stimulation for the treatment of refractory seizures. *Epilepsia* **41**(9), 1195–1200.

Devous, M. D., Husain, M., Harris, T. S., Rush, A. J. (2002). Effects of VNS on regional cerebral blood flow in depressed subjects. *Poster presented at the 42nd Annual New Clinical Drug Evaluation Unit Meeting*, Boca Raton, FL, June 10–13, 2002.

Dodrill, C. B., Batzel, L. W., Queisser, H. R., Temkin, N. R. (1980). An objective method for the assessment of psychological and social problems among epileptics. *Epilepsia* **21**, 123–135.

Edeh, J., Boone, B. K., Corney, R. H. (1990). Epilepsy, psychiatric morbidity, and social dysfunction in general practice. Comparison between clinic patients and clinic nonattenders. *Neuropsychiatry Neuropsychol Behav Neurol* **3**, 180–192.

Elger, G., Hoppe, C., Falkai, P., Rush, A. J., Elger, C. (2000). Vagus nerve stimulation is associated with mood improvements in epilepsy patients. *Epilepsy Res* **42**, 203–210.

George, R., Salinsky, M., Kuzniecky, R., Rosenfeld, W., Bergen, D., Tarver, W. B., Wernicke, J. F. (1994). First International Vagus Nerve Stimulation Study Group. Vagus nerve stimulation for treatment of partial seizures: 3. Long-term follow-up on first 67 patients exiting a controlled study. *Epilepsia* **35**(3), 637–643.

George, M. S., Nahas, Z., Bohning, D. E., Kozel, F. A., Anderson, B., Chae, J.-H., Lomarev, M., Denslow, S., Li, X., Mu, C. (2002). Vagus nerve stimulation therapy: a research update. *Neurology* **59**(Suppl 4), S56–S61.

Groves, D. A., Brown, V. J. (2005). Vagal nerve stimulation: a review of its applications and potential mechanisms that mediate its clinical effects. *Neurosci Biobehav Rev* **29**, 493–500.

Gureje, O. (1991). Interictal psychopathology in epilepsy: prevalence and pattern in a Nigerian clinic. *Br J Psychiatry* **158**, 700–705.

Hallböök, T., Lundgren, J., Stjernqvist, K., Blennow, G., Strömblad, L.-G., Rosén, I. (2005). Vagus nerve stimulation in 15 children with therapy resistant epilepsy; its impact on cognition, quality of life, behavior and mood. *Seizure* **14**, 504–513.

Hamilton, M. (1959). The assessment of anxiety states by rating. *Br J Med Psychol* **32**, 50–55.

Hamilton, M. (1960). A rating scale for depression. *J Neurol Neurosurg Psychiatry* **12**, 371–379.

Hamilton, M. (1967). Development of a rating scale for primary depressive illness. *Br J Soc Clin Psychol* **6**, 278–296.

Handforth, A., DeGiorgio, C. M., Schachter, S. C., Uthman, B. M., Naritoku, D. K., Tecoma, E. S., Henry, T. R., Collins, S. D., Vaughn, B. V., Gilmartin, R. C., Labar, D. R., Morris, G. L., Salinsky, M. C., Osorio, I., Ristanovic, R. K., Labiner, D. M., Jones, J. C., Murphy, J. V., Ney, G. C., Wheless, J. W. (1998). Vagus nerve stimulation therapy for partial-onset seizures: a randomized active-control trial. *Neurology* **51**, 48–55.

Harden, C. L., Pulver, M. C., Ravdin, L. D., Nikolov, B., Halper, J. P., Labar, D. R. (2000). A pilot study of mood in epilepsy patients treated with vagus nerve stimulation. *Epilepsy Behav* **1**, 93–99.

Harden, C. L. (2002). The co-morbidity of depression and epilepsy. *Neurology* **59**(Suppl 4), S48–S55.

Helmstaedter, C., Elger, C. E. (1994). Behavioral psychosocial scales for epilepsy. *J Epilepsy* **7**, 13–23.

Hermann, B. P., Seidenberg, M., Haltiner, A., Wyler, A. R. (1991). Mood state in unilateral temporal lobe epilepsy. *Biol Psychiatry* **30**, 1205–1218.

Hermanowicz, N. (2003). Cranial nerves IX (glossopharyngeal) and X (vagus). In *Textbook of Clinical Neurology* (C. G. Goetz, ed.), pp. 211–222. Philadelphia, PA: Saunders.

Hoppe, C., Helmstaedter, C., Scherrmann, J., Elger, C. E. (2001). Self reported mood changes following 6 months of vagus nerve stimulation in epilepsy patients. *Epilepsy Behav* **2**, 335–342.

Indaco, A., Carrieri, P. B., Nappi, C., Gentile, S., Striano, S. (1992). Interictal depression in epilepsy. *Epilepsy Res* **12**, 45–50.

Jacoby, A., Baker, G. A., Steen, N., Potts, P., Chadwick, D. W. (1996). The clinical course of epilepsy and its psychosocial correlates: findings from a UK community study. *Epilepsia* **37**, 148–161.

Kling, M. A., Loyd, D., Sansbury, N., Ren, K., Murphy, A. Z. (2003). Effects of short-term VNS therapy on Fos expression in rat brain nuclei. *Poster presented at the 58th Annual Scientific Convention of the Society of Biological Psychiatry*, San Francisco, CA, May 15–17, 2003.

Krahl, S. E., Clark, K. B., Smith, D. C., Browning, R. A. (1998). Locus coeruleus lesions suppress the seizure-attenuating effects of vagus nerve stimulation. *Epilepsia* **39**, 709–714.

Krahl, S. E., Senanayake, S. S., Handforth, A. (2001). Destruction of the peripheral C-fibers does not alter subsequent vagus nerve stimulation induced seizure suppression in rats. *Epilepsia* **42**, 586–589.

Labiner, D. M., Ahern, G. L. (2007). Vagus nerve stimulation therapy in depression and epilepsy: therapeutic parameter settings. *Acta Neurol Scand* **115**, 23–33.

Landolt, H. (1953). Serial electroencephalographical correlations in epileptic psychoses (twilight states). *Electroenchephalogr Clin Neurophysiol* **5**, 121–122.

Landolt, H. (1958). Serial electroencephalographic investigations during psychotic episodes in epileptic patients and during schizophrenic attacks. In *Lectures on Epilepsy* (A. M. Lorentz de Hass, ed.), pp. 91–133. Amsterdam: Elsevier.

Lanska, D. J. (2002). Corning and vagal nerve stimulation for seizures in the 1880s. *Neurology* **58**, 452–459.

Marangell, L. B., Rush, A. J., George, M. S., Sackeim, H. A., Johnson, C. R., Husain, M. M., Nahas, Z., Lisanby, S. H. (2002). Vagus nerve stimulation (VNS) for major depressive episodes: one year outcomes. *Biol Psychiatry* **51**, 280–287.

Mason, B. J., Kocsis, J. H., Leon, A. C., Thompson, S., Frances, A. J., Morgan, R. O., Parides, M. K. (1993). Measurement of severity and treatment response in dysthymia. *Psychiatry Ann* **23**, 625–631.

Mason, B. J., Kocsis, J. H., Leon, A. C. (1993). Measurement of severity and treatment response in dysthymia. *Psychiatry Ann* **23**, 625–631.

Mendez, M. F., Cummings, J. L., Benson, D. F. (1986). Depression in epilepsy: significance and phenomenology. *Arch Neurol* **43**, 766–770.

Mendez, M. F., Taylor, J. L., Doss, R. C., Salguero, P. (1994). Depression in secondary epilepsy: relation to lesion laterality. *J Neurol Neurosurg Psychiatry* **57**, 232–233.

Montgomery, S. A., Åsberg, M. (1979). A new depression scale designed to be sensitive to change. *Br J Psychiatry* **134**, 382–389.

Nahas, Z., Marangell, L. B., Husain, M. M., Rush, A. J., Sackeim, H. A., Lisanby, S. H., Martinez, J. M., George, M. S. (2005). Two-year outcome of vagus nerve stimulation (VNS) for treatment of major depressive episodes. *J Clin Psychiatry* **66**(9), 1097–1104.

Nemeroff, C. B., Mayberg, H. S., Krahl, S. E., McNamara, J., Frazer, A., Henry, T. R., George, M. S., Charney, D. S., Brannan, S. K. (2006). VNS therapy in treatment-resistant depression: clinical evidence and putative neurobiological mechanisms. *Neuropsychopharmacol* **31**, 1345–1355.

Penry, J. K., Dean, J. C. (1990). Prevention of intractable partial seizures by intermittent vagal nerve stimulation in humans: preliminary results. *Epilepsia* **31**(Suppl 2), S40–S43.

Perini, G., Mendius, R. (1984). Depression and anxiety in complex partial seizures. *J Nerv Ment Dis* **172**, 287–290.

Piazzini, A., Canevini, M. P., Maggiori, G., Canger, R. (2001). Depression and anxiety in patients with epilepsy. *Epilepsy Behav* **2**, 481–489.

Ramsay, R. E., Uthman, B. M., Augustinsson, L. E., Upton, A. R. M., Naritoku, D., Willis, J., Treig, T., Barolat, G., Wernicke, J. F. (1994). The First International Vagus Nerve Stimulation Study Group. Vagus nerve stimulation for treatment of partial seizures: 2. Safety, side effects, and tolerability. *Epilepsia* **35**(3), 627–636.

Reid, S. A. (1990). Surgical technique for implantation of the neurocybernetic prosthesis. *Epilepsia* **31**(suppl 2), S38–S39.

Robertson, M. M., Trimble, M. R., Townsend, H. R. (1987). Phenomenology of depression in epilepsy. *Epilepsia* **28**, 364–372.

Robertson, M. M., Channon, S., Baker, J. (1994). Depressive symptomatology in a general hospital sample of outpatients with temporal lobe epilepsy: a controlled study. *Epilepsia* **35**, 771–777.

Rush, A. J., George, M. S., Sackeim, H. A., Marangell, L. B., Husain, M. M., Giller, C., Nahas, Z., Haines, S., Simpson, R. K., Jr., Goodman, R. (2000). Vagus nerve stimulation (VNS) for treatment-resistant depression: a multicenter study. *Biol Psychiatry* **47**, 276–286.

Rush, A. J., Marangell, L. B., Sackeim, H. A., George, M. S., Brannan, S. K., Davis, S. M., Howland, R., Kling, M. A., Rittberg, B. R., Burke, W. J., Rapaport, M. H., Zajecka, J., Nierenberg, A. A., Husain, M. M., Ginsberg, D., Cooke, R. G. (2005a). Vagus nerve stimulation for treatment-resistant depression: a randomized, controlled acute phase trial. *Biol Psychiatry* **58**(5), 347–354.

Rush, A. J., Sackeim, H. A., Marangell, L. B., George, M. S., Brannan, S. K., Davis, S. M., Lavori, P., Howland, R., Kling, M. A., Rittberg, B., Carpenter, L., Ninan, P., Moreno, F., Schwartz, T., Conway, C., Burke, M., Barry, J. J. (2005b). Effects of 12 months of vagus nerve stimulation in treatment-resistant depression: a naturalistic study. *Biol Psychiatry* **58**(5), 355–363.

Schmitz, E. B., Moriarty, J., Costa, D. C., Ring, H. A., Ell, P. J., Trimble, M. R. (1997). Psychiatric profiles and patterns of cerebral blood flow in focal epilepsy: interactions between depression, obsessionality, and perfusion related to the laterality of epilepsy. *J Neurol Neurosurg Psychiatry* **62**, 458–463.

Seidenberg, M., Hermann, B., Noe, A., Wyler, A. R. (1995). Depression in temporal lobe epilepsy: interaction between laterality of lesion and Wisconsin Card Sort Performance. *Neuropsychiatry Neuropsychol Behav Neurol* **8**, 81–87.

Strauss, E., Wada, J., Moll, A. (1992). Depression in male and female subjects with complex partial seizures. *Arch Neurol* **49**, 391–392.

Van Bockstaele, E. J., Peoples, J., Telegan, P. (1999). Efferent projections of the nucleus of the solitary tract to peri-locus coeruleus dendrites in rat brain: evidence for a monosynaptic pathway. *J Comp Neurol* **412**, 410–428.

Victoroff, J. I., Benson, F., Grafton, S. T., Engel, J., Jr., Mazziotta, J. C. (1994). Depression in complex partial seizures: electroencephalography and cerebral metabolic correlates. *Arch Neurol* **51**, 155–163.

Zabara, J. (1985a). Peripheral control of hypersynchronous discharge in epilepsy. *Electroencephalogr Clin Neurophysiol* **61**, S162.

Zabara, J. (1985b). Time course of seizure control to brief, repetitive stimuli. *Epilepsia* **26**, 518.

Zabara, J. (1992). Inhibition of experimental seizures in canines by repetitive vagal stimulation. *Epilepsia* **33**, 1005–1012.

Zersen, D. V., Koeller, D. M., Rey, E. R. (1970). Die Befindlichkeits-Skala Bf-S-ein einfaches Instrument zur Objektivierung von Befindlichkeits-Störungen insbesondere im Rahmen van Längsschnitt-Untersuchungen. *Arzneimittelforschung* **20**, 915–918.

Zung, W. W. K. (1971). A rating instrument for anxiety disorders. *Psychosomatics* **12**, 371–379.

CHAPTER 20

Do Psychological Therapies Alleviate Epileptic Seizures?

G. A. Baker and J. Eatock

INTRODUCTION

The continual development of newer antiepileptic drugs designed to maximize seizure control and minimize side effects has enabled the majority of individuals with epilepsy to have a better quality of life. However, a percentage of people continue to experience seizures regardless of medication and for certain sections of the world's population the full spectrum of antiepileptic drugs (AEDs) are not easily accessible or affordable (Engel, 2005). Seizures can also occur despite optimal pharmacological therapy and can affect an individual's quality of life and sense of control due to their sudden onset. Various anecdotal accounts report on the complex relationship between certain psychological factors and seizure occurrence (Mostofsky and Balaschak, 1977). This has understandably led to patients attempting to self-develop behaviors that can prevent a seizure occurring (Fenwick, 1992, 1995).

While medication and lifestyle changes are part of the long-term strategy of dealing with seizures it is apparent that people with epilepsy additionally aspire to maintain as much control over their condition as possible by being able to apply techniques that can not only warn of a seizure but hopefully prevent it from happening (Fenwick, 1995). Epilepsy patients have consistently reported factors that seem to play a part in causing seizures and in some cases, they have even been able to induce their seizures (Ng, 2002). Nonadherence to medication, alcohol, sleep deprivation and stress can all be seen to contribute (Aird, 1983) but at the root of these causes there are various psychological and behavioral processes operating. Therefore it seems logical to look beyond simple dispensing of AEDs (Fenwick, 1995) and examine how a psychological approach taking into account individual circumstances may be of benefit in controlling seizures (Wolf, 2002).

EVOLUTION OF EPILEPSY TREATMENT

Epilepsy and neurology in general stemmed from the psychiatric domain but with the "medicalization" of epilepsy in the 20th century (Brown and Muller, 2001) psychological aspects of seizures have been relegated to the background with the exception of treating psychiatric comorbidity (Mostofsky and Balaschak, 1977). Advances in diagnostic techniques and the acceptance of organic causes of seizures has recategorized epilepsy as a medical condition rather than a mental illness (Betts, 1981). In his comprehensive history of epilepsy, Temkin (1971) describes how by the end of the 4th century treatment of epilepsy was mostly "...dietetic, surgical and pharmacological" (p. 67). This could equally be stated as the case for the 21st century.

Present day epilepsy management largely relies on the use of AEDs in which an estimated 70% of individuals will have good seizure control, a lesser number of people undergo surgery and a smaller section of the population who modify their diet (e.g. ketogenic diet, vitamin supplements). The main objective for the clinician is to establish optimal drug therapy to minimize seizure occurrence. However, as will be outlined, pharmacotherapy does not control seizures in all people with epilepsy and for certain sections of the population they are not available or desirable.

Pharmacotherapy

As pharmacological treatment has emerged as the mainstay there has been an emphasis on developing newer formulations in pursuit of achieving the balance between seizure freedom and fewer unwanted side effects (e.g. Schachter, 1999; Sander, 2004). Up until 1994 there were six commonly used AEDs (Nadkarni et al., 2005) but today 18 drugs are licensed for use in epilepsy in the UK with 1 appearing per year over the last 10 years (Kelso and Cock, 2005). Despite the wider choice of AEDs available to the clinician, seizures in an estimated 30% of patients become refractory and as yet there is no evidence that newer AEDs reduce this percentage (Nadkarni et al., 2005). Recently it has been suggested that this 30% may be reduced through a more thorough and systematic trial of AEDs. Luciano and Shorvon (2007) found that 28% of their patients with refractory seizures studied became seizure free when swapped to a previously untried AED. However, there were still a percentage who do not respond to pharmacotherapy or who object to, or cannot afford, medication. Additionally, it cannot be assumed that the 70% classified as having treatment success are content with their treatment. It has long been recognized that from the patient perspective, quality of life is dependent on more than seizure control (Sander, 2005).

Problems of AEDs

A key difficulty in taking AEDs involves the various and well documented side effects associated with individual drugs, dosage and combinations. Older AEDs such as valproate, carbamazepine and phenytoin commonly have side effects such as headache, dizziness, and fatigue (Brodie and Dichter, 1996); newer AEDs additionally produce similar side effects (Dichter and Brodie, 1996). The age of the patient can also create difficulties. Whilst there has been a large body of literature focusing on AEDs and child development, prescription of AEDs can also be problematic when dealing with the elderly. The elderly are more likely to be receiving medication for other medical conditions which may interact with AEDs causing side effects which may be of a serious nature (Schachter, 1999). Newer AEDs may limit side effects from drug interactions but for the elderly population who do not receive free treatment the cost may prove prohibitive (Nadkarni et al., 2005).

Various studies have shown that AEDs adversely affect cognitive function (e.g. Aldenkamp et al., 2004). Discontinuation of AEDs has shown to improve complex cognitive processes and attention (Hessen et al., 2006). Cognitive effects are of particular concern in the case of children with epilepsy where parents have the difficult decision in balancing control of seizures with possible negative effects on their children's learning and education (Loring and Meador, 2004; Vinayan, 2006). However, while the effect of AEDs on cognitive function is still being investigated it is important to emphasize the detrimental effects uncontrolled seizures can have on brain development and cognitive processes; therefore the clinician has to consider both these factors in developing a treatment regimen (Marsh et al., 2006).

The treatment gap

Throughout any discussion concerning treatment for seizures there is a tendency to adopt a western world perspective and assume that all treatment options are available for each patient. The International League Against Epilepsy (ILAE) defines the treatment gap as "The difference between the number of people with active epilepsy and the number whose seizures are being appropriately treated in a given population at a given point of time, expressed as a percentage" (Meinardi et al., 2001). The definition covers the problem of affording or accessing treatment, it also recognizes that patients need to have an epilepsy treatment that is appropriate to the individual and their circumstances. While epilepsy affects all sections of the population regardless of race or class, not every individual has the same access to treatment, with an estimated 3 out of every 4 people receiving no treatment at all (WHO, 2001). This is particularly true with developing countries. Dua et al. (2006) cataloged how lower income countries had limited diagnosis tools and treatment available to epilepsy patients – epilepsy specialists were only available in 55% of these countries. The pharmaceutical industry concentrates on the minority (20%) of people with epilepsy-those living in the developed world (Engel, 2005).

In addition, costs to both health provider and patient prohibit the use of newer AEDs. Phenobarbital, a drug used since 1912, is a first line AED in 95% of the countries surveyed; this is not based on clinical effectiveness alone but the fact that it is considerably cheaper compared to other drugs (Engel, 2005). Additionally, the spectrum of AEDs available in the western world is unavailable in undeveloped countries resulting in more individuals having refractory epilepsy (Engel, 2005). As outlined earlier, the development of newer AEDs has been promoted in order to limit the type of side effects experienced with older AEDs; this effectively means that there will be an epilepsy population who may have seizure control but at the expense of experiencing undesirable side effects. As a consequence of side effects, treatment failure, and the treatment gap, alternative methods for controlling seizures have been pursued by the patient and their family.

The psychological approach

Psychological problems associated with epilepsy

A review about psychological therapies for alleviating seizures must acknowledge the psychological problems experienced by some individuals with epilepsy (see other chapters for more detailed discussions). Various studies examining the incidence of psychological problems have been conducted but have arrived at different conclusions. Estimates of incidence of psychological problems varies; Fiordelli et al. (1993) concluded that there was no significant difference in the number of psychiatric problems encountered with their 100 epilepsy patients in comparison to controls with only 19 out of the 100 identified as having psychiatric problems. In stark contrast, a recent community study of 5,834 patients recorded that 41% had a diagnosis of a psychiatric disorder during the study period of 3 years (Gaitatzis et al., 2004).

Psychological comorbidity is also likely to be underreported-an individual may be experiencing difficulties but this is not made evident to the clinician; O'Donoghue et al. (1999) found that only a third of those who could be classified as having definite or borderline anxiety or depression (as determined by Hospital Anxiety and Depression (HAD) scores) had symptoms reported in their medical notes. Smith et al. (1986) was able to study 622 epilepsy patients throughout the United States and found that they scored significantly higher in measures of tension and depression.

Anxiety has been reported at a rate of 25–33% and depression 10–15% in epilepsy patients (Smith et al., 1991; Baker et al., 1996). Seizure frequency has been shown to increase a patient's level of anxiety and depression (Jacoby et al., 1996). Smith et al. (1991) emphasize that how the patient perceives their seizure severity can affect the depression experienced. Using the concept of "attributional style" from Seligman's learned helplessness model (Seligman and Maier, 1967), Hermann *et al.* (1996) noted that those individuals who believed difficulties encountered in their everyday lives were associated with having the diagnosis of epilepsy were more at risk of depression regardless of seizure severity. Both factors (seizure frequency and seizure severity) disproportionately affect the refractory epilepsy population and this is the group predominately used in examining the effectiveness of psychological therapies in controlling seizures.

Seizure precipitants

Aird (1983) classifies causes of seizures in terms of seizure inducing and seizure triggering factors. The inducing factors are related to the environment of the individual and lower seizure threshold while the physiological or chemical actions that create a seizure are classified as seizure-triggering. Wolf (2005) divides precipitants into nonspecific facilitation (e.g. stress, alcohol) and specific reflex triggers (e.g. photosensitivity). Fenwick (1995) argues that brain activity involved in creating seizures can be directly related to behavior and therefore it follows that seizures can be generated or inhibited by certain behaviors. Whatever the mechanism involved it is agreed that certain external factors do contribute to seizures.

While it has been documented that some seizures are precipitated by specific cognitive tasks or external stimuli such as flashing lights, this only applies to a subset of people with epilepsy. More general variables seem to play a part in producing seizures and patients seem to be aware of them. Antebi and Bird (1993) reported that 83 out of their 100 patients could recognize factors associated with seizures. Informants could also list seizure causes in 80% of the patients. Spector *et al.* (2000) found that 91 of their 100 patients could identify at least one or more precipitant of their seizures. Nakken *et al.* (2005) reported that 53% of their 1,677 subjects noted at least one factor. In the Nakken *et al.* study there was a marked difference in percentage of those reporting seizure triggers dependent on whether their epilepsy was currently active or not (73% of the active epilepsy group compared to 42% of the inactive group). There also appeared to be a genetic influence in that monozygotic twins had a higher concordance rate in selecting the same precipitants compared to dizygotic twins (0.8 for MZ, 0.33 for DZ).

Amongst the results of various studies, stress has emerged as one of the main causes of seizures (Frucht *et al.*, 2000; Spector *et al.*, 2000; Haut *et al.*, 2003; Nakken *et al.*, 2005). Stress is problematic to define and is multilayered; emotional stress can cause sleep difficulties which may result in increased alcohol consumption all of which can trigger seizures (Nakken *et al.*, 2005; Schramke and Kelly, 2005). Neugebauer *et al.* (1994) found that sleep duration or interrupted sleep was correlated with seizures. However, in Haut *et al.*'s (2003) study they found that stress was experienced as a precipitating factor even when sleep problems were not associated. Stress has also been identified as a factor in the increased standardized mortality rate (SMR) associated with people with epilepsy – even in those who are seizure free – this suggests that seizures alone are not enough to explain levels of stress experienced by people with epilepsy (Yuen *et al.*, 2007). Interestingly, reporting of stress as a factor seems to have a cultural dimension in that 13% of patients in Pakistan believe stress to play a contributing role compared to 34% in Austria (Frucht *et al.*, 2000).

The underreported population of patients with epilepsy who actively seek to bring about a seizure – Fenwick (2005) noted a rate of 23% of the Maudsley Clinic admitted to self-induction, which is higher than most studies – lends weight to the evidence of psychological factors contributing to seizures. Realistic figures of those who self induce seizures are difficult to obtain; depending on the research methodology used, most patients are unlikely to disclose this to a clinician who is looking to enable seizure minimization for their patient (Spector *et al.*, 2000). Most self-induced seizures are related to photosensitive epilepsy and can be triggered by such actions as blinking or staring at a television set and it is mostly children who make up the cases (Ng, 2002). However, as Fenwick (1995) has noted, self induction of seizures is not restricted to external stimuli. A patient can adopt a mental state that has previously brought about seizures; this can be feelings of sadness, attempting to clear the mind or excitement (Fenwick, 1995). While there are varying motivations in bringing about a seizure – such as avoidance of stressful situations or attention seeking, it also represents an individual's desire for control in that they can decide when to have a seizure which may then allow seizure freedom for a longer period of time (Ng, 2002). Anecdotal evidence based on patient's reports whilst attending clinics demonstrates that many can frequently cite various precipitants they feel to be responsible for their seizures (Table 20.1).

TABLE 20.1   Seizure precipitants – as viewed from a patient perspective

- Stress
- Fatigue
- Sleeping patterns
- Alcohol
- Flashing lights/strobes
- Depression
- Anxiety
- Confrontation
- Heat
- Working with computers
- Noise

Source: Baker, Jacoby, Eatock (2007) (unpublished)

THE ROLE OF PSYCHOLOGICAL THERAPIES IN CONTROLLING SEIZURES

Psychological therapies designed to tackle seizures themselves and not psychological problems associated with living with epilepsy come under the umbrella of "alternative" treatment. Despite advances in the treatment of epilepsy even the most industrialized and wealthy nations in the world are reluctant to dismiss other treatments for epilepsy. Fisher *et al.* (2000) found that 17% of their 1,023 patients in the United States had been using alternative therapy although psychological therapies were less commonly used (biofeedback representing 2%, breathing/relaxation techniques 0.4%, mind control 0.1%). Sirven *et al.* (2003) reported that 44% of their sample (167 from 379) had used alternative therapies with stress reduction and yoga viewed as the most beneficial. Haut *et al.* (2003) found that just over half of their patients sampled were willing to embark on stress reduction training. This illustrates that there is a demand for alternative treatment in preference for solely relying on taking AEDs; however just as one drug cannot be used for all people with epilepsy, the individual psychological therapy may also depend on a number of factors.

Lifestyle behaviors

Wolf (2002) describes how nonpharmacological treatment of seizures fall into three domains; nonspecific treatment, specific prevention of seizures and arrest of seizures. Nonspecific treatment of seizures includes dealing with sleep-related problems, alcohol, and stress. Nakken *et al.* (2005) remark how for some patients with epilepsy a slight adaptation of lifestyle behaviors has brought about the same degree of success in controlling seizures as the use of new AEDs. While moderating alcohol intake or maintaining regular sleep patterns can be viewed as a conscious decision, several other factors such as avoiding or coping with stress or recognition of what triggers seizures can also be employed to minimize their occurrence.

Relaxation therapy

Stress as a seizure inducing factor has been widely reported by patients but has often been neglected in establishing its link with seizures and has led to a call for more systematic research (Betts, 1992). When discussing stress management it is unavoidable to omit relaxation therapy, which is often the main component of stress management programs. Within relaxation therapy, a number of techniques have been developed including muscle tension/relaxation, visualization and controlled breathing, often with many of these methods used concurrently. Another method which is gaining popularity is autogenic training, a form of self hypnosis designed to enable the individual to exert self control over certain physiological actions (Berger, 2005). Autogenic training can reduce levels of stress and has been shown to lessen anxiety and depression in other medical conditions such as cancer (Hidderley and Holt, 2004).

Relaxation therapy involves a certain amount of sessions with a trainer who teaches the techniques to an individual or group who are then encouraged to practice them at home for 20-minute sessions twice daily. Tapes are given to patients to follow, in some cases these are recordings done by the original therapist (Rousseau et al., 1985). Relaxation therapy is relatively inexpensive in that groups of people can be taught in each session, it requires no specialized equipment such as electroencephalograms (EEGs) and can seemingly be taught to most age ranges.

Puskarich et al. (1992) used Bernstein and Borkavec's (1973) muscle relaxation training program in order to train 24 people with epilepsy. Out of the twenty four, thirteen were in the treatment group while 11 represented a placebo group (quiet sitting). By the end of the 8-week follow-up, seizure frequency was reduced significantly in the treatment group when compared to baseline. Rousseau et al. (1985) divided their eight patients into two groups; one of which underwent sham treatment followed by progressive relaxation treatment while the other group undertook the real treatment only. They found a significant decrease in seizure frequency which could not be attributed to AEDs as the drug regimens were unaltered throughout the test period. Of equal importance, when interviews were conducted patients reported that they had better quality-of-life outcomes with improved sleep and increased perceived control of their seizures.

Dahl et al. (1987) used three treatment groups: contingent relaxation, attention control and waiting list control with their group of 18 adults. The group had no changes to AED regimens for at least a month before the start of therapy and throughout the therapy. Significant reduction of seizure frequency only occurred in the relaxation group, the attention control group – which involved discussion about epilepsy and seizures in general – actually reported an increased frequency of seizures. In view of their results it appeared unlikely that increased attention as a potential placebo effect could be responsible for reducing seizure frequency. Patients in the relaxation group were surveyed after a 10-week follow-up period and 80% stated that they believed they were definitely helped by the treatment and these positive responses were mirrored in a reduction of seizures (Dahl et al., 1987).

Berger (2001) used a variety of relaxation techniques including diaphragmatic breathing, imagery, progressive muscle relaxation and autogenic training and compared results with a control group. Quality-of-life scores were measured alongside seizure frequency; the results showed that only the treatment group could demonstrate significant improvements in both QoL and seizure frequency. Cabral and Scott (1976) studied three patients in a biofeedback relaxation crossover study and noted that 2 out of the 3 showed improved EEG patterns with the remaining patient more responsive to biofeedback. This provides further evidence for Wolf's contention that certain stimuli can both inhibit or precipitate seizures depending on the individual (Wolf, 2005).

Biofeedback

Biofeedback is the ability of an individual to identify and alter their physiological state.

Neurofeedback is a form of biofeedback but involves the process of operant conditioning in relation to the ability to alter EEG patterns (Sterman, 2005). In particular a 12–20 Hz rhythm is desired and reinforced (the Sensorimotor Rhythm or SMR) as this seems to have antiepileptic effects (Roth et al., 1967). A second method of biofeedback also uses the EEG. Using similar techniques to SMR training, patients are encouraged to regulate slow cortical potentials (SCPs), reducing the levels of cortical negativity, which is believed to play a role in seizure genesis (Monderer et al., 2002). Both forms of feedback involve patients receiving information about the direction in which they are altering the EEG; feedback can be through graphics or animations on a computer screen (Kotchoubey et al., 1996; Nagai et al., 2004a, b) or they may receive auditory feedback with different sounds informing them of how they are altering the EEG (Tozzo et al., 1988). Through processes of feedback and reward the desired rhythm on the EEG becomes a learned response with the hope that once the visual or auditory feedback is removed the patient has been conditioned to reproduce the desired effects on the EEG and therefore reduce seizure susceptibility.

Fried et al. (1984), testing the notion that seizure occurrence can be linked to chronic hyperventilation, used a biofeedback process with a group of epilepsy patients undergoing training in diaphragmatic breathing. Feedback could be obtained by showing the patients the effect of their particular respiration technique on the EEG. Finally, a new method of biofeedback has been utilized by Nagai et al. (2004a, b) who measured galvanic skin response (GSR) via electrodes on the participants' fingers and found that changes in GSR were inversely related to cortical excitation. If this association can be further qualified it would have important implications in treating epilepsy in that GSR measurements could be used to predict whether the individual's current physiological state was making them vulnerable to a seizure occurring.

Sterman (2000), in a review of various biofeedback studies, calculated that 82% of 174 patients with uncontrolled seizures had improved their seizure control. Tozzo et al. (1988), with their group of six adults with refractory epilepsy, noted both a reduction in seizure rate and in severity by using SMR training. Similarly, Kotchoubey et al. (1996) found that the average severity (frequency and strength) of seizures decreased with their group of 12 patients with drug-resistant seizures. Fried et al. (1984) successfully used diaphragmatic breathing training to reduce seizures with some patients who previously had convulsive seizures having none throughout the 7-month period of the study.

In both the Fried et al. (1984) and Kotchoubey et al. (1996) studies, the improvements could not be attributed to changes in AED regimens as these remained unchanged during the biofeedback training; in the Kotchoubey study there were no changes in the preceding 5 months. In an attempt to rule out the possibility of a placebo effect, Nagai *et al.* divided their patients into a biofeedback training group and a sham treatment group. The treatment group underwent bio-feedback training based on GSR and had a significant decrease in seizure frequency in comparison to the sham treatment group (Nagai *et al.*, 2004a, b).

Predictors of which patients may benefit the most from biofeedback have been examined. Strehl *et al.* (2005) identified seizure focus, level of cortical excitability and personality variables such as life satisfaction and negative response to stressful situations. Andrews and Schonfeld (1992), who reported that 69 out of their 83 patients had total control of their complex partial seizures by the end of the train-ing, suggested that frequency of seizures was the only significant predictor in their study.

Educational programs

As part of an overall aim of raising awareness of epilepsy, its psychosocial conse-quences and treatment, various educational programs have been developed. This has subsequently been expanded to include additional aspects of self-management including identifying seizure precipitants and seizure control techniques (May and Pfäfflin, 2002). A recent Cochrane review examining self-management education defines self management as "...the range of actions that need to be taken by people with epilepsy to improve their quality of life" (Shaw *et al.*, 2007, p. 2). Therefore, educational programs have a broader remit in that they are not solely used to impart knowledge of epilepsy but also attempt to equip an individual with the ability to become an expert of their individual epilepsy (May and Pfäfflin, 2002).

The MOSES (Modular Service Package Epilepsy) (May and Pfäfflin, 2002) and the SEE (Sepulveda Epilepsy Education program) (Helgeson *et al.*, 1990) are two examples of multi-component training that have been used. The MOSES program consists of nine units covering issues such as basic knowledge of epilepsy, diagnosis, self-control, psychosocial aspects and prognosis. The SEE divides train-ing into two main categories of medical aspects of epilepsy and social/emotional aspects of epilepsy. Within each are various topics such as what epilepsy is, treat-ment options, seizure control, psychological problems and social difficulties.

Both programs have been assessed using a randomized controlled study design with training taking place over a 2-day schedule. May and Pfäfflin (2002) and Helgeson *et al.*'s 1990 study reported an increase in participant's levels of knowl-edge about epilepsy and a decrease in seizure frequency – reaching statistical significance in the May 2002 study. A small pilot study using five patients with stress-induced seizures also found a reduction in seizure frequency following a course of training (Oosterhuis, 1994). Unusually, the SEE and the MOSES pro-grams did not significantly improve levels of depression, anxiety and other emo-tional problems associated with living with epilepsy. This is in stark comparison to Olley *et al.* (2001) who reported how levels of depression and anxiety were reduced as a consequence of their training program. This may reflect a cultural bias in that Olley *et al.* (2001) were investigating individuals with epilepsy in Nigeria, a country where there is limited knowledge about epilepsy, where it is viewed as contagious and is shrouded in superstition. In these circumstances improving knowledge of epilepsy may alleviate some of the associated emotional difficulties.

Cognitive behavior therapy

A number of behavioral approaches have been classified under the general heading of cognitive behavior therapy (CBT). It is difficult to clearly distinguish whether a therapy can be described as CBT. A broad definition encompasses investigating how an individual interacts with their environment, the development of positive, reinforcing behaviors, analysis of the "core" of the problem from which behaviors (symptoms) are based, identification of maladaptive thoughts and behaviors and subsequent replacement with more appropriate or useful behaviors and coping strategies (Hazlett-Stevens and Craske, 2005). The application of CBT in alleviating seizures has involved both dealing with the negative thought processes associated with having seizures and attempts to equip individuals with more control over their seizures. As with other psychological therapies there can be a great deal of overlap between methods with some CBT programs incorporating elements of relaxation and/or biofeedback and with some educational programs borrowing heavily from aspects of CBT.

Self control of seizures

Tackling the problem of control of seizures could be of benefit in reducing levels of stress and depression experienced which in turn may reduce seizures (Benak, 2001). The method by which seizures can be arrested differs for each individual as it depends on the situation or behaviors that typically cause seizures (Wolf, 2002). A technique which may appear benign and help prevent seizures in one person may actually induce seizures in another (Wolf, 2005). The more successful the attempts at seizure arrest become over time could also have longer term implications with reports that seizures can disappear completely; the mechanism for this is unclear but it seems to be different to how AEDs work (Wolf, 2002) and is of special interest when it is considered that these strategies are being employed by a population with refractory seizures.

The ability to arrest a seizure allows an individual to regain some level of perceived control over their condition. Individuals experiencing seizures can feel that they are not in control of their life in general or that anything positive which happens in their life can ultimately be ruined by an unexpected seizure (Gehlert, 1994). Spector *et al.* (2001) found that while their sample of people with intractable seizures believed they were able to exert a certain level of control, they could split the group into high controllers and low controllers. High controllers were more successful in being able to identify seizure precipitants and in avoiding situations likely to increase seizure occurrence but were not different to the low controllers in terms of anxiety, depression or self esteem experienced. However, there were differences between the groups in their perceived level of control with low controllers scoring highly on external health locus of control beliefs (Spector et al., 2001). This has to be considered a factor in being able to successfully equip the individual with successful seizure arrest techniques.

Countermeasures

As has been outlined earlier in this chapter, patients can readily identify factors that increase the likelihood of a seizure. The next logical step is to develop a behavior which can be used to abort a seizure; a countermeasure. Fenwick (1995, p. S48) describes how a countermeasure is "...a behaviour performed either in a seizure-prone situation or at the arrest of a seizure, with the aim of inhibiting a seizure completely or halting its spread." Wolf (1997) states how countermeasures can be seen as one of the oldest therapies: reports exist dating back to the 1st and 2nd centuries of how certain individuals on sensing a seizure would attempt to apply a ligature to the affected limb.

Various countermeasures have been self reported by patients including concentration, relaxation and muscle tension (Antebi and Bird, 1993). The development of a successful countermeasure is dependent on how much warning (if any) an individual has, whether the warning signs can be recognized and whether the countermeasure is judged as effective. Understandably, it has been shown that patients who feel a countermeasure is ineffective simply stop applying them (Lee and No, 2005). Seizure type can also affect whether people with epilepsy experience premonitory symptoms–Hughes et al. (1993) found that in their sample those who could identify symptoms all had partial epilepsy.

The effectiveness of CBT amongst people with epilepsy appears to vary according to what is judged as a successful outcome from the patient perspective. While Au et al. (2003) found a slight reduction in seizure frequency in their group, their study along with Goldstein et al. (2003) and Tan and Bruni (1986) failed to reach significant levels. However, all three studies reported an improvement in various psychosocial factors such as stress management and anxiety (Au et al., 2003), depression (Goldstein et al., 2003) and therapist's rating of psychological adjustment (Tan and Bruni, 1986). Conversely, Spector et al.'s (1999) study of seven adults with intractable seizures noted significant seizure reduction but no improvement in psychosocial status. Martinovic (2001) assessed young adults with juvenile myoclonic epilepsy and found improvements for both seizure control and psychological problems such as depression and anxiety in the CBT treatment group when compared to their antistress program group (Table 20.2).

TABLE 20.2   Summary of psychological therapies cited

	Principal intervention (as defined by authors)	Country
Relaxation Therapy		
Dahl et al. (1987)	Contingent relaxation treatment (progressive muscle relaxation)	Sweden
Puskarich et al. (1992)	Progressive relaxation therapy	USA
Rousseau et al. (1985)	Progressive relaxation therapy	USA
Biofeedback		
Fried et al. (1984)	Diaphragmatic respiration training and CO_2 biofeedback	USA
Kotchoubey et al. (1996, 1999)	Slow cortical potential (SCP) biofeedback	Germany
Nagai et al. (2004)	Galvanic skin response (GSR) biofeedback	UK
Behavioral Therapies		
Au et al. (2003)	Cognitive behavioral therapy including relaxation training, self control strategies	Hong Kong
Dahl et al. (1985, 1992)	Analysis of seizure behavior, seizure control techniques	Sweden
Goldstein et al. (2003)	Cognitive behavioral – epilepsy-related problems, psychopathology, seizure reduction strategies	UK
Schmid-Schönbein (1998)	Self control techniques	Germany
Spector et al. (1999)	Cogntive behavioral group intervention/self–control	UK
Tan and Bruni (1986)	Group cognitive behavioral therapy	Canada
Educational Programs		
Gillham (1990)	Self control of seizures/alleviation of psychological disorders	UK
Helgeson et al. (1990)	Psycho-educational program (SEE)	India
May and Pfäfflin (2002)	Educational program (MOSES)	Germany
Olley et al. (2001)	Psycho-educational program	Nigeria
Oosterhuis (1994)	Psycho-educational program	The Netherlands
Combined Therapies		
Andrews and Schonfeld (1992)	EEG biofeedback/counseling/seizure triggers	USA
Cabral and Scott (1976)	EEG biofeedback and relaxation	UK
Martinovic (2001)	Antistress program/Individual cognitive behavior therapy	Serbia
Reiter and Andrews (2000)	Neurobehavioral treatment including EEG biofeedback	
Strehl et al. (2005)	Slow cortical potential (SCP) feedback and behavioral self management program	Germany
Tozzo et al. (1988)	Biofeedback and relaxation	USA

EVIDENCE OF THE EFFECTIVENESS OF PSYCHOLOGICAL THERAPIES

Theoretical issues

Explanations for how seizures are generated largely rely on the medical model, focusing on abnormal electrical discharges taking place within the brain (Fenwick, 1992). As Fenwick (1992) argues, changes in brain activity alone cannot predict at any given time whether a seizure will fully develop or be arrested. Behaviors can affect relative levels of neuron excitation or inhibition and this can determine whether a seizure will develop fully. While a medical model approach may look at effective doses of AEDs or investigate adherence to medication, a behavioral approach to seizures could involve keeping a diary of the situations in which seizures occurred, emotions at the time preceding the seizure and any warning signals experienced and descriptions of the postictal experience. Fenwick (1995) used this technique with the ABC charts (Antecedents, Behavior, Consequences) in order to tailor specific behavioral interventions with individual patients.

The Andrews/Reiter Program (Reiter et al., 1987) similarly moves away from the medical model of seizures and examines each individual's patterns of behavior, emotions and external factors which may precipitate seizures with the aim of developing positive behaviors to halt seizures. Using a behavioral approach to alleviating seizures poses some difficulties in being able to accurately ascertain effectiveness of the techniques. While a drug may gradually be increased until it is judged as reaching a therapeutic level, positive behavior adjustments may take longer to achieve and are more problematic to monitor.

Methodological issues

Dahl (1999) discusses how psychological therapies used in epilepsy have their efficacy judged within a "mechanistic paradigm" of medicine that is preoccupied with measuring physiological changes. Dahl calls for a shift toward using a behavioral medical model rather than assessing all treatment for epilepsy within this traditional medical model (Dahl, 1999). However, while the research outlined in this chapter has focused on psychological therapies, the outcome measures have predominately been similar to those used in drug trials – frequency of seizures, number of seizures, time seizure free. Frequency of seizures can be misleading in assessing psychological interventions that may be capable of preventing a full seizure by recognizing preseizure symptoms or auras, but for neurology an aura would be defined as a seizure (Dahl et al., 1987). Arriving at a methodology that is judged as valid and acceptable to both behavioral and medical perspectives once again illustrates how epilepsy "lies in the borderland" of psychiatry and neurology (Betts, 1993).

A recent systematic review of randomized controlled trials (RCTs) assessing psychological therapies (Ramaratnam *et al.*, 2006) included many of the studies highlighted here and found that, overall, evidence for the effectiveness of the various therapies was limited and hindered by a number of methodological factors. In particular, there is a dearth of well-designed RCTs, the number of patients being investigated is small, AED regimens are not kept constant or are not made explicit and seizure types are not always reported (Ramaratnam *et al.*, 2006). The authors conclude that these methodological flaws need to be addressed in order to determine the effectiveness of the various treatments.

The "active" component

As Mostofsky and Balaschak (1977) state, psychological approaches used in controlling seizures are not mutually exclusive in terms of specific methods employed; there is a blurring of methods and techniques used which impede any attempt to judge effectiveness. Biofeedback studies frequently use relaxation and/or breathing techniques as well as elements of CBT such as identifying warnings of a seizure and developing countermeasures. Some CBTs employ biofeedback as part of the training process and educational programs may also contain elements of CBT. Tozzo *et al.* (1988), in a study involving biofeedback, comment on the difficulty in establishing the "active ingredient". In Kaplan's (1975) biofeedback study, a successful reduction in seizures was attributed to the relaxation techniques rather than changes on the EEG.

The role of relaxation training in preventing seizures has also been questioned. Dahl *et al.* (1987) discuss how relaxation techniques for some people may serve to distract an individual from whatever stimuli is triggering seizures – the "distraction hypothesis" – and therefore it would seem useful to concentrate on developing specific behaviors that could halt the seizure instead. Relaxation therapy has been shown to have a possible beneficial effect (Ramaratnam *et al.*, 2006) but this is not conclusive due to the various methodological issues already outlined previously; additionally there are some studies which have shown that some patients had an increase in seizure frequency instead (Puskarich *et al.*, 1992; Goldstein *et al.*, 2003). However, due to the small sample sizes being used, this negative effect may be atypical and it is difficult to generalize these findings to the general epilepsy population.

Gillham (1990) found that the two behavioral treatments used (self control of seizures, alleviation of psychological disorders) appeared to either have a common effective ingredient or were essentially the same treatment. She concludes that both treatments may have equipped the individuals with new coping skills that could improve both seizure control and psychopathology associated with their epilepsy (Gillham, 1990). Goldstein's (1997) review of psychological therapies asks whether studies investigating the effects of relaxation therapy involve training in breathing patterns. This is of crucial importance as Fried *et al.*'s (1984) study illustrated how diaphragmatic respiration training could reduce seizures even after feedback was withdrawn.

Goldstein (1997) suggests that instead of investigating the effectiveness of singular therapies it may be more useful to undertake large scale RCTs comparing cognitive/behavioral therapies (relaxation, countermeasures) against neurophysiological therapies (e.g. biofeedback). However, even this dichotomy appears to be problematic as there can still be a crossover between the two when the previous example of relaxation and breathing techniques is considered. A further complication in ascertaining effectiveness is that some authors consider biofeedback to be based on the same psychological principles as behavioral therapy; Kotchoubey *et al.* (1996) describe how both use operant learning theory.

Placebo effects

In the quest to determine which elements of psychological therapies are alleviating seizures, the potential for placebo effects to occur is worthy of examination. Quy *et al.* (1979) hypothesized that in their biofeedback study the placebo effects and/or the relaxation training brought about the reduction in seizure rates rather than the biofeedback process. Mostofsky and Balaschak (1977) remark that the by-product of increased attention and/or the therapist's personality may provide unintentional reinforcement. It has been documented that attention can be a powerful reinforcer for some patients. Spector *et al.* (2000) report how 15% of their patient group admitted to inducing seizures at some point (mostly in childhood as a means of avoiding school exams in order to receive increased parental attention). In order to address these issues, studies have been designed to use control or sham treatment groups alongside the therapy being assessed (Rousseau *et al.*, 1985; Dahl *et al.*, 1987; Nagai *et al.*, 2004a, b). Aird (1983) argues that where placebo effects are present they are generally short lived and disappear over longer term follow-up especially if the patient is being seen by the original referring clinician and not the consultant. While the avoidance of placebo effects is desirable it is also interesting to note that placebo effects themselves are psychological, therefore if they do reduce seizures how is this being achieved (Betts, 1992)?

Heterogeneity of individuals with epilepsy

As a rule the individuals undergoing psychological therapies for seizures are atypical of the general epilepsy population with the majority being those whose seizures are refractory to AED treatment (Mostofsky and Balaschak, 1977) or who are unwilling to use AEDs (Wolf and Okujava, 1999). Sterman and Egner (2006) emphasize that biofeedback has been deemed as the last resort in treating seizures in that it is predominately used with patients whose seizures have proved drug resistant, therefore success rates should be judged in this context. The age of the individual patient and the age at which epilepsy was diagnosed may also be complications in using psychological therapies; biofeedback, which requires high levels of concentration and can be time intensive, may not be useful to younger children (Kneen and Appleton, 2006). Muller (2001) reported that behavioral treatment programs were less effective amongst young children and individuals with learning difficulties.

As the patients studied tend to be those for whom AED treatment has failed investigators are left with small number of patients with various seizure types. This factor may explain differences in levels of success across the various therapies and further hinders the ability to use treatment appropriate to an individual (Benak, 2001). Patients with mixed seizure types and no obvious precipitating factors for their seizures are more difficult to treat with behavioral strategies (Wolf and Okujava, 1999). Ironically, patients with poorly controlled seizures can have a higher success rate at stopping seizures simply because their increased seizure frequency allows them more opportunities to practice and develop their arrest methods (Lee and No, 2005).

Practical considerations

Clear judgment about the effectiveness of psychological therapies is confounded when they are being applied alongside an AED regimen. Few studies are available where AED usage has remained unchanged and there have been failures in reporting changes to drug or dosage (Mostofsky and Balaschak, 1977; Monderer *et al.*, 2002). AEDs themselves may negatively interact with therapies, in particular they may affect the aura experienced (Muller, 2001). Muller (2001), in his meta-analysis of nocturnal seizures, found that seizure freedom increased as AEDs were gradually withdrawn under close supervision. Schmid-Schönbein (1998) equally found that certain changes to AED drug or dosage could enable patients to successfully use self control techniques.

A patient's willingness to adhere to psychological therapies also contributes to the degree of success in alleviating seizures (Wolf and Okujava, 1999). Therapies are more time consuming with sessions being held in a clinic setting and then patients being encouraged to further practice techniques at home. Practicing techniques regularly takes up more time than simply taking medication and patients may be resistant to altering behaviors (Engel, 2005). It is also likely that a patient's belief in their ability to alter seizure behavior determines how effective the psychological therapy is (Goldstein, 1997).

Another practical consideration in the use of psychological therapies is the financial cost to both the health care system and the patient. Sterman (2005) estimates that the cost of one biofeedback session is approximately $100, with one to three sessions needed per week over a period of 3 months. CBT has been shown to be effective in a group setting for Au *et al.* (2003) and Spector *et al.* (1999) but not in

Tan and Bruni's (1986) study, which was based on individual patient sessions. The number of sessions needed has also been debated with some researchers suggesting that more beneficial outcomes would be achieved via additional sessions (Goldstein et al., 2003). This obviously has implications for staff resources and related costs. However, certain components of behavioral training such as identifying and avoiding seizure inducing stimuli could be viewed as inexpensive and particularly useful in areas of the world where AEDs are not readily available (Engel, 2005).

CONCLUSION

In concluding this chapter we believe that there are a number of issues that still need to be resolved. In particular:

- Are psychological therapies effective in preventing seizures and how can effectiveness be evaluated?
- If certain therapies are shown to be effective, which ones should be provided, for whom in particular and in what circumstances?
- What are the costs associated with providing treatment – the number of psychologists available is limited, if specialist nurses can be used who will be the training providers and at what financial cost to the health care system?

What is clear is that there is a demand for alternative treatments (Fisher et al., 2000; Sirven et al., 2003) and there is a subset of patients who are not achieving seizure control through AEDs or do not wish to take them (Wolf and Okujava, 1999). Psychological therapies allow a patient control over their epilepsy to a certain extent (Gehlert, 1994; Spector et al., 2001). It has been established that behaviors, stress and other external stimuli can either trigger or predispose individuals to seizures (Wolf, 2005) but equally there is a growing body of evidence to show that behaviors can also be used to prevent them. What is problematic to ascertain is whether psychological therapies are tackling seizures directly or are alleviating the psychopathology associated with epilepsy. It would seem prudent to incorporate psychological therapies into everyday routine care for people with epilepsy but can this presently be justified when evidence of effectiveness is inconclusive? These issues need to be resolved by the undertaking of high quality research using well-designed RCTs with clear pragmatic outcomes. Until then psychological therapies, while no longer viewed as a last resort, will realistically remain an adjunct to pharmacotherapy rather than a replacement.

REFERENCES

Aird, R. B. (1983). The importance of seizure-inducing factors in the control of refractory forms of epilepsy. *Epilepsia* **24**, 567–583.

Aldenkamp, A. P., Baker, G. A., Meador, K. J. (2004). The neuropsychology of epilepsy: What are the factors involved? *Epilepsy Behav* **5**(Suppl 1), 1–2.

Andrews, D. J., Schonfeld, W. H. (1992). Predictive factors for controlling seizures using a behavioural approach. *Seizure* **1**, 111–116.

Antebi, D., Bird, J. (1993). The facilitation and evocation of seizures. A questionnaire study of awareness and control. *Br J Psychiatry* **162**, 759–764.

Au, A., Chan, F., Li, K., et al. (2003). Cognitive-behavioral group treatment program for adults with epilepsy in Hong Kong. *Epilepsy Behav* **4**, 441–446.

Baker, G. A., Jacoby, A., Chadwick, D. W. (1996). The associations of psychopathology in epilepsy. A community study. *Epilepsy Res* **25**, 29–39.

Benak, J. C. (2001). To quantify experience? Methodological issues in the behavioural/psychological treatment of complex partial seizures. *Seizure* **10**, 48–55.

Berger, N. M. (2001). The effect of a stress management training program on epilepsy. AES Proceedings. *Epilepsia* **42**(Suppl 7), 245.

Berger, N. (2005). Autogenic training. In *Complementary and Alternative Therapies for Epilepsy* (O. Devinsky, S. C. Schachter, S. Pacia, eds.). New York: Demos Medical Publishing, Ch 7, pp. 57–63.

Bernstein, D. A., Borkovec, T. D. (1973). *Progressive Relaxation: A Training Manual for Therapists.* Champaign, IL: Research Press.

Betts, T. A. (1981). Epilepsy and the mental hospital. In *Epilepsy and Psychiatry* (E. H. Reynolds, M. R., Trimble, eds.). Edinburgh: Churchill Livingstone, pp. 175–184.

Betts, T. A. (1992). Epilepsy and stress (editorial). *Br Med J* **305**, 378–379.

Betts, T. A. (1993). Neuropsychiatry. In *A Textbook of Epilepsy* (J. Laidlaw, A. Richens, D. W. Chadwick, eds.). Edinburgh: Churchill Livingstone, pp. 445–448.

Brodie, M. J., Dichter, M. A. (1996). Antiepileptic drugs. *New Engl J Med* **334**, 168–175.

Brown, S., Muller, B. (2001). The third major pillar of epilepsy treatment – using psychological approaches. *Seizure* **10**, 1–2.

Cabral, R. J., Scott, D. F. (1976). Effects of two desensitization techniques, biofeedback and relaxation, on intractable epilepsy: follow-up study. *J Neurol Neurosurg Psychiatry* **39**, 504–507.

Dahl, J., Melin, L., Lund, L. (1987). Effects of a contingent relaxation treatment program on adults with refractory epileptic seizures. *Epilepsia* **28**(2), 125–132.

Dahl, J., Brorson, L. O., Melin, L. (1992). Effects of a broad-spectrum behavioral medicine treatment program on children with refractory epileptic seizures: an 8 year follow-up. *Epilepsia* **33**(1), 98–102.

Dahl, J. (1999). A behaviour medicine approach to epilepsy – time for a paradigm shift? *Scan J Behav Ther* **28**(3), 97–114.

Dichter, M. A., Brodie, M. J. (1996). New antiepileptic drugs. *New Engl J Med* **334**, 1583–1590.

Dua, T., de Boer, H. M., Prilipko, L. L., et al. (2006). Epilepsy care in the world: results of an ILAE/IBE/WHO global campaign against epilepsy survey. *Epilepsia* **47**(7), 1225–1231.

Engel, J., Jr. (2005). Alternative treatments. *Epilepsy: Global Issues for the Practising Neurologist.* New York, USA: Demos Medical Publishing.

Fenwick, P. B. C. (1992). The relationship between mind, brain and seizures. *Epilepsia* **33**(Suppl 6), S1–S6.

Fenwick, P. (1995). The basis of behavioral treatments in seizure control. *Epilepsia* **36**(Suppl 1), S46–S50.

Fenwick, P. (2005). Seizure generation. In *Complementary and Alternative Therapies for Epilepsy* (O. Devinsky, S. C. Schachter, S. Pacia, eds.). New York: Demos Medical Publishing, Ch 5, pp. 43–52.

Fiordelli, E., Beghi, E., Bogliun, G., Crespi, V. (1993). Epilepsy and psychiatric disturbance: a cross-sectional study. *Br J Psychiatry* **163**, 446–450.

Fisher, R. S., Vickrey, B. G., Gibson, P., et al. (2000). The impact of epilepsy from the patient's perspective II: views about therapy and health care. *Epilepsy Res* **41**, 53–61.

Fried, R., Rubin, S. R., Carlton, R. M., et al. (1984). Behavioral control of intractable idiopathic seizures: self-regulation of end-tidal carbon dioxide. *Psychosom Med* **46**(4), 315–331.

Frucht, M. M., Quigg, M., Schwaner, C., et al. (2000). Distribution of seizure precipitants among epilepsy syndromes. *Epilepsia* **41**(12), 1534–1539.

Gaitatzis, A., Carroll, K., Majeed, A., et al. (2004). The epidemiology of the comorbidity of epilepsy in the general population. *Epilepsia* **45**(12), 1613–1622.

Gehlert, S. (1994). Perceptions of control in adults with epilepsy. *Epilepsia* **35**(1), 81–88.

Gillham, R. A. (1990). Refractory epilepsy: an evaluation of psychological methods in out-patient management. *Epilepsia* **31**, 427–432.

Goldstein, L. H. (1997). Effectiveness of psychological interventions for people with poorly controlled epilepsy. *J Neurol Neurosurg Psychiatry* **63**, 137–142.

Goldstein, L. H., McAlpine, M., Deale, A., et al. (2003). Cognitive behaviour therapy with adults with intractable epilepsy and psychiatric co-morbidity: preliminary observations on changes in psychological state and seizure frequency. *Behav Res Ther* **41**, 447–460.

Haut, S. R., Vouyiouklis, M., Shinnar, S. (2003). Stress and epilepsy: a patient perception survey. *Epilepsy Behav* **4**, 511–514.

Hazlett-Stevens, H., Craske, M. G. (2005). Brief cognitive-behavioral therapy: definitions and scientific foundations. In *Handbook of Brief Cognitive Behaviour Therapy* (F. W. Bond, ed.). Hoboken NJ, USA: John Wiley & Sons Inc., Ch 1, pp. 1–20.

Helgeson, D. C., Mittan, R., Tan, S. Y., et al. (1990). Sepulveda epilepsy education: the efficacy of a psychoeducational treatment program in treating medical and psychosocial aspects of epilepsy. *Epilepsia* **31**(1), 75–82.

Hermann, B. P., Trenery, M. R., Colligan, R. C. (1996). Learned helplessness, attributional style and depression in epilepsy. *Epilepsia* **37**(7), 680–686.

Hessen, E., Lossius, M. I., Reinvang, I., et al. (2006). Influence of major antiepileptic drugs on attention, reaction time, and speed of information processing: results from a randomized, double-blind, placebo-controlled withdrawal study of seizure-free epilepsy patients receiving monotherapy. *Epilepsia* **47**(12), 2038–2045.

Hidderley, M., Holt, M. (2004). A pilot randomized trial assessing the effects of autogenic training in early stage cancer patients in relation to psychological status and immune system responses. *Eur J Oncol Nurs* **8**, 61–65.

Hughes, J., Devinsky, O., Feldmann, E., et al. (1993). Premonitory symptoms in epilepsy. *Seizure* **2**, 201–203.

Jacoby, A., Baker, G. A., Steen, A., et al. (1996). The clinical course of epilepsy and its psychosocial correlates: findings from a UK community study. *Epilepsia* **37**, 148–161.

Kaplan, B. J. (1975). Biofeedback in epileptics: equivalent relationship of reinforced EEG frequency to seizure reduction. *Epilepsia* **16**, 477–485.

Kelso, A. R. C., Cock, H. R. (2005). Advances in epilepsy. *Br Med Bull* **72**, 135–148.

Kneen, R., Appleton, R. E. (2006). Alternative approaches to conventional antiepileptic drugs in the management of paediatric epilepsy. *Arch Dis Child* **91**, 936–941.

Kotchoubey, B., Busch, S., Birbaumer, N. (1999). Changes in EEG Power spectra during biofeedback of slow cortical potentials in epilepsy. *Appl Psychophysiol Biofeedback* **24**(4), 213–233.

Kotchoubey, B., Schneider, D., Schleichert, H., et al. (1996). Self-regulation of slow cortical potentials in epilepsy: a retrial with analysis of influencing factors. *Epilepsy Res* **25**, 269–276.

Lee, S. A., No, Y. J. (2005). Perceived self-control of seizures in patients with uncontrolled partial epilepsy. *Seizure* **14**, 100–105.

Loring, D. W., Meador, K. J. (2004). Cognitive side effects of antiepileptic drugs in children. *Neurology* **62**, 872–877.

Luciano, A. L., Shorvon, S. D. (2007). Results of treatment changes in patients with apparently drug resistant chronic epilepsy. *Ann Neurol* **62**(4), 375–381.

Marsh, E. D., Brooks-Kayal, A. R., Porter, B. E. (2006). Seizures and antiepileptic drugs: Does exposure alter normal brain development? *Epilepsia* **47**(12), 1999–2010.

Martinovic, Z. (2001). Adjunctive behavioural treatment in adolescents and young adults with juvenile myoclonic epilepsy. *Seizure* **10**, 42–47.

May, T. W., Pfäfflin, M. (2002). The efficacy of an educational treatment program for patients with epilepsy (MOSES): results of a controlled, randomized study. *Epilepsia* **43**(5), 539–549.

Meinardi, H., *et al.* (2001). On behalf of ILAE Commission. *Epilepsia* **42**, 136–139.

Monderer, R. S., Harrison, D. M., Haut, S. R. (2002). Neurofeedback and epilepsy. *Epilepsy Behav* **3**, 214–218.

Mostofsky, D. I., Balaschak, B. A. (1977). Psychobiological control of seizures. *Psychol Bull* **84**(4), 723–750.

Muller, B. (2001). Psychological approaches to the prevention and inhibition of nocturnal epileptic seizures: a meta analysis of 70 case studies. *Seizure* **10**, 13–33.

Nadkarni, S., LaJoie, J., Devinsky, O. (2005). Current treatments of epilepsy. *Neurology* **64**(Suppl 3), S2–S11.

Nagai, Y., Goldstein, L. H., Critchley, H. D., Fenwick, P. (2004a). Influence of sympathetic autonomic arousal on cortical arousal: implications for a therapeutic behavioural intervention in epilepsy. *Epilepsy Res* **58**, 185–193.

Nagai, Y., Goldstein, L. H., Fenwick, P. B. C., *et al.* (2004b). Clinical efficacy of galvanic skin response biofeedback training in reducing seizures in adult epilepsy: a preliminary randomized controlled study. *Epilepsy Behav* **5**, 216–223.

Nakken, K. O., Solaas, M. H., Kjeldsen, J., *et al.* (2005). Which seizure-precipitating factors do patients with epilepsy most frequently report? *Epilepsy Behav* **6**, 85–89.

Neugebauer, R., Paik, M., Hauser, W. A., Nadel, E., Leppik, I., Susser, M. (1994). Stressful life events and seizure frequency in patients with epilepsy. *Epilepsia* **35**(2), 336–343.

Ng, B. Y. (2002). Psychiatric aspects of self-induced epileptic seizures. *Aust New Zeal J Psychiatry* **36**, 534–543.

O'Donoghue, M. F., Goodridge, D. M. G., Redhead, K., *et al.* (1999). Assessing the psychosocial consequences of epilepsy: a community-based study. *Br J Gen Pract* **49**, 211–214.

Olley, B. O., Osinowo, H. O., Brieger, W. R. (2001). Psycho-educational therapy among Nigerian adult patients with epilepsy: a controlled outcome study. *Patient Educ Counsel* **42**, 25–33.

Oosterhuis, A. (1994). A psycho-educational approach to epilepsy. *Seizure* **3**, 23–24.

Puskarich, C. A., Whitman, S., Dell, J., *et al.* (1992). Controlled examination of effects of progressive relaxation training on seizure reduction. *Epilepsia* **33**, 675–680.

Quy, R. J., Hutt, S. J., Forrest, S. (1979). Sensorimotor rhythm feedback training and epilepsy: some methodological and conceptual issues. *Biol Psychol* **9**, 129–149.

Ramaratnam, S., Baker, G. A., Goldstein, L. H. (2006). Psychological treatments for epilepsy. *Cochrane Database Syst Rev* (4). Art.No:CD002029

Reiter, J. M., Andrews, D. J., Janis, C. (1987). *Taking Control of Your Epilepsy: A Workbook for Patients and Professionals*. Santa Rosa, CA: Andrews/Reiter Research Program.

Reiter, J. M., Andrews, D. J. (2000). A neurobehavioral approach for treatment of complex partial epilepsy: efficacy. *Seizure* **9**, 198–203.

Roth, S. R., Sterman, M. B., Clemente, C. C. (1967). Comparison of EEG correlates of reinforcement, internal inhibition and sleep. *Electroencephalogr Clin Neurophysiol* **23**, 509–520.

Rousseau, A., Hermann, B., Whitman, S. (1985). Effects of progressive relaxation on epilepsy: analysis of a series of cases. *Psychol Rep* **57**, 1203–1212.

Sander, J. W. (2004). The use of antiepileptic drugs-principles and practice. *Epilepsia* **45**(Suppl 6), 28–34.

Sander, J. W. (2005). Ultimate success in epilepsy – the patient's perspective. *Eur J Neurol* **12**(Suppl 4), 3–11.

Schachter, S. C. (1999). Antiepileptic drug therapy: general treatment principles and application for special patient populations. *Epilepsia* **40**(Suppl 9), S20–S25.

Schmid-Schonbein, C. (1998). Improvement of seizure control by psychological methods in patients with intractable epilepsies. *Seizure* **7**, 261–270.

Schramke, C. J., Kelly, K. M. (2005). Stress and epilepsy. In *Complementary and Alternative Therapies for Epilepsy* (O. Devinsky, S. C. Schachter, S. Pacia, eds.). New York: Demos Medical Publishing, Ch 3, pp. 25–32.

Seligman, M. E., Maier, S. F. (1967). Failure to escape traumatic shock. *J Exp Psychol* **74**(1), 1–9.

Shaw, E. J., Stokes, T., Camosso-Stefinovic, J. *et al.* (2007). Self-management education for adults with epilepsy. *Cochrane database of Syst Rev* (2). Art No:CD004723.

Sirven, J. I., Drazkowski, J. F., Zimmerman, R. S., *et al.* (2003). Complementary/alternative medicine for epilepsy in Arizona. *Neurology* **61**, 576–577.

Smith, D. B., Craft, B. R., Collins, J., *et al.* (1986). Behavioral characteristics of epilepsy patients compared with normal controls. *Epilepsia* **27**, 760–768.

Smith, D. F., Baker, G. A., Dewey, M., *et al.* (1991). Seizure frequency, patient perceived seizure severity and the psychosocial consequences of intractable epilepsy. *Epilepsy Res* **9**, 231–241.

Spector, S., Trannah, A., Cull, C., *et al.* (1999). Reduction in seizure frequency following a short term group intervention for adults with epilepsy. *Seizure* **8**, 297–303.

Spector, S., Cull, C., Holdstein, L. H. (2000). Seizure precipitants and perceived self-control of seizures in adults with poorly-controlled epilepsy. *Epilepsy Res* **38**, 207–216.

Spector, S., Cull, C., Goldstein, L. H. (2001). High and low perceived self-control of epileptic seizures. *Epilepsia* **42**(4), 556–564.

Sterman, M. B. (2000). Basic concepts and clinical findings in the treatment of seizure disorders with EEG operant conditioning. *Clin Electroencephal* **31**(10), 45–55.

Sterman, M. B. (2005). Neurofeedback therapy. In *Complementary and Alternative Therapies for Epilepsy* (O. Devinsky, S. C. Schachter, S. Pacia, eds.). New York: Demos Medical Publishing, Ch 6, pp. 53–56.

Sterman, M. B., Egner, T. (2006). Foundation and practice of neurofeedback for the treatment of epilepsy. *Appl Psychophysiol Biofeedback* **31**(1), 21–35.

Strehl, U., Kotchoubey, B., Trevorrow, T., *et al.* (2005). Predictors of seizure reduction after self regulation of slow cortical potentials as a treatment of drug-resistant epilepsy. *Epilepsy Behav* **6**, 156–166.

Tan, S. Y., Bruni, J. (1986). Cognitive-behaviour therapy with adult patients with epilepsy: a controlled outcome study. *Epilepsia* **27**(3), 225–233.

Temkin, O. (1971). *The Falling Sickness – A History of Epilepsy from the Greeks to the Beginning of Modern Neurology* 2nd edition. Baltimore & London: Johns Hopkins Press.

Tozzo, C. A., Elfner, L. F., May, J. G., Jr. (1988). EEG Biofeedback and relaxation training in the control of epileptic seizures. *Int J Psychophysiol* **6**, 185–194.

Vinayan, K. P. (2006). Epilepsy, antiepileptic drugs and educational problems. *Indian Paediatr* **43**, 786–794.

Wolf, P. (1997). Behavioral therapy. In *Epilepsy: A Comprehensive Textbook* (J. Engel, Jr., T. A. Pedley, eds). Philadelphia, PA: Lippincott-Raven Publishers.

Wolf, P. (2002). The role of nonpharmaceutic conservative interventions in the treatment and secondary prevention of epilepsy. *Epilepsia* **43**(Suppl 9), 2–5.

Wolf, P. (2005). From precipitation to inhibition of seizures: rationale of a therapeutic paradigm. *Epilepsia* **46**(Suppl 1), 15–16.

Wolf, P., Okujava, N. (1999). Possibilities of non-pharmacological conservative treatment of epilepsy. *Seizure* **8**, 45–52.

World Health Organization (WHO) (2001). Epilepsy: aetiology, epidemiology and prognosis, Fact Sheet No **165**. Revised February 2001, Geneva.

Yuen, A. W. C., Thompson, P. J., Flugel, D., *et al.* (2007). Mortality and morbidity rates are increased in people with epilepsy: Is stress part of the equation? *Epilepsy Behav* **10**, 1–7.

Index

Printed and bound by CPI Group (UK) Ltd, Croydon, CR0 4YY

03/10/2024

01040414-0005